# So You Want to Be a Chef?

## Your Guide to Culinary Careers

......................

**Lisa M. Brefere**

CEC, AAC

**Karen Eich Drummond**

EdD, FMP, RD

**Brad Barnes**

CMC, CCA, AAC

......................

WILEY

John Wiley & Sons, Inc.

*Library of Congress Cataloging-in-Publication Data:*

Brefere, Lisa M.
    So you want to be a chef? : your guide to culinary careers / Lisa Brefere,
Karen Eich Drummond, Brad Barnes.
        p. cm.
    Includes bibliographical references and index.
    ISBN 0-471-64691-1 (paper w/CD-ROM)
    1. Cooks—Juvenile literature. 2. Cookery—Vocational guidance—
Juvenile literature. I. Drummond, Karen Eich. II. Barnes, Brad. III.
Title.
    TX652.5.B715 2005
    641.5'023—dc22
                                                    2004023267

Printed in the United States of America

10 9 8 7 6 5 4 3 2

I NEVER PROPERLY THANKED THE PEOPLE IN MY LIFE who helped me nurture my abilities, fuel my passions, and evaluate my flaws. I am grateful to have the opportunity to dedicate this book to those who have been part of my culinary journey. The impact of these people on my career can never be properly measured or reciprocated.

I dedicate this book, which celebrates the enthusiasm and energy of our profession, to:

**Ed Esposito,** for giving a young woman with short checked pants and red socks an opportunity to step confidently into a man's kitchen with respect and equality.

**Rene Mettler,** for recognizing my natural talent, teaching me the art of dedication, and meticulously refining my skills.

**Fritz Sonnenschmidt,** for mentoring my path, cultivating my vision, and always finding the time to talk.

**Brad Barnes,** my partner, my friend, for keeping the bar high, never compromising on quality, and for giving me the desire to strive for individual perfection.

**To my family: My beloved mother,** for teaching me to balance my life, to persist in reaching for my dreams, and to remember the words "never give up." **My dad,** for my work ethic, my inventive mind, and my obsession for perfection. **My sister Leslie**, who fostered my childhood, teaching me love for family traditions, guiding my many choices, supporting my decisions, and giving countless hours of counseling. **My brother, Edward,** for urging me to stay focused, develop leadership qualities, and follow my dreams. **My sister Lauren,** for cherishing her baby sister with the love and care of a mother and coaching me with a good-natured heart and a nonjudging character, giving me the aptitude to excel in my career. **My children, Joey Jr., Julia, Johnny, and Jeremy,** for keeping me young at heart, whole in spirit, and openminded in thinking. And to **Joe, my husband,** with thanks for the decades of support, the years of encouragement, the months of advice, and the lifetime of friendship. Your level, calming nature has taught me patience and respect and showed me opportunities I never would have found on my own. You showed me how to take time to pick the tomatoes.

My inspiration for this book is a direct result of the wonderful people in my life. They have given me the strength and perseverance to encourage others following a similar career path. Thank you all for allowing me to give back.

LISA BREFERE, CEC, AAC

THIS BOOK IS DEDICATED TO OUR INDUSTRY, to the camaraderie of chefs and cooks. It is dedicated to the love for food we all share and our fascination with the effect heat has on ingredients. This book was written because our industry, although sincere and rewarding, can be an undescribed, directionless, intimidating maze, and over the years I have seen too many young cooks with fear and curiosity in their hearts about the uncertainty of just where this career will lead them. It is my greatest hope that this book will enlighten its audience, remove the mystery, and help all interested find their way through the fantastic journey of love, craft, art, and science we all take for granted as we enjoy each plate of food put in front of us.

Most of all, this book is dedicated to the brigade of incredibly dedicated people who have helped me see my way in my chosen career, the men and women I recognize as my mentors. I hope I have taken their teachings and formed them into a tool that will help many more culinarians than I could ever touch otherwise.

BRAD BARNES, CMC, CCA, AAC

# Contents

# Introduction

---

## Welcome

by Lisa Brefere, CEC, AAC

### "SO WHY DID YOU WANT TO BECOME A CHEF?"

After 27 years as a professional Culinarian and Chef, this is still the question I am most frequently asked. When I, a woman, picked this career 28 years ago, family, friends, and relatives dubbed me certifiably crazy. "Why would you want to be a Chef?" they asked. "To flip hamburgers the rest of your life?" Truthfully, no one thing lured me into this profession. I graduated from a well-known culinary school and come from an Italian-American background; thus, my career, and its inspiration, has centered on wholesome foods, natural flavors, and the freshest possible ingredients.

Early memories of my childhood revolve around large family meals, with wonderful smells of traditional dishes permeating the house. At least monthly we would get together with family and friends at a huge buffet table overflowing with delicacies like Aunt Columbia's crab corn bread, Mrs. Armando's orange butter cake, and Aunt Cindy's veal cutlet with olives and fresh garden tomatoes. I vividly remember coming home from nursery school and finding a big paper bag of something moving in the refrigerator. When I pushed the bag, something snapped back at me. That was my first experience with lobsters, the king of the sea, as my mother called them, when she taught me at the tender age of five the art of cleaning and cooking this treasure. It wasn't until some years later that I realized my early dining experiences were not the norm for the 1960s, when TV dinners, frozen vegetables, and convenience foods were becoming popular.

In 1975, I was introduced to the retail aspect of the culinary business at an obscure little store in Hayward, California. It was a drive-in dairy concept where customers drove into the middle of the store, where all the products were visible. The customers called out orders as I ran around collecting their groceries from a large selection of local dairy products, canned goods, homemade tortillas, bakery items, and fresh seasonal produce. I was 15 and thought the whole thing was fascinating. At the end of the day, the owner gave me expiring dairy, wilted produce, and day-old bread to take home, where I lived with my big sister. These I cooked, along with a limited amount of secondary cuts of meat and poultry. Although this was the most primitive phase of my career, it was the most innocent and adventurous. I didn't have a reputation to uphold; I just did whatever I wanted, and everyone in the house was thrilled to be catered to as long as they didn't have to help or clean. I was so taken by the experience of cooking that I never minded the work, which was always followed by praise.

My next experience was at an old-fashioned coffee shop, gourmet restaurant, and airplane catering company located at the Westchester County Airport. This is where I was introduced to my first real kitchen arrangement, complete with a hot/cold station, expediter, meal period rushes, beautifully displayed foods catered to corporate jets, and the adrenaline buzz of a busy environment. There was a magic about the kitchen, a sense of passion, commitment, love, and enthusiasm. You are there as a Chef by choice, pursuing a profession that grows from within, an occupation that reflects history, tradition, honor, and pride. Either you love it or you hate it. The heat, smoke, work, tireless hours just go with the territory. It's magnetic, exciting, ever-changing, and rewarding. You get up every day and love what you know, what you do, and what you get paid for.

### DO YOU THINK YOU ARE BORN TO BE A CHEF?

Perhaps you are, perhaps not, but what I do know is you are born with work ethics, pride, dedication, and perseverance. You were born to lead, born to teach, born to mentor, and, most important, born to create. I didn't know what I wanted either. I loved the scene of the kitchen, the fast pace, the routine that called for immediate attention, instant results, split-second decisions, long-term goals, with many short-term conclusions. I knew early on I needed to work with my hands, but at the same time I needed to use my intelligence to be a motivator. I was convinced from the beginning of my education that my hands would work in tandem with my mind and cooking would be the driving force of both.

### SO YOU WANT TO BECOME A CHEF?

You don't become a Chef; you strive, you excel, you achieve, you commit, and you train. Your heart throbs, expediting a busy night, watching your creations sail out the door with your signature attached. I can't remember the most dramatic and wonderful experience in my career that quantified my decision. There have been so many over the years, and so many more are to come. I can select examples to reflect on that are not only highlights but events I am extremely proud of: as a young apprentice, at my first culinary competition, being critiqued by the great Herman Rush, as he gave me his dignified nod of approval; when Rene Mettler invited me to carve ice in Japan; when Fritz Sonnenschmidt welcomed me to the 1988 Big Apple Chapter Olympic Team; four years ago, when my longtime colleague and friend, Brad Barnes, CMC, and I teamed up to open our own culinary concepts company, B&B Solutions.

There have been honors and accomplishments that both reinforced my choice and reenergized my desire to be a chef.

The motivation for this book is to share the journey of a profession and the wisdom, knowledge, and guidance of today's industry leaders. As a new culinarian, it is essential to map a career path, to understand your many choices so you avoid taking unnecessary turns that hinder your success. You have chosen a wonderfully rewarding field that can take you to places you never imagined possible. Now, keep these fundamentals in your pocket for the passage that lies ahead: Listen. Remember. Write Down. Ask Questions. Practice. Learn the Basics. Develop a Style. Stay Humble. Keep Listening. Continue Your Education. And most important: Give Back to Others.

In Food we Trust!

## Welcome

### by Brad Barnes, CMC, CCA, AAC

It really all started with a love for food. At a very young age, I remember being amazed that the things grownups called fruits could be so incredibly tasty and even grow on trees. The things we could eat were infinite and all had a life of their own; nothing tasted or felt quite like anything else. One of my earliest gourmet meal experiences was at the Officer's Club in Fort Benning, Georgia, at the age of five, when I dined on escargot and frogs' legs. It was of course not until later that I realized, in my grandmother's kitchen, there was a real plan, a special way to do things, so that food might taste so good. I was puzzled and intrigued by the way hot oil made meat or potatoes turn brown, and that brown by itself tasted different from the inside of the food. I wondered why my grandmother put milk on top of the fridge with old biscuits and corn bread. She told me so the milk would curdle and the bread would get hard. But then she mixed it all up in a couple of days with sugar and eggs and baked it into an indescribably delicious product she called bread pudding. It all seemed quite magical.

I don't think I ever really accepted that it was easy to make food taste good; it seemed to be so cared for by anyone cooking. Cooking or "fixing" a meal appeared to be a process and always done in very specific ways for each item to be prepared. Of course I now know that the methodology behind this application of heat to edible goods in order to render them palatable is far more complicated than my childhood mind could have ever guessed. But I felt it was all in the way it was done, and not much else could affect the outcome. This notion was reinforced by watching Julia Child and the Galloping Gourmet along with the Three Stooges or Bugs Bunny, all of which had a profound influence on my being as a chef and person. By my early teens I was preparing foods myself and trying to refine the methods I used to control the product's behavior. I read different sources on the same subject, drew my conclusions, and made my own adjustments to the printed cooking methods.

When I needed to get a job, I discovered I could actually be paid to cook! So I got a job at the age of 15 in a high-volume Italian-style restaurant. I was in charge of the salad bar for a bit, and then I moved through the kitchen at each station, loving every step. When I went off to college, my need for money increased, and so did my interest in cooking. I secured a job as a pantry cook and again moved through the stations. At some point (this is not a career path recommendation) I decided that cooking was much better than college, and I quit school. Inside of a year, I had become a reasonably integral part of a restaurant team and was asked to help open a new project in the suburbs. I started as a supervisor and within six months the Chef and Sous had bolted and I was made Chef. Talk about a rude awakening! I toiled as never before to succeed at this task, and stayed for five years.

At this point I attained my culinary education, developed goals, and began to move forward with my career. I became a competitor and began to garner medals at competitions with my sights on the Culinary Olympics in Germany. I went, won perfect gold, and continued working toward my goal of becoming a Certified Master Chef, which I attained in 1996. Another eight years have passed, along with three more Olympics, and I sit writing to you about the fantastic trip I have experienced, all in the name of food.

I must convey to you that this book was primarily written to outline the many possibilities in our industry and to be an aid in planning your career. You should realize, however, that no matter what, to be really good, there is a secret requirement, a special skill you will need, an ability held dear by those who possess it: You must be a good cook.

I do not want to try to quantify why this business held my heart so strongly, but I can tell you that few things in this life have grasped my soul as the knowledge of how to make people (and myself) feel good through the art of cooking. Of course, the science involved is intriguing, and the art of the process and all its infinite parts form a never-ending challenge, but I still think that the people we affect are the true reward of mastering this craft. When you can gather a group of friends to enjoy your work or serve someone confined to a hospital bed whose only pleasure is your meals, then you know what the real satisfaction of this business is. To demonstrate to a class of young cooks the way egg whites become the body of a perfectly risen soufflé, or how the roux changes to a nutty brown before it's used to build a silky espagnole sauce, will help you realize how precious our craft is. As you watch the bride and groom sit in their special places enjoying your heartfelt labor and realize you are a big part of so many special times in so many people's lives, then you will know what you have done with your life. And as you sit back and remember the fragrance of your grandmother's home just before dinner, when all you did was play and eat, or the smell of breakfast when you rolled out of bed, it will truly be undeniable that to be a good cook you must be one part artist, one part scientist, and one part magician.

## The Culinary Professional

People who work as Chefs and in other positions in the culinary field are considered culinary professionals. A *professional* is someone who works in a specific type of occupation referred to as a *profession*. Fields such as teaching and nursing are also referred to as professions. A profession has the following characteristics.

- It requires mastery of a specialized body of knowledge and an extensive period of training.
- It provides a service to society.
- It maintains high standards of conduct and competency for its members.
- It offers its members opportunities to become certified or licensed and to be active in professional organizations.
- It asks members to be committed to lifelong learning or continued study.

As you can see in Figure I-1, the culinary profession meets these requirements by being service-oriented and having a body of knowledge, code of conduct (Table I-1, from the American Culinary Federation), certification programs, and lifelong learning.

The American Culinary Federation (ACF) is the largest and most prestigious organization dedicated to professional chefs in the United States today. Their mission is as follows.

It is our goal to make a positive difference for culinarians through education, apprenticeship, and certification, while creating a fraternal bond of respect and integrity among culinarians everywhere.

**Figure I-1**
The Culinary Profession

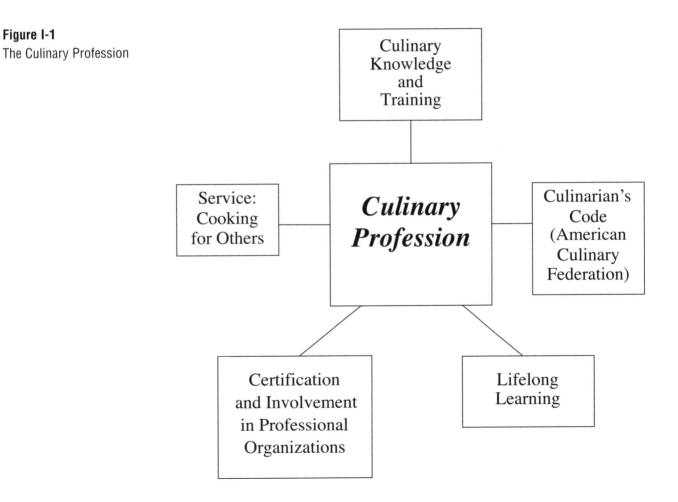

| **Table I-1 Culinarian's Code (American Culinary Federation)** |
| --- |
| I pledge my professional knowledge and skill to the advancement of our profession and to pass it on to those that are to follow. |
| I shall foster a spirit of courteous consideration and fraternal cooperation within our profession. |
| I shall place honor and the standing of our profession before personal advantage. |
| I shall not use unfair means to effect my professional advancement or to injure the chances of another colleague to secure and hold employment. |
| I shall be fair, courteous, and considerate in my dealings with fellow colleagues. |
| I shall conduct any necessary comment on, or criticism of, the work of fellow colleagues with careful regard for the good name and dignity of the culinary profession, and will scrupulously refrain from criticism to gain personal advantage. |
| I shall never expect anyone to subject himself to risks which I would not be willing to assume myself. |
| I shall help to protect all members against one another from within our profession. |
| I shall be just as enthusiastic about the success of others as I am about my own. |
| I shall be too big for worry, too noble for anger, too strong for fear, and too happy to permit the pressure of business to hurt anyone, within or without the profession. |

Adopted by the American Culinary Federation, Inc., at its national convention in Chicago, August 1957. Copyright © 2004 American Culinary Federation. All rights reserved. Reprinted with permission.

Getting certified in the culinary field is a tremendous asset and can help you advance your career. ACF offers the only comprehensive certification program for Chefs in the United States. ACF certification is a valuable credential awarded to Cooks, Chefs, Pastry Cooks, and Pastry Chefs after a rigorous evaluation of industrial experience, professional education, and detailed testing. Table 18-1 (page 355) lists ACF certifications such as Certified Executive Chef.

ACF also offers its members many opportunities to keep up to date with the latest in knowledge and skills through its monthly publications (*The National Culinary Review* and *Center of the Plate*), seminars, workshops, national conventions, and regional conferences. ACF sanctions U.S. culinary competitions and oversees international competitions that take place in the United States. ACF also accredits culinary programs at the secondary and postsecondary levels. Local chapters of ACF offer members opportunities to network with nearby culinary professionals. The ACF chef apprenticeship program is described in the next section.

Culinary professionals work in the foodservice industry, which is a part of the hospitality industry. Hospitality can defined as giving guests a generous and cordial reception, and then offering them a pleasant and sustaining environment. The hospitality industry includes hotels, restaurants, and therefore any kind of institution that offers shelter, food, and often additional services to people away from their homes. The hospitality professions are among the oldest in the world, for cooks and innkeepers alike have a long history of making guests feel at home.

Inns and taverns have played an important role in history from the ancient Greek taverns, to the European inns during the Crusades, the old English inn, and eventually the early American tavern, inn, and hotels.

## Culinary Training Programs

Formal training in the culinary field is imperative for anyone who wants to work as a chef. In the past, some chefs learned their trade through long-term apprenticeships. However, today's professional culinary programs do a lot more than just cooking and baking. Most culinary programs now offer more business courses (financial management, marketing, and so on) as well as computer training to better prepare chefs to assume leadership and managerial roles in the industry and to manage large, complex foodservice operations. Culinary training also has adapted to reflect changing food trends and eating habits. For example, chefs and cooks must know a wide variety of food preparation techniques and cooking styles. They also must know how to prepare foods to accommodate various dietary restrictions to satisfy health-conscious eating styles and to meet the needs of an increasingly international clientele. Chefs and cooks also must be creative and know how to inspire other kitchen staff to develop new dishes and inventive recipes.

Professional culinary courses are available at high schools, culinary schools, two-year colleges, four-year colleges and universities, hotel and restaurant associations, and trade unions. The Armed Forces are another source of training and experience in foodservice work. Satisfactory completion of coursework can result in the award of a certificate or diploma, depending on the program.

Certificate programs usually last from several weeks to as long as 12 or 18 months. These programs offer certificates (not degrees) upon completion. Many types of certificate programs are available for individuals looking to start a career in the culinary field and also for experienced culinary professionals looking to expand and update their skills. You can find certificate programs at many schools, colleges, and universities.

Culinary degree programs at culinary schools, colleges, and universities usually offer associate degrees, but more are also offering bachelor's degrees. An associate degree requires two academic years of study. The most common associate degree for the culinary field is the associate of occupational science (AOS) degree. Other associate degrees include the associate of arts (AA), associate of science (AS), and associate of applied science (AAS).

Obtaining a bachelor's degree requires four years of study. Depending on the program, culinary students pursuing a bachelor's degree take courses in management, marketing, finance, communications, purchasing, and cultural cooking as well as cooking and baking. If you plan to complete an associate degree first, and then pursue a bachelor's degree at another college, chances are some, but not all, of your credits for your associate degree will transfer to your bachelor degree. Try to find out early which credits are transferable.

Many culinary schools and colleges operate year-round, admitting students every few weeks. To apply, you must have a high school diploma and, sometimes, standardized test scores (SAT or ACT) and work experience (though not necessarily in cooking). Many schools have three culinary programs: culinary skills, baking and pastry, and management. Students often combine programs.

To find culinary schools in your area, you can use the Internet and the Yellow Pages. A number of online sites list culinary schools, but the only site that lists all possible culinary schools is a federal website, http://nces.ed.gov/ipeds/cool/. IPEDS College Opportunities On-Line is your direct link to nearly 7,000 schools, colleges, and universities in the United States. This website is sponsored by the National Center for Education Statistics in the U.S. Department of Education, and it is designed to help college students and future students understand the differences between schools and colleges and how much it costs to attend them. You can find out about a specific school, if you have one in mind, or you can search for a school based on location, program, or degree offerings. On this site, you can search specifically for Personal and Culinary Services programs.

Another way to become trained as a Chef is through an apprenticeship program. The ACF sponsors more than 80 culinary apprenticeship programs in the United States. ACF apprentices are required to work for three years under the supervision of a Chef and to attend school part-time throughout the apprenticeship. Apprentices gain work experience in all the kitchen stations and learn about recipe and menu development, purchasing, financial management, sanitation, and safety. Information about ACF apprenticeships is on their website (www.acfchefs.org).

Table I-2 lists questions to consider when choosing a culinary program. One of the questions to ask is whether or not the program is accredited. Accreditation is a status granted to an educational institution or a program found to meet or exceed written standards of educational quality. In the United States, accreditation is not a requirement, but most schools (beyond high school) are accredited or in the process of becoming accredited. Accreditation ensures the school or program offers a high-quality education to its students.

The Council for Higher Education Accreditation (CHEA) recognizes the following regional accrediting agencies for colleges and universities in the United States.

- Middle States Association of Colleges and Schools
- New England Association of Schools and Colleges
- North Central Association of Colleges and Schools
- Northwest Association of Schools and Colleges
- Southern Association of Colleges and Schools
- Western Association of Schools and Colleges

In addition to these regional accrediting agencies, several national agencies, including the American Culinary Federation Educational Institute Accrediting Commission, are active. Accreditation by ACF requires that curriculum, faculty, facilities, resources, support staff, and organizational structure all substantially meet the standards set by the Accrediting Commission. These standards were set by industry leaders in both culinary arts and culinary arts education. They are monitored constantly by the Accrediting Commission for accuracy in the culinary profession. The standards are the centerpiece of what constitutes a viable program.

## Table I-2 Questions to Consider When Choosing a Culinary School

Is the school in a convenient location?

Do you like the campus?

When are classes scheduled? Does the program offer a flexible schedule: day, evening, and on-line options?

What are the admission requirements?

What courses are required to complete the program?

How long will the program take to complete?

Do the students like the program?

Are the kitchens and other facilities sufficient and up-to-date?

Does the school run a restaurant, bakery, or similar operation?

Does the school offer internships or externships where you work in a kitchen for credit?

Will you get a certificate or a degree? If you get a degree, what type of degree?

Is the school accredited?

Do the instructors seem knowledgeable and approachable?

What are the qualifications of the faculty?

What is the instructor-student ratio?

How many contact hours do the students receive in the kitchen labs or other lab components of the program?

What is the job placement office like, and what are their job placement rates? What is the average starting salary for graduates?

What types of financial aid are available?

How does the school select financial aid recipients and determine your financial need?

How much debt may you incur to attend school, and how does it compare to what you expect to earn once you are working?

What percentage of students complete the school's programs? What percentage transfer out?

Who are the alumni and what kind of positions do they hold in the industry?

Culinary or foodservice programs that are ACF-accredited have been judged against the published standards. They undergo a self-evaluation and report their findings to the Accrediting Commission. The commission then authorizes a fact-finding team to visit the school to verify compliance to standards. The self-evaluation, the report of the fact-finding team, and the program response are studied by the commission, which then grants accreditation to deserving programs. Check the ACF website (www.acfchefs.org) to see if the program you are interested in is accredited by ACF. Another national agency is the Accreditation Commission for Programs in Hospitality Administration (ACPHA). ACPHA is the accrediting arm of the Council on Hotel, Restaurant, and Institutional Education (CHRIE).

Once you have started in a culinary program, make the most of it by working hard. You can also start your career off right by doing any of the following while you are in school.

◆ Get involved as a student member of ACF.

◆ Join any other culinary student organization and become an active member.

◆ Enter an ACF-sponsored student culinary competition.

◆ Go to local trade shows and conferences. Students can usually attend for greatly reduced fee.

◆ Get work experience in the field, either as part of an externship/internship or on your own.

◆ Start building a portfolio with photographs, certificates, class projects, recipes that you have tested, and so on. Document everything you do that could be useful when you are looking for a job or internship.

# How to Use This Book

*So You Want to Be a Chef?* is organized in five parts, as shown here. The first three parts discuss culinary careers in over 20 industry segments.

### PART 1. FEEDING FRENZY (CHAPTERS 1–6)

These chapters discuss culinary careers in mostly traditional and always busy settings including hotels, restaurants, cruise lines, clubs, catering, and supermarkets. At the beginning of Part 1 is a discussion of possible career paths and education path advice.

### PART 2. FEEDING THE MASSES (CHAPTERS 7–10)

These chapters tell you about careers in on-site foodservices such as business and industry, schools and universities, hospitals, retirement communities, and the military. In on-site foodservices, the foodservice is offered to employees, students, patients, residents, and the military as a needed service. At the beginning of Part 2 is a discussion of managed service companies, career paths, and education path advice.

### PART 3. HOME ON AND OFF THE RANGE (CHAPTERS 11–12)

The two chapters in Part 3 discuss culinary careers that may or may not involve as much cooking as found in the jobs discussed in Parts 1 and 2.

Learn about working as a Culinary Educator, Research and Development Chef, Private or Personal Chef, Purchasing Director, Food Stylist or Food Photographer, Food Editor, Food Writer and Author, Public Relations Manager, and even a Celebrity Chef.

### PART 4. ADVICE FROM INDUSTRY LEADERS (CHAPTER 13)

The culinary field boasts many renowned individuals who have excelled and contributed to the culinary arts profession. This chapter comprises interviews with four professionals whose stories are fascinating to read and who offer useful advice for people starting out in this exciting field.

### PART 5. MANAGING YOUR CAREER (CHAPTERS 14–18)

Once you have read about lots of jobs and listened to some of the industry's top professionals, it's time to think about your own career. Part 5 guides you, step by step, through writing your resume, putting together a job search portfolio, locating jobs, interviewing, handling job offers, and advancing your career.

This book will help you decide what type of jobs you might like (see Table I-3), what type of operations you would want to work in (see Table I-4), and how to get job offers so that one day you will land your dream job.

To help you get the most out of chapters 1 to 12, which discuss culinary careers in many settings (also referred to as segments), each chapter was organized using seven headings: The Feel, A Day in the Life, The Reality, Earnings and Outlook, Professional Organizations, Interview, Organization and Job Description.

## Table I-3 Culinary Jobs

**Baker**

Responsible for the bakeshop in a foodservice establishment. Ensures that the products produced in the pastry shop meet the quality standards established in conjunction with the Pastry Chef and Executive Chef.

**Banquet Chef**

Coordinates and supervises the cooking, plating, and setup of buffet tables for all banquet functions to ensure clients' specifications are adhered to and that the function runs smoothly and efficiently.

**Banquet Cook**

Prepares, cooks, and plates food for banquets.

**Broiler/Grill Cook**

Responsible for all grilled, broiled, and roasted items prepared in the kitchen of a foodservice establishment. Portions food items prior to cooking, such as meats and fish. Other duties include preparation of à la carte mise en place, plating and garnishing cooked items, and preparing appropriate garnishes for menu items. Responsible for maintaining a sanitary kitchen workstation.

**Chef**

Supervises and coordinates activities concerning all back-of-the-house operations and personnel, including food preparation, kitchen, and storeroom areas. Supervises cooks and food preparation personnel to ensure food adheres to standards of quality, including sanitation standards. Purchases food items, supplies, and equipment. Plans menus and food production. Controls costs. Hires, discharges, schedules, trains, and evaluates back-of-the-house staff. Meets with clients to plan special menus.

**Catering Chef**

Coordinates and supervises the cooking, plating, and setup of buffet tables for all catered functions to ensure clients' specifications are adhered to and that the function runs smoothly and efficiently.

**Executive Chef**

The department head responsible for any and all kitchens in a foodservice establishment. Ensures that all kitchens provide nutritious, safe, eye-appealing, properly prepared and seasoned food. Maintains a safe and sanitary work environment for all employees. Other duties include menu planning, budgets, forecasting control of payroll, food cost, and other culinary financials. Specific duties involve food preparation and the establishment of SOPs (Standard Operating Procedures), training employees in cooking methods, presentation techniques, portion control, and nutritional balance.

**Garde Manger**

Prepares cold foods, including salads and dressings, cold hors d'oeuvre (and sometimes hot), sandwiches, plating, desserts. Creates specialty items such as terrines, pâtés, platters for banquets and buffets, and ice carvings.

**Pantry Cook**

Responsible for all cold food items prepared in the kitchen of a foodservice establishment. Portions and prepares cold food items such as salads, cold appetizers, desserts, sandwiches, salad dressings, and cold banquet platters. Responsible for maintaining a sanitary kitchen workstation.

**Pastry Chef**

Responsible for the pastry shop in a foodservice establishment. Ensures that the products produced in the pastry shop meet the quality standards in conjunction with the Executive Chef. In an establishment, the pastry chef is usually responsible for pastries, cakes, plated desserts, friandise, breads, and bakery items. The Pastry Chef may also be responsible for decorative centerpieces such as marzipan figures, pastillage, chocolate sculptures, blown or pulled sugar. Develops recipes and prepares desserts, including cakes, pies, cookies, sauces, glazes, and custards. In large establishments, there may be an Executive Pastry Chef.

**Personal Chef**

Responsible for the preparation, cooking, serving, and sorting of foods on a cook-for-hire basis. Responsible for menu planning and development, marketing, financial management, and operational decisions of private business; provides cooking services to a variety of clients; possesses a thorough knowledge of food safety, sanitation, and culinary nutrition.

## Table I-3 Culinary Jobs (continued)

### Preparation Cook

Entry-level job working under the direction of a supervisor or Chef. Prepare ingredients by cleaning, peeling, cutting, measuring, etc.

### Roundsman

Also called Relief Cook, Swing Cook, or Tournant. Replaces other cooks as needed.

### Sauté Cook

Responsible for all sautéed items prepared in the kitchen of a foodservice establishment. Portions and prepares food items prior to cooking, such as fish, seafood, meats, and poultry. Other duties include preparing batters, sauces, soups, plating and garnishing cooked items for sautéed foods. May also prepare fried items. Responsible for maintaining a sanitary kitchen workstation.

### Saucier/Sauce Chef

Responsible for all soups, sauces, and condiments prepared in the kitchen of a foodservice establishment. Prepares stock, thickening agents, soups, and soup garnishes for a large production kitchen. Responsible for maintaining a sanitary kitchen workstation.

### Sous Chef/Executive Sous Chef

The Sous Chef is second-in-command in the kitchen, directing and managing cooks and other kitchen workers, and taking over in the absence of the executive chef. In a large establishment, the Sous Chef may be in charge of food production for one kitchen. In a smaller operation, the Sous Chef ensures that all food production workers are performing their duties as prescribed by the quality standards established by the Executive Chef. The Sous Chef assumes all the duties of the Executive Chef in the Chef's absence.

Some descriptions are adapted from *Model Position Descriptions for the Restaurant Industry.* National Restaurant Association, 2000. Reprinted with permission.

## Table I-4 Where Culinary Professionals Work

| | |
|---|---|
| Hotels, resorts, and other lodging | Recreation (such as sports arenas) |
| Cutting-edge dining | Hospitals |
| Other full-service restaurants | Nursing homes |
| Limited-service restaurants | Continuing care retirement communities |
| Bars and taverns | Community (senior) centers |
| Cruise lines | Military dining |
| Country clubs | Prisons |
| Catering companies | Bakery and pastry shops |
| Food markets | Industrial baking plants |
| Manufacturing and industrial plants | Managed service companies |
| Commercial and office buildings | Food industry |
| Schools, colleges, and universities | Television, magazines, and newspapers |
| Transportation (such as airlines and railways) | Private venues |

### THE FEEL

This section is designed to give you the big picture: who your customers are, how you serve them, who you work with. This section is written, like everything else in this book, by people who have firsthand experience.

## A DAY IN THE LIFE

Ever wonder what a typical day is like for a Chef? Well, first of all, "typical" days are rare! Every day brings different customers, new situations, unique challenges. Even so, you will read firsthand accounts of Chefs describe how they interact with others, do their work, handle problems, enjoy the camaraderie of their coworkers, and more.

## THE REALITY

Being a Chef has its challenges, and this is where we talk about them in a no-nonsense way. You know the kitchen can often be hot and steamy, the work environment like a pressure cooker, and you're required to work many hours. But we also talk about the unique realities of each segment: dealing with the executive committee in a hotel, working with hospital patients, and feeding troops in a war zone.

## EARNINGS AND OUTLOOK

For each segment, we discuss what Chefs are earning as well as the occupational outlook. For example, employment of Chefs and Cooks who prepare meals-to-go in supermarkets should increase faster than the average as people continue to demand quality meals and convenience.

In general, job openings for Chefs and Cooks are expected to be plentiful through 2012. Competition for jobs in the top kitchens of higher-end restaurants should be keen. Overall employment of Chefs and Cooks is expected to increase about as fast as the average for all occupations over the 2002–2012 period. Employment growth will be spurred by increases in population, household income, and leisure time that will allow people to dine out and take vacations more often. In addition, growth in the number of two-income households will lead more families to opt for the convenience of dining out.

Projected employment growth, however, varies by segment. The number of higher-skilled Chefs and Cooks working in full-service restaurants is expected to increase about as fast as the average. Much of the increase in this segment, however, will come from more casual rather than upscale full-service restaurants. Dining trends suggest increasing numbers of meals eaten away from home, growth in family dining restaurants, and greater limits on expense-account meals.

Chef salaries vary greatly according to region of the country and type of food-service establishment. Salaries are usually highest in elegant restaurants and hotels, where many Executive Chefs are employed, and in major metropolitan areas. Salaries also rise as the sales volume of the operation increases. Table I-5 shows the average hourly and annual wage for Chefs and Head Cooks in various industries and locations as of 2003.

## PROFESSIONAL ORGANIZATIONS

In addition to the ACF are many other culinary professional associations. Table I-6 lists over 35 more associations with their website addresses. For more specific information on each association, see the Appendix.

**INTERVIEW**

For each segment of the foodservice industry, you can read an interview with a seasoned professional who describes his or her job, work experience, and career path. They give you advice on the many ways you can achieve the goals of being a successful businessperson in the foodservice industry today. These interviews will give you insight into their real-life experiences—their many obstacles and successes, and the direction they used to achieve their career goals. These interviews should be used as a tool to let you identify your objectives for your education and employment opportunities. Use this knowledge as a guide to avoid unnecessary turns that will hinder your success and waste precious time in a career or education path.

**ORGANIZATION AND JOB DESCRIPTION**

Finally, each chapter includes an example of an organizational chart along with actual job descriptions of various culinary positions so you can see what a chef does as well as the position requirements.

**Table I-5   2003 Mean (Average) Hourly and Annual Wages for Chefs and Head Cooks**

| | INDUSTRIES | |
| --- | --- | --- |
| | Mean (Average) Hourly Wage | Mean (Average) Annual Wage |
| All Industries | $15.68 | $32,620 |
| Full-Service Restaurants | $14.80 | $30,780 |
| Hotels and Lodging | $19.04 | $39,610 |
| Cruises | $17.97 | $37,390 |
| Elementary and Secondary Schools | $12.65 | $26,320 |
| Colleges, Universities, and Professional Schools | $18.28 | $38,010 |
| Hospitals | $18.43 | $38,320 |
| Nursing Care Facilities | $16.70 | $34,740 |

| Top-Paying States for Chefs and Head Cooks | Top-Paying Metropolitan Areas for Chefs and Head Cooks |
| --- | --- |
| New Jersey | Stamford-Norwalk, Connecticut |
| District of Columbia | York, Pennsylvania |
| Washington | Monmouth-Ocean, New Jersey |
| Hawaii | Newark, New Jersey |
| Connecticut | Detroit, Michigan |

Source: U.S. Department of Labor, Bureau of Labor Statistics, 2003 National Occupational Employment and Wage Estimates.

In addition to these seven sections, we offer advice about career path and education for the Feeding Frenzy chapters (1–6) on pages 2–10. Career path and education advice for the Feeding the Masses chapters (7–10) are located on pages 122–126.

## Table I-6  Professional Associations

American Culinary Federation (ACF), www.acfchefs.org

American Dietetic Association (ADA), www.eatright.org

American Hotel and Lodging Association (AHLA), www.ahla.com

American Institute of Baking (AIB), www.aibonline.org

American Institute of Wine and Food (AIWF), www.aiwf.org

American Personal Chef Association (APCA), www.personalchef.com

American Society for Healthcare Food Service Administrators (ASHFSA), www.ashfsa.org

Association for Career and Technical Education, www.acteonline.org

Black Culinarian Alliance (BCA), www.blackculinarians.com

Bread Baker's Guild of America, www.bbga.org

Club Managers Association of America (CMAA), www.cmaa.org

Confrérie de la Chaîne des Rôtisseurs, www.chaineus.org

Dietary Managers Association (DMA), www.dmaonline.org

Foodservice Consultants Society International (FCSI), www.fcsi.org

Foodservice Educators Network International (FENI), www.feni.org

Institute of Food Technologists (IFT), www.ift.org

International Association of Culinary Professionals (IACP), www.iacp.com

International Caterers Association, www.icacater.org

International Council of Cruise Lines, www.iccl.org

International Council on Hotel and Restaurant Institutional Education (I-CHRIE), www.chrie.org

International Food Service Executives Association (IFSEA), www.ifsea.com

International Foodservice Manufacturers Association (IFMA), www.ifmaworld.com

International Inflight Food Service Association (IFSA), www.ifsanet.com

Les Dames d'Escoffier International, www.ldei.org

Military Hospitality Alliance (MHA), www.mhaifsea.com

National Association of College and University Foodservice (NACUFS), www.nacufs.org

National Association of Foodservice Equipment Manufacturers (NAFEM), www.nafem.org

National Association for the Specialty Food Trade (NASFT), www.fancyfoodshows.com

National Food Processors Association, www.nfpa-food.org

National Ice Carving Association (NICA), www.nica.org

National Restaurant Association, www.restaurant.org

National Society for Healthcare Foodservice Management (HFM), www.hfm.org

Research Chefs Association, www.culinology.com

Retailer's Bakery Association (RBA), www.rbanet.com

School Nutrition Association (SNA), www.schoolnutrition.org

Société Culinaire Philanthropique, www.societeculinaire.com

Society for Foodservice Management (SFM), www.sfm-online.org

United States Personal Chef Association (USPCA), www.uspca.com

Women's Foodservice Forum (WFF), www.womensfoodserviceforum.com

Women Chefs and Restaurateurs, www.womenchefs.org

## CD-ROM

This book is accompanied by a CD-ROM, which includes a resume worksheet and checklist, resume templates, portfolio worksheet and checklist, job log, templates for cover letters and thank-you letters, and goal-setting chart.

## Resources for Instructors

To assist the instructor, an *Instructor's Manual* (0-471-69602-1) has been developed. It includes learning objectives, a chapter outline, class activities, and test questions for each chapter in the book. In addition, PowerPoint® slides and student worksheets are available for downloading at **www.wiley.com/college.**

## Acknowledgments

We would like to thank the following people, who are not only consummate professionals, but close friends who have supported us in our journey in creating *So You Want to Be a Chef?*

**Roland Schaeffer, C.E.C, A.A.C.; Retired Corporate Chef of Heinz Foods**
For his years of knowledge and his insights into the manufacturing, research, and development fields of the foodservice profession.

**Steve Goldstein; President, Food / Thinque**
For his marketing, sales, and food marketing input in Part 3, Home On and Off the Range. His contribution to the industry has been a driving force in the field of research and development.

**David St. John-Grubb C.E.C. C.C.E. C.-I.E. A.A.C.; Corporate Chef, The JM Smucker Company**
For his charming comments on cruise line Chefs, which gave us the ability to express in writing the journey of a young apprentice aboard a grand vessel.

We are grateful to all of the interviewees for giving their time, expertise, and patience in presenting to you a taste of their career and education paths. Their commitment to our profession and their desire to give back what they have learned is an inspiration to all.

We also appreciate of the feedback of the reviewers who read the initial proposal for this book as the manuscript. They are:

Sarah Gorham, Art Institute of Atlanta

Joseph Renfroe, York Technical Institute

David Weir, Orlando Culinary Academy/Le Cordon Bleu

# Part

# 1

# FEEDING FRENZY

Here's a glimpse of what to expect in Part 1, "Feeding Frenzy."

Just 20 minutes to showtime. You've been cooking and tasting all day in your restaurant. You scan a quick panorama of your kitchen: Production is cooled and put away; cook's mise en place filled, ready, and displaying their magnificent natural colors; the cooks are hanging, telling war stories of other busy nights; sauté pans are neatly stacked like a Doric column only an arm's length away from the stove; refrigerators stocked with seasoned proteins awaiting the first click of the POS; preparations on the cold station lined up perfectly like soldiers in battle formation waiting their first command. It's a feeling you must live, a feeling mixed with pride, obsession, perfection, compulsive actions that delight your guest with satisfaction. We strive to be the best we can, and we try damn hard.

Part 1 of this book discusses culinary careers and what working as a chef is like in:

1. restaurants
2. hotels
3. cruise lines
4. clubs
5. catering
6. supermarkets

1

For many chefs, these foodservices are the heart of the industry. They are busy, noisy, unpredictable, and full of activity. Being a chef in any of these venues is a passion, an art, an obsession that must be preserved and nurtured daily.

## FEEDING FRENZY CAREER PATH GUIDE

Figure P1-1 is the Career Path Guide for culinary positions in restaurants, hotels, cruise lines, clubs, catering, and supermarkets. It is only a *guide* for you to get an idea of what type of work experience you need to move forward in your career. The Career Path Guide indicates the number of months in various positions that ideally will prepare you for an Executive Chef position and others noted at the bottom of the chart.

**Figure P1-1**
Feeding Frenzy
Career Path

## On the Way

| Nonmanagement positions* | | | | | | | |
|---|---|---|---|---|---|---|---|
| Prep cook | 8 | 12 | 12 | 8 | 8 | 8 | 8 |
| Pantry cook | 6 | 6 | 6 | 6 | 6 | 6 | 6 |
| Grill cook | 12 | 12 | 12 | 12 | 12 | 12 | 12 |
| Sauté cook | 12 | 12 | 12 | 12 | 12 | 12 | 12 |
| Expediter | 12 | 12 | 12 | 12 | 12 | 12 | 12 |
| Breakfast cook | 6 | 4 | 4 | | 8 | 8 | 6 |
| Banquet cook | 12 | 8 | 24 | 8 | 12 | 12 | 12 |
| Butcher | 4 | 12 | 6 | 6 | 6 | 6 | 6 |
| Roundsman | 8 | 8 | 8 | 12 | 8 | 8 | 8 |
| Retail counter cook | | | 8 | | 12 | | |
| Garde manger | 12 | 12 | 8 | 8 | 12 | 12 | 12 |
| Pastry cook | 8 | 12 | 8 | 8 | 12 | 8 | 12 |
| F. H. service staff | 8 | 8 | 8 | 8 | | 8 | 8 |
| **Management positions\*\*** | | | | | | | |
| Off-premises chef | | | 12 | | | | |
| Sous chef | 12 | 24 | 12 | 12 | 12 | 12 | 6 |
| Banquet chef | 12 | | 12 | 8 | 12 | 12 | 12 |
| Restaurant chef | 12 | | | 24 | 12 | 12 | 12 |
| Executive sous | 24 | | 12 | | | 12 | 12 |
| **Finish as** | Executive Chef Hotel | Cutting-Edge Chef | Executive Chef Catering | Corporate Chef Multi-Unit Dining | Executive Chef Supermarket | Executive Chef Country Club | Executive Chef Cruise Ship |

*Number of months in the position at 40 hours a week.
**Number of months in the position at 55 hours a week.

The Career Path Guide will help you understand the time commitment needed to reach many of the top management positions in the foodservice industry. By no means do we claim that these paths are the only way or that the time frames are precise. Some of you will advance much more quickly than we indicate, and some will need more time. The length of your workweek, your dedication, and your commitment are all factors in the time frame needed to obtain your final goal. Also, you may be able to hold two positions at once. For example, you could probably get experience as an expediter while working as a grill, sauté, or pantry cook. Similarly, you could be a roundsman, butcher, and garde manger at the same time.

## FEEDING FRENZY EDUCATIONAL PATH ADVICE

Figures P1-2 to P1-8 are the Education Path Advice charts for restaurants (cutting-edge and multi-unit), hotels, cruise lines, clubs, catering, and supermarkets. The Education Path Guide displays the level of importance (through the height of the balls) of formal education, knowledge, and competencies needed for positions in different settings.

When you plan your educational path, the key is to remember that education is an investment in the future on which you will draw over your entire career. You may want to go to a two-year intensive cooking degree program, then work in the kitchen to perfect your skills, and return to school for another two years or so to get a bachelors degree. Make a plan to maximize your time and your education.

The foodservice, hospitality, and culinary professions are first and foremost hands-on professions, but you should understand that they present minds-on challenges as well. As your hands carry out the precision tasks of common kitchen activities, your mind must retain, compute, and comprehend these important skills you are perfecting. As your mind absorbs the academic aspects of your profession—for example, financial reporting, computer techniques, menu costing, and menu development—the hands-on aspects make these skills come alive. The hands-on, which is the passionate and creative part of your profession, complements and motivates all facets of the work. Did you ever notice that when you are committed to a topic, you find that learning about it is easy? You're not studying, you're absorbing; you're not being tested, you're remembering.

Don't be intimidated by the Education Path Advice. Keep in mind that any academic program you need to complete will be filled with examples and situations related to the hospitality field that is your first love. Take one of the subjects in the education path that may sound threatening. For instance, food and beverage (F&B) financials may be scary because of the word *financials*, which makes it sound like the class is all math problems. This is true—except the math problems all represent real-life situations you will need to know how to handle as your career unfolds. Financials and math problems are not as scary when you encounter them in the business you are committed to and passionate about.

Continue to advance your abilities with classes that will be an asset to your career. You could learn Spanish, which is spoken every day in many kitchens. Or you could take a public speaking course to help you do cooking demonstrations or other group presentations. These investments in your future will definitely pay off.

The educational tools we have mapped out are a matrix of the many directions available. We hope you use them to inspire, plan, and evaluate a successful journey toward a rewarding culinary career.

**Figure P1-2**
Cutting-Edge Dining—
Education Path Advice

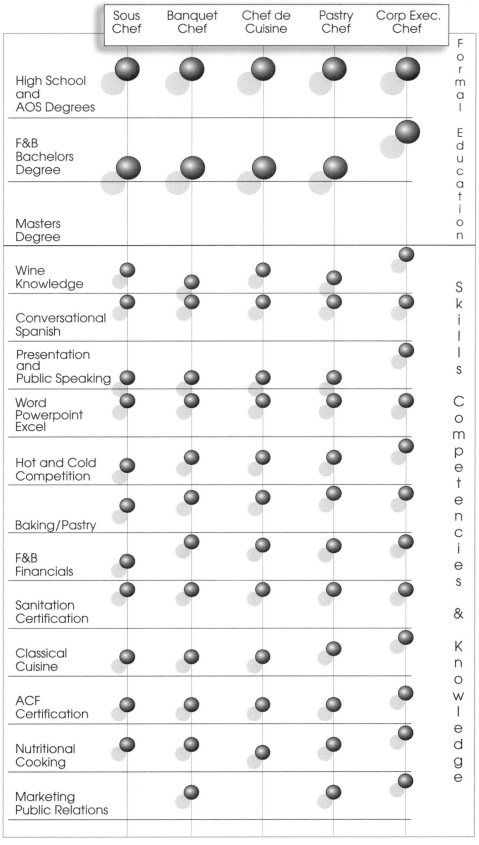

Height of ball indicates importance.

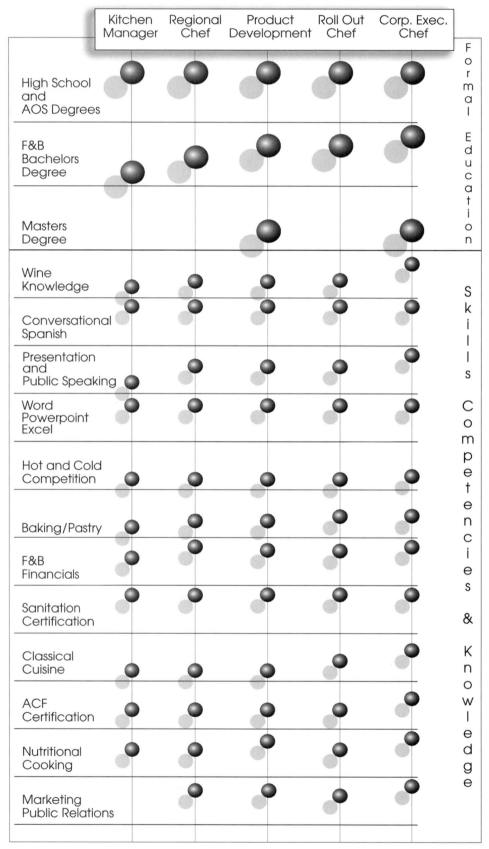

**Figure P1-3**
Multi-Unit—
Education Path
Advice

Height of ball indicates importance.

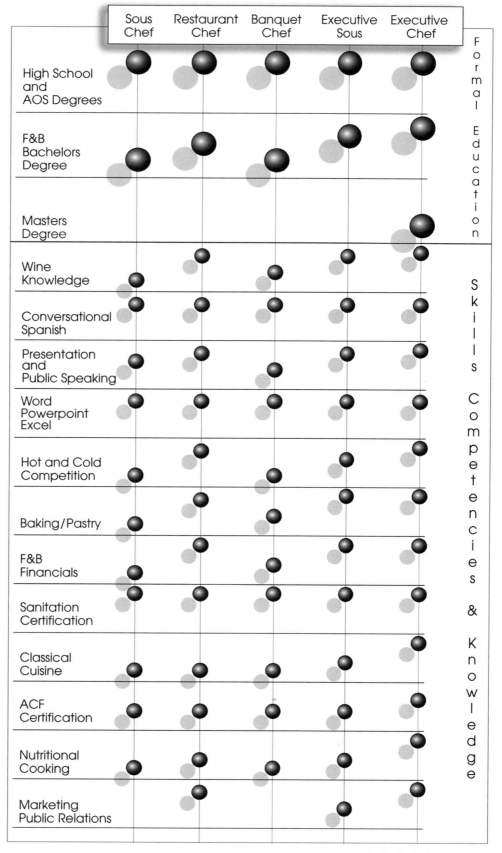

**Figure P1-4**
Hotels—Education
Path Advice

Height of ball indicates importance.

**Figure P1-5**
Cruise Lines—
Education Path
Advice

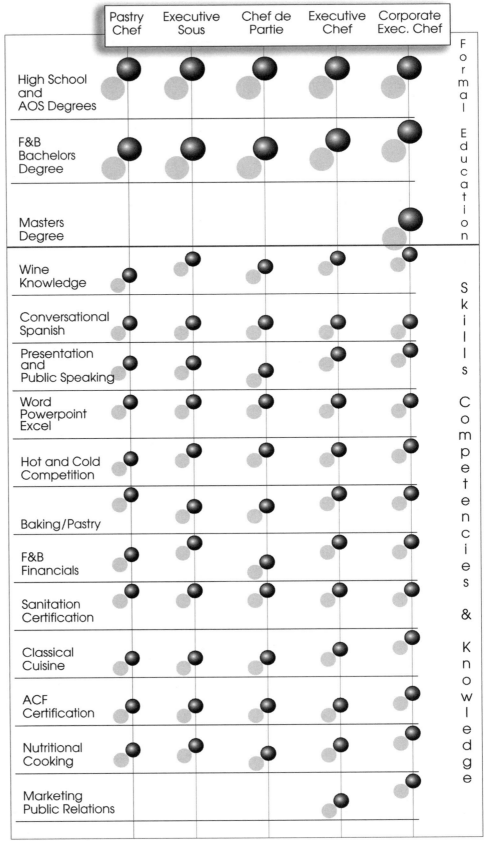

Height of ball indicates importance.

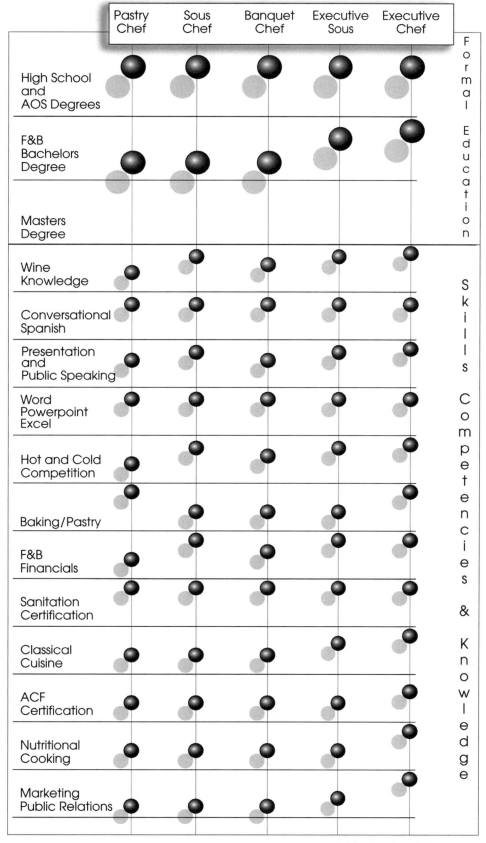

**Figure P1-6**
Clubs—Education Path Advice

Height of ball indicates importance.

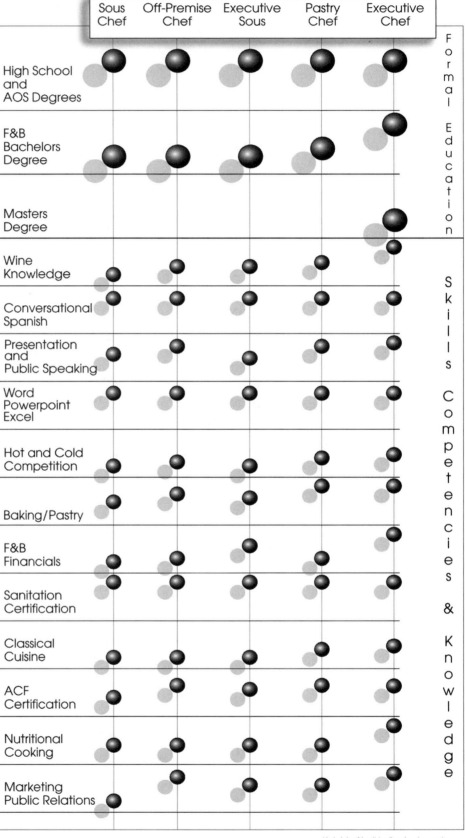

**Figure P1-7**
On- and Off-premises
Catering—Education
Path Advice

Height of ball indicates importance.

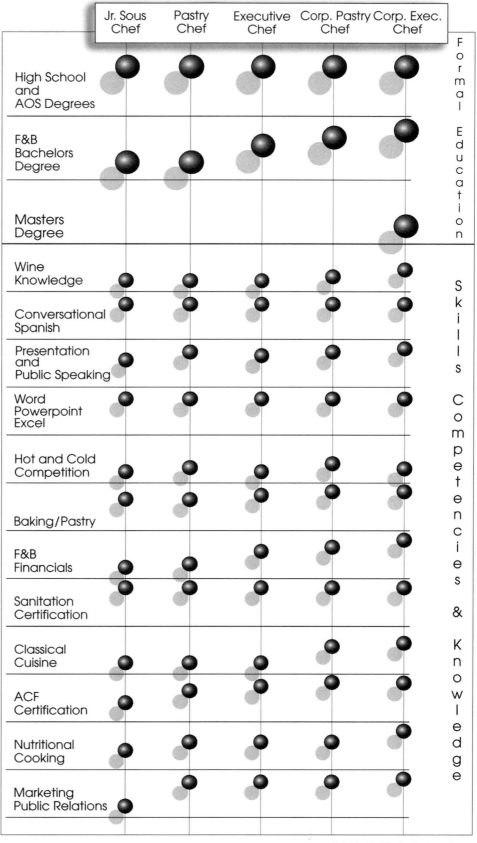

**Figure P1-8**
Supermarket—
Education Path
Advice

Height of ball indicates importance.

# Culinary Careers in Restaurants

## Introduction

**NOT LONG AGO, EATING OUT WAS RESERVED** for special occasions and celebrations. Times have changed. With more dual-income families, fast and convenient meals are a must. Restaurants are now an essential part of the American lifestyle, with Americans spending 47 cents out of every food dollar to dine out. With an annual economic impact of more than $1 trillion, the restaurant industry is huge, and it employs over 11 million people—more than any other private-sector industry.

The restaurant industry boasts over 800,000 restaurants in large cities, small towns, rural areas, and every place in between, and it presents consumers with more menu choices than ever before. About 45% of restaurants are limited-service eating places such as fast-food restaurants and cafeterias that primarily serve guests who select items and pay before eating. Full-service restaurants account for about 39%; these cater to patrons who order and are served while seated, then pay after eating.

National chains are a growing segment of full-service restaurants. These restaurants usually offer efficient table service, well-priced familiar menu items prepared by moderately skilled culinary employees, and a substantially nicer physical setting than limited-service establishments. By contrast, customers at upscale or cutting-edge dining places tend to seek a relaxed and elegant atmosphere in which to enjoy skillfully prepared food and leisurely but professional service. Cost-conscious and time-strapped guests increasingly eat at midscale or family restaurants rather than at elegant dining establishments.

You can categorize restaurants by level of service, such as limited service or full service. You can also use two culinary perspectives to roughly group restaurants. First, you can examine how much cooking is done from scratch and how much cooking utilizes convenience or premade products such as soup bases or frozen prepared foods. Second, you can look at who develops the menu and recipes. Are the menus and recipes mostly the product of corporate decisions, or is a chef solely responsible for menus and recipes?

Using the culinary perspective, most restaurants fall into one of these four categories.

1. Cutting-edge restaurants: Also called fine dining, these restaurants feature cooking from scratch almost exclusively, with the chef/owner developing the menus and recipes.

2. Upscale casual restaurants: These restaurants make most of their own menu items, and the chefs make most of their own menus and recipes.

3. Value-driven, multi-unit restaurants: The food in these restaurants is the result of corporate decision making and recipes, but it is still cooked mostly from scratch and of high quality. Examples include Houston's Restaurants and the Cheesecake Factory.

4. Chain restaurants: The restaurants in this category do little cooking from scratch, and most, if not all, of the menu is corporate driven.

As you can imagine, chefs at cutting-edge, upscale casual, and value-driven restaurants encounter more culinary challenges and tasks on a daily basis. In chain restaurants, chefs are more often used in research and the development of new products, an exciting area discussed in the "Interview" section of this chapter and in chapter 12.

# The Feel

## OPENING SCENE

The scene is a dust-filled space with raw walls and exposed ceilings. In the middle of the room is a giant unrefined board covered by crisp white architectural plans. The precise tiny blue lines represent the master plan of doorways, kitchen equipment, seating arrangements, and workflow. A voice from across the table interrupts my thinking. "Everything looks in order, Chef?" At first glance, these clean, crisp drawings would awe anyone with their professional appearance and military precision, but as I painfully understand each carefully planned line, I find the hidden mistakes most kitchens take to their graves. This was my first newly designed kitchen, and I planned to recognize those flaws with my years of working these disastrous kitchens. "No," I quickly answer without hesitation. "I need to walk the space and plan each station as if we were cooking our first meal."

| The Feel | A Day in the Life | The Reality | Earnings and Outlook | Professional Organizations | Interviews | Organization and Job Descriptions |

**15**

## THE RESTAURANT SCENE

The opening of a restaurant is like the birth of a child. You have nurtured your dreams, shaped your experience, and perfected your cuisine. Your interest in the project includes the dining room décor, color scheme, carpeting patterns, seating arrangements, bar and lounge atmosphere, not to mention the style of china, silver, and glassware. The list of kitchen smallwares grows with every breath of creativity and menu design. The months before were filled with sleepless nights when you mentally prepared menus over and over again. The final cuts were meticulously executed in a peer's kitchen or your own home, with friends and family as taste testers. Many items are classic dishes you've cooked hundreds of times—with a new twist of excitement, of course. It seems like a lifetime to get to this point—countless hours, immeasurable months, and vaporizing years—to finally hear the words of stature: "The Chef." I'm only 28, yet I've matured in a practice filled with mistakes and successes. I cooked day and night, weekends and holidays, and, usually, my days off. If I wasn't cooking, I was talking about cooking. Now, today, I have an owner who believes in my spirit, my style, my ability—and, most important, my food. Every inch of the restaurant must be perfect, the concept exact, and the menu inviting and realistic. We need to please, surprise, and satisfy with every bite. We must create an experience and supply a value so when our clients leave we hear the musical words "We'll definitely be back."

## A Day in the Life

8 AM: The phone is ringing and I feel like I'm saying hello, but I'm dreaming. The phone continues to ring. In my half-conscious sleep where the mind is willing but the body feels last night's exhausting covers, I grab the phone and say hello, wondering who dared to call this early. Static, and a voice saying, "Yo, Chef, I got the gold." I can't make out who it is or what is being said in my semi-awake state. "Who is this?" I ask. "It's Jimmy from J.J.'s, the winters are in, and you're the first to know." The sleeping chef in me comes to life, wide-eyed and fantasizing about my next creation with these morsels of black magic. "Get me a pound," I eagerly reply. I laugh as I hang up. "Truffles," I sigh, eyes open with anticipation of the day ahead. My husband lies next to me, unaffected by the episode. He is my Sous Chef, equally committed to the passion that drives us in our profession. I'll get up and let him sleep in.

The pitter-patter of little feet comes running across the house. "Mommy, Mommy, are you UP?" It's my five-year-old son, bright-eyed and bushy-tailed, getting ready with the housekeeper to go to school. I make it a point to see him off on the bus every day or let him sleep in with us a few times a month. The life of a restaurant chef isn't conducive to the stereotypical American family schedule. You eat dinner many nights at 9:30 or 10 PM; you work nights and weekends, and with luck you grab a slow Sunday off, depending on your restaurant business. You miss many family gatherings unless you can carefully plan with coverage for the Chef, which is difficult. Many restaurants close to give the Chef and crew time off for vacation. The lifestyle is addicting. Your routine, even with children, revolves around your work schedule and allows minimal compromise. So begins another power-filled day. The home is secured, the boy is off to school, and my mind races with newly imagined dishes.

## MORNING PRODUCTION

I arrive at the restaurant at 9:15 AM; the buzz of the morning's production is well underway. The scent of veal stock fills my nose with the aftertones of Lobster Bisque. Just then another scent disturbs these priceless aromas. "What's burning?" I blurt out with fear that a hard-earned sauce or soup has been trashed. "Just a burnt pot of coffee," replies a waiter. In a split second, with a dark roast in my hand, I'm creeping through every inch of mise en place—or *meeze*, as it's known in the industry—for the lunch menu. I'm called downstairs for every order arriving to weigh, touch, smell, taste, and return if needed. It's Thursday, and the weekend prep must begin. The setup for lunch is stocked and ready for service to begin. The maître d' stops to review the lunch specials, carefully taking notes and asking questions so he can instruct his well-trained staff.

## AFTERNOON RECAP

Next, a quick recap of yesterday's sales, purchases, and expenses together with another flavorful java. Price, cost, waste, utilization, and profitability are real words you must live by daily. Indulging in underutilization, excess, and extravagance doesn't work in our occupation. I love the best—but still we need to make money on every dollar earned. We are here for all the above and more. We are magicians in our field, taking simple goods and creating masterpieces. My train of thought is interrupted by the hum of lunch service. The typical business crowd demands value, quality, and atmosphere. In our 70-seat restaurant, we work diligently to please the masses by constructing cool sandwiches, innovative appetizers, assorted organic greens topped with goodies, entrées with sparkle, and desserts to die for. We want guests in the seats, not dust on the tables. The key is to understand what customers want to eat and then figure out how to entice them in a style that pleases you and suits the kitchen's philosophy. Don't ever make the mistake of thinking this is easy!

The turn from lunch to dinner is quick. There is a mellowing silence before the next meal period hits. The kitchen is madly prepping all the final details of a wonderfully planned day. We are anxious and excited, anticipating the rush of 150 reservations already on the books.

| The Feel | A Day in the Life | The Reality | Earnings and Outlook | Professional Organizations | Interviews | Organization and Job Descriptions |

**17**

## DINNER

Dinner is elegant and trendy yet comfortable, with style and class. The bar will rock with guests dining on a wide range of interesting appetizers that express the culinary path of their creator. Experience is something a chef can always fall back on, an archive preparation from years past, a specialty from a previous job. Knowledge is the secret that keeps the adrenaline and enthusiasm flowing from day to day, year to year.

The staff prepares a casual and tasty family meal. All sit and enjoy the last supper before the wave of the evening's reservations. Headwaiter Andy walks around collecting a buck from staff members predicting the evening's covers. One hundred fifty now on the books, a 20-top with a set menu at the same time as the bulk of the reservations. You get used to everyone coming at once. It's important for the staff to keep cool, calm, and collected so the dinner chaos in the back goes unnoticed out front. I review the specials with the waitstaff and do a plate presentation and tasting. The front-of-the-house staff is my link to the customer. They must be educated on my style and philosophy as well as my passion for pleasing.

Just 20 minutes to showtime, you've been cooking and tasting all day. You scan a quick panorama of your kitchen: Production is cooled and put away; cook's mise en place filled, ready, and displaying their magnificent natural colors; the cooks are hanging, telling war stories of other busy nights; sauté pans are neatly stacked like a Doric column only an arm's length away from the stove; refrigerators are stocked with seasoned proteins awaiting the first click of the POS; preparations on the cold station are lined up perfectly like soldiers in battle formation waiting their first command. It's a feeling you must live, a feeling that mixes pride, perfection, obsession, and the need to delight your guests. We strive to be the best we can, and we try damn hard.

## EVENING SERVICE

There is a flow through the evening's slam; we work meticulously to expedite each diner's request. Like a conductor in front of an orchestra, the chef controls the tempo, dictates the beat, and balances the rhythm. The line fills with dupes, table 6 appetizers for 7 people, table 8 pick up entrées for 4, table 15 VVIP special amuse, table 5 rush an extra mushroom flan, table 12 steak black and blue, Bar needs 2 appetizer samplers. When it's on, it's on, sending a cool rush of accomplishment through your entire body. The last of the dinner tickets is completed. The cooks' challenge all evening is to keep their station as organized and neat as possible. This shows their skill level and intellectual preparation. When the last ticket is out, we all high-five each other. "That was awesome," I comment. We all feel pretty high from the great success, but there is still dessert and cleanup. The hum of the dining room is still in full swing with last-minute requests, wine service, and chitchat. I make my way among the tables, greeting, smiling, and making small talk with those guests who are looking for my attention. You want to be available, approachable, sincere, but not annoying to your clientele.

It's 10:25 PM. I wonder how my son is doing. He's at Grandma's tonight, I quickly remember. I make a mental note that on my day off we will take him to the zoo. What a great kid he is. My thinking is interrupted by the maître d'. "Great job, Chef! All is well!" "Thanks! Why don't you buy my staff a beer? They deserve it," I reply. I personally walk around and thank everyone for a great evening. Dishwashers first, for they are my backbone; many of them grow into prep and cook positions. Tools are sharpened and put away, ordering finished and called in, cleanup started, prep sheets put in order for tomorrow's production. The cooks are organizing, wrapping, labeling, dating, icing fish, and carefully putting away their preparations. All seems calm; I think it's time to leave. It was a great day, but tomorrow is just around the corner. "Wait, Chef," calls out Craig, my Sous Chef. "You won the pool with 142 exact. Twenty bucks is yours." From halfway down the parking lot I yell back, "Keep it, Cheshu [his nickname]! You deserve it! See you tomorrow."

## The Reality

This segment is probably the most fun, at least for this author—a restaurant chef for about 20 years. The level of creativity and thought applied to your cooking is challenging at every turn and drives the energy of the entire staff. You set the rules regarding style, culture, and philosophy. Of course, you need to make money, so there are plenty of business concerns for which you are also responsible. Even so, the ever-present motivator is to create and serve great cutting-edge food to your guests. You are the artist, the mad chemist behind the dishes that fill your customers with pleasure and force all other matters to the back of your mind. The stove is your workshop. Life is good when it is full and the sting of the heat hits your face.

Naturally, the romance is governed by the fact that you are here to make money and so is everyone else employed by the restaurant. You must constantly keep in mind that if your art does not sell, the people who are your support and co-conspirators will go hungry. In this lies the challenge of being cutting-edge and meeting a need for a share of the market that ultimately must be substantial enough to support your work. All too often, this little detail is overlooked until it is too late and the ship goes down. The people who come through the door do so for many reasons and have any number of expectations; you as a chef must understand this well enough to hit the mark and drive sales through your heartfelt efforts.

| The Feel | A Day in the Life | The Reality | Earnings and Outlook | Professional Organizations | Interviews | Organization and Job Descriptions |

**19**

## THE GRIND

You are in by 9 or 10 AM, depending on the work and the way you organize the staff to perform your prescribed tasks and make ready for lunch service. You probably chat with the lunch sous chef and the cooks and begin to taste their work—all the sauces, marinades, dressings, and other preparations—to make sure they will represent you in a fashion you are happy with. The mise en place must be inspected, because the freshness and good condition of your groceries are paramount. Today you will work on the stove for six or seven hours, stopping to review sales reports and the day's purchases, and then glide into the blood pressure–raising rush we call service. There's not much time to waste on rest or breaks. No matter; you probably won't even know when the sun sets. You are happy to be surrounded by the group you know as "the family" and still are not sure why anyone would call this work.

## PHYSICAL ISSUES

Steam, hot oil, and particularly bubbling butter are very hot and will instantly create the pain with which you are all too familiar. Your knives are sharp; you wield them quickly, and even with the years of practice, sometimes a julienne goes astray and you slice a digit or two. It is bound to happen over and over, especially in the early years, between the rush and the nerves. It is part and parcel of your job, and after a couple hundred burns they don't really hurt so much (or that part of your brain refuses to feel them anymore!). This is a really physical job. Your hands are in ice-cold meat for 30 minutes, then over the stove in the blazing heat; next you yank the 40-pound stockpot off the stove and carry it to the sink, then bend too far over and strain your back. The body-wrenching positions never seem to end. This goes on day after day and year after year, until one day you realize you are not so young and invincible anymore, so you decide that maybe the young cook you are so diligently teaching should help you pick up that pot. Understanding the physical strains and urging your staff to pay attention to them is your responsibility. You probably should have gotten smarter sooner, but when passion is the force behind what you do, it is easy to be blind.

## THE BENEFITS FOR YOUR CAREER

The things you do stimulate all the senses in your body; your work brings pleasure to your customers and satisfaction to your staff. You are the person responsible for this business, this place where people come to nourish themselves and be part of your family for a shift. You set the tone for the staff, who work together as brothers and sisters, forgetting the occasional dispute and keeping close the success they achieve every night after service when you, the chef, gather them all, have a cold one or a glass of wine, share some food, and thank them for their attention to the night's work. By the time we all say good night, it is late, and tomorrow will be here soon. The ability to do this type of work day in and day out is testament to your capabilities and talents, which may take you elsewhere some day.

## YOUR LIFE OUTSIDE THE JOB

As in any career, you must build a balance. You have a family outside the restaurant family, and it is important that your management skills are strong enough for you to build a ship that can float without you. It is tough in a restaurant because, to varying degrees, it is built around you, your thoughts, and your abilities. If you spend time creating and teaching the culture while grooming your staff and sous chef, then you will be able to get away with minimal damage. It is ultimately necessary to focus on this aspect of your work, as the time from the job is nearly as important as the time on it.

# Earnings and Outlook

The earnings of chefs vary greatly by region and by type of restaurant. Salaries are usually highest in elegant restaurants and in major metropolitan areas, and full-service restaurants pay more than limited-service restaurants. Salaries also rise as the sales volume of the restaurant increases.

| 2003 Earnings in Restaurants for Chefs and Head Cooks | | |
|---|---|---|
| | **WAGE ESTIMATES** | |
| | **Mean (Average) Hourly Wage** | **Mean (Average) Annual Wage** |
| Full-Service Restaurants | $14.80 | $30,780 |
| Limited-Service Restaurants | $12.66 | $26,330 |

**SOURCE:** *2003 OES National Industry-Specific Occupational Employment and Wage Estimates*, Bureau of Labor Statistics, 2003.

For more recent statistics, go to a website such as **http://salary.monster.com**, and click on "Restaurant and Food Services." Then you can get salary information for chef positions throughout the United States.

Job opportunities in restaurants should be plentiful. Employment in occupations concentrated in full-service restaurants, including skilled chefs, is expected to grow slightly faster than overall employment in the foodservices and drinking places industry. Qualified chefs should be in demand.

Increases in population, dual-income families, and dining sophistication contribute to job growth in restaurants. Consumer demand for convenience and ready-to-heat meal options adds to the variety of employment settings in which chefs may work. Moderate-priced restaurants that offer table service will afford increasing job opportunities as these businesses expand to accommodate both an older and more mobile population and families with young children. Fine dining establishments, which appeal to affluent, often older customers, also should grow as the 45-and-older population increases rapidly. The number of limited-service and fast-food restaurants that appeal to younger diners should increase more slowly than in the past.

# Professional Organizations

General culinary associations, such as the American Culinary Federation, are discussed in Appendix A. This section addresses associations for culinary and foodservice professionals working in restaurants.

## NATIONAL RESTAURANT ASSOCIATION AND THE NATIONAL RESTAURANT ASSOCIATION EDUCATIONAL FOUNDATION

### *Who Are They?*

The National Restaurant Association is the leading business association for the restaurant industry. Together with the National Restaurant Association Educational Foundation (NRAEF), the Association's mission is to represent, promote, and educate (through reports, publications, research, training materials, networking, etc.) the restaurant/foodservice industry.

The NRAEF is the not-for-profit organization dedicated to fulfilling the educational mission of the National Restaurant Association. The NRAEF is the premier provider of education resources, materials, and programs to attract, develop, and retain the industry's workforce. Examples of NRAEF programs include ServSafe® food safety certification, the Prostart® School-to-Career program, the ProMgmt.® certificate program, and the Foodservice Management Professional (FMP) certification program.

### *Where Are They?*

National Restaurant Association
1200 17th Street NW
Washington, DC 20036
202-331-5900
**www.restaurant.org**

National Restaurant Association Educational Foundation
175 West Jackson Boulevard, Suite 1500
Chicago, IL 60604-2814
800-765-2122
312-715-1010 (in Chicago)
**www.nraef.org**

### *Publication (through the National Restaurant Association)*

Monthly online magazine: *Restaurants USA Online*

### *Certification (through the National Restaurant Association Educational Foundation)*

Foodservice Management Professional (FMP)

## WOMEN CHEFS AND RESTAURATEURS (WCR)

### *Who Are They?*

The Women Chefs and Restaurateurs (WCR) promote the education and advancement of women in the restaurant industry. Their goals are exchange, education, enhancement, equality, empowerment, entitlement, environment, and excellence. WCR offers a variety of networking, professional, and support services.

### *Where Are They?*

304 West Liberty Street, Suite 201
Louisville, KY 40202
877-927-7787
**www.womenchefs.org**

### *Publication*

Quarterly newsletter: *Entrez!*

## Interviews

### WALDY MALOUF, Chef/Co-owner, Beacon Restaurant

#### Q / Can you tell me a little bit about the history of Beacon?

*A: We opened the first Beacon five years ago in Manhattan. I was the Director of Operations and Executive Chef at The Rainbow Room with the Baum Emil Group at the time when we decided not to renew our lease for various reasons. Mostly it was just too expensive. Joe Baum and the group came to me and I told them I wanted to open two restaurants, a casual one, and then later on, a more formal one.*

*The first restaurant, Beacon, would be casual and based on the sort of foods that I enjoyed and had experienced in my travels to places in the Mediterranean and South America. I had this concept of high-quality ingredients, simply prepared yet still sophisticated, and cooked with open-fire grills, wood-burning ovens, and rotisseries. At the time, the market seemed to warrant something more like that as opposed to a more formal, expensive restaurant. The Baum Group agreed, and I became a partner with them. We opened Beacon to excellent reviews. We've enjoyed the good fortune of success. We just got over the five-year hump and I know it will continue to be successful.*

### Q / I know you're involved in more than Beacon. Tell us what else you do.

*A: In addition to operating Beacon, I do a lot of other things. I'm in the process of opening a small pizzeria on Sixth Avenue. This is a prototype pizzeria using a wood-burning oven.*

*I've written two cookbooks, and I'm in the process of writing a third one. The first cookbook,* Hudson River Cookbook, *was based on my cuisine at the Hudson River Club, where I used products and produce from the Hudson Valley.* High Heat *is about grilling and roasting, the common cooking methods at Beacon Restaurant. It is geared toward the home cook more than the first book. The publishers have asked me to do another grill cookbook, in a dictionary format listed by ingredient. Lastly, I am also working on* Slow Burn, *which is about braising and slow cooking.*

*I have a consulting company called Waldy Malouf Hospitality Concepts. I work on independent consulting jobs for food companies, restaurants, hotel chains, resorts, airlines, grocery stores, and food production. I am consulting for a wine bar/restaurant in Chicago and Northwest Airlines. I am involved in numerous organizations, Windows of Hope being one of my personal favorites. I'm involved with a number of schools and charities. I am an advocate of giving back and being part of the community. It's something that you do for more than just the networking or the publicity.*

### Q / What steered you into this business?

*A: When I was 13, I wanted to buy a motorcycle, so I went to work in a pancake house as a dishwasher. I quickly changed positions when I found out I could make $1.00 more an hour if I worked the pancake grills. I liked working the grill.*

*I have always been interested in food. I had a Sicilian grandmother and a Lebanese grandmother and grandfather. My mother is a New England farm girl. I grew up on a farm my family owned in Massachusetts. We later moved to Florida. We kept the farm, and visited every summer. I had a natural affinity with the farm, animals, and the food my family cooked there. I enjoyed cooking, but at the time it was not considered a serious profession.*

*When I was 15, I worked as a dishwasher at a large country club that opened near our home in Florida. The kitchen was run by a French chef and a German chef who had trained in France. They had recently come to the United States, and were looking for an apprentice. They offered me the apprenticeship.*

*It paid more than washing dishes, so I took the job. They took me under their wings, and I enjoyed it. I would work the breakfast shift before school and then I would come back and work an early dinner service after school. I did this all through high school and never thought anything of it. It was a way of earning money. I continued to explore different positions within the kitchen. I enjoyed the camaraderie of the kitchen, the work, and the work ethic.*

*Although I did well in school, I was never the studious student type. When I was about 17, a new clubhouse at the country club opened up and I was put in charge of about 30 people. It was only open for breakfast, lunch, and dinner on the weekend. It was in this position that I realized I was good with people. I had learned from other people how to supervise, work with people, and get people to do what you needed them to do.*

*Then I graduated high school, and went to college with the idea that I was going to follow in my father's footsteps. I went into pre-law but continued to work in restaurants. I moved to Tallahassee to attend Florida State for my second year. I went to work for another country club that had a classical chef. One night we hosted a dinner for the Confrérie de la Chaîne des Rôtisseurs (a professional association). A representative of the Culinary Institute of America was there drumming up the students. He came into the kitchen and talked us into coming to visit the school. I visited the school out of curiosity. I loved the facility and program so much that I decided to enroll. My parents basically disowned me because I quit college and went into a career that at the time they did not believe was acceptable. I had a strong attraction to culinary arts. Thinking back now, I believe it was the discipline that attracted me.*

**Q / What did you do after graduation?**

*A:* *I graduated in 1975 and took a temporary job as a Banquet Chef at the St. Regis Hotel in New York City. Then I worked as the Assistant Saucier at the Four Seasons, and became the Saucier a couple of months later after the Saucier quit. That was one of the greatest jobs I ever had. The Four Seasons was a very professionally run restaurant and kitchen. Everything was prepared from scratch, and menus were changed four times a year and were ingredient driven. The food and service were serious business. I worked with a high-level professional team of people. It was here that I realized that as much as the cooking was important, so was the front of the house, service, marketing, promotion, and advertising. The food is a major part of a restaurant, but it is still just a part of the restaurant. I also saw that anybody that was very successful, no matter what their position was in the restaurant, always acted like they were owners. They didn't treat it as if it were just a job. It was a lifestyle.*

*After working at the Four Seasons, I was offered a position to open a restaurant in Key West, Florida, as the Executive Chef of a big restaurant at the Pier House Hotel. I put together a kitchen, the menus, and got involved in the design, construction, and marketing. From Florida, I went back to New York to work as the Executive Sous Chef at the newly opened King Cole Room at the St. Regis Hotel.*

**Q / It's interesting that you haven't said that you ever interviewed for a job. You never had to look for a job. You had a network and connections, so you fell into jobs.**

*A: I've interviewed but not applied for jobs.*

*I stayed at the St. Regis for about a year and then became the Chef of a new restaurant, Christopher's on East 63rd Street. From there, I worked as the morning Sous Chef at Le Côte Basque, one of the oldest, best-known, and finest New York City restaurants. It was an extremely competitive work environment.*

*I took a position as Chef in a small 45-seat restaurant by the United Nations that had no written menu. Basically, the menu changed daily and consisted of four appetizers, main courses, salad, and desserts. We offered a four-course lunch and four-course dinner. The kitchen staff consisted of a dishwasher and myself. It was truly an interesting experience. I worked there for about a year, basically by myself. It was this solitary time that helped develop my cuisine. There were no menu restrictions other than it had to be good. There was always a fish, bird, and meat on the menu. You did everything yourself including desserts and butchering. There was no one there to help you so discipline was crucial. I developed my own style and perfected kitchen organization.*

*A year later, I was offered a position at La Crémaillère, a French country restaurant in Banksville, New York. I stayed there for seven years.*

**Q / Did you wish you had done your bachelor's degree?**

*A: I still wish today I'd done a bachelor's degree. I have taken business courses at New York University, the New School, and art classes at the Japanese Institute and the Japanese Society. I have educated myself, but I probably would have been able to get into ownership sooner if I had taken the business courses earlier. It has taken me a really long time to get where I am today.*

**Q / What kind of advice would you give young culinarians?**

*A: Education is incredibly important, as is your ability to work with other people. As much as the chef is involved in the cooking, he/she also has to acquire the skill of working well with people.*

*Knowledge of history, finance, cuisines, and people from around the world has to be much more today than in years past. I would get a solid education before venturing on my career path. There are so many things that I never thought would be part of my job, from architectural design to colors to music, finances, and loans. I could not have accomplished anything that I have without the cooperation and assistance of hundreds if not thousands of people during my career. The ability to lead people along will get people to work with you. Your success is almost guaranteed because you can count on other people to help you develop a great team. You can't do it by yourself. So develop leadership. Those leadership abilities are something that should be taken seriously as part of your training and part of your education.*

### Q / What trends do you see developing?

*I think one of the things that didn't exist that much before was how much the public and consumers look to the food industry to influence what people eat and what they cook and eat at home. The reason we have better restaurants in the United States today is mostly because customers want better restaurants and are demanding better.*

*There is a thirst for knowledge about food that didn't exist 20 years ago. I think it is ready to blow across the country. Just look at the Food Network, magazines, the higher profile of American chefs. People want to know where their food is coming from, who is cooking it, and why it is being cooked a certain way. People across the board are taking more interest in what they are eating and where it comes from. And I think that is going to overall have an increased impact on how we are eating 10 years from now.*

### CAMERON MITCHELL, President, Cameron Mitchell Restaurants

### Q / Tell me about your company, Cameron Mitchell Restaurants.

*We are a $100,000,000-a-year multiconcept, full-service company that operates in six states in the Midwest. We have 10 different concepts, 27 different restaurants, and a full-service, off-premises catering company. We also manage a small restaurant group including four restaurants.*

### Q / How did you get started doing this?

*I was the Director of Operations for a small restaurant company before I started on my own in 1993. I left my operations job in July of 1992. It took 14 months to put the first restaurant together. I was 30 years old at the time when we opened our first one. I started with partners who I needed to raise capital. Today we have over 180 partners, including 30 operating partners within the company. We raised money on several different occasions throughout the course of the company's history.*

### Q / How did you get into this business?

*I've always said I got fired from my first two jobs, which were mowing lawns and delivering newspapers, and I started washing dishes in high school. It was the only job I could hold. So I stuck with it. I came from a traditional suburban high school here in Columbus, where 93% of the kids went off to college. Both my brothers were doctors, but I wasn't ready to go off to a four-year college. So I lived at home with my mom, and I worked for a large regional casual dining chain based here in Columbus, Ohio. I was with them for about a year when I almost got fired. I was put on suspension for three days and probation for 30 days for being late to work. I couldn't get myself out of bed. I was just working for beer money and was kind of a wayward kid at the time.*

*After my suspension, I came back to work and I was late again. Luckily the manager on duty turned the other cheek. That was a Tuesday and thank goodness he did, because on Friday, I was working a double shift as a line cook during the day and a host at night, and during shift change it was just pandemonium. All of a sudden, time froze, and I had an epiphany. I looked across to the hot line and said, "This is what I want to do with the rest of my life." I went home that night and wrote out my goals. I wanted to go to the Culinary Institute of America, become Executive Chef at 23, General Manager at 24, a Regional Director at 26, Vice President of Operations at 30, and President of a restaurant company at 35. That was it. On Friday, I was just going to work, but the next day I was working for a career. My whole world changed. I went from the bottom of the basement to becoming an Opening Team Cordinator for them until finally I was Kitchen Manager.*

**Q / What triggered that decision?**

**A:** *I love what I am doing. This is what I'm going to do for the rest of my life. I had one more epiphany that was kind of fundamental as well. If we fast-forward a few years, I graduated from culinary school, came back to Columbus, and worked for a local restaurant company. We had one restaurant when I joined the company. I started as a Sous Chef opening up our second restaurant. I became Sous Chef for this company and then six months later I was promoted to executive chef, so I became Executive Chef at 23, just like my goal was. And I became General Manager at 24 and then became their Regional Director at 26, and then Operations Director. We had six restaurants then. As I got closer to the center of the organization, I had some fundamental problems with the company.*

*I decided then to start my own restaurant company, Cameron Mitchell Restaurants. I wanted to build a special company that had a culture and foundation that was second to none, with people that loved to work there. And that is what we are still doing today.*

*So I started with a legal pad and pen and really no money. I had a little money to live off of, but I had no money to start a business. It was a lot of hard work and perseverance and courage to go through those times. I worked for about six months on a restaurant project downtown in Columbus. I was just getting ready to finalize the lease and the plans for starting construction. I raised all the money through partnerships for this project, and the landlord went bankrupt at the last minute. The bank repossessed the building, so everything fell through. I was totally broke. I sent my partners their checks back. I explained to them what happened and sent them my new business plan for a smaller project. We were finally able to get it financed. It was a long 14 months to get the first one open. It's been a struggle, and it's still a struggle today. Our cost of goods is through the roof right now, and profitability is down from last year. When you build a restaurant you expect to pull a profit, but it doesn't always happen like that.*

**Q / Can you talk about the management philosophy of Cameron Mitchell Restaurants?**

**A:** *At Cameron Mitchell Restaurants, we believe in our associates. The foundation of our company is based upon integrity—each individual adhering to a* *code of core values that determines behavior and interactions. Treating one another in accordance with these philosophies establishes strong communication and an enjoyable work environment. Ultimately, we believe we will experience continued success by maintaining this company culture.*

*We have five questions that we ask ourselves.*

*1. What do we want to be? An extraordinary restaurant company.*

*2. Who are we? Great people delivering genuine hospitality.*

*3. What is our role? To make raving fans of our associates, guests, purveyors, partners, and communities.*

*4. What is our mission? To continue to thrive, driven by our culture and fiscal responsibilities.*

*5. What is our goal? To be better today than we were yesterday and better tomorrow than we are today.*

*Our people are the foundation of this organization. When the company puts its people first, the results are spectacular. The tools we utilize and the theories by which we operate all stem from this belief. Superior service comes from the heart. We realize that our guests will have a wonderful experience only when our associates are truly happy.*

*The value of an individual is never held higher than the value of the team. For the team to function at its greatest potential, all individuals on the team must work in harmony. It is important that no individual disrupts the positive chemistry of the team.*

*To preserve our great work environment, we hire only upbeat and positive people from a variety of cultures and backgrounds. A positive attitude is an approach to a way of life. It is a conscious choice and the driving force behind exceptional service. Everyone is responsible for fostering an atmosphere that encourages positive attitudes.*

*Work should be fun. We have an exciting work environment that is filled with laughter and smiles. Guaranteed fun equals guaranteed success.*

*Quality is built up front and permeates everything we do. We believe that there is no room for mediocrity. If we are not better than the rest, we become a commodity and are chosen only for price. We measure our quality constantly. What we choose to do, we choose to do best. Our quality is 100% guaranteed. We do it right the first time.*

*We foster open and honest communication. Communication breaks down the barriers to success. We uphold an open-door policy. Feedback creates learning, understanding, and growth. The only bad idea is the one not communicated. We communicate with respect and eloquent language. When we are all informed, we can move forward together.*

*We believe in the creative process. There is art in everything that we do, and art is important because some of the most memorable aspects of service are creative ones. Pride should be taken in even the simplest tasks. Time, people, ideas, artistic beauty, and togetherness all define a creative process, which is inherent to our success and to our organization.*

The | A Day | The | Earnings | Professional | | Organization and
Feel | in the Life | Reality | and Outlook | Organizations | Interviews | Job Descriptions

**29**

*We are committed to the growth of all our associates, our company, and our community. We believe that without the growth of our associates, we too will become stagnant. Everything changes around us and we intend to change with it, not to be left behind and forgotten. We are committed to the educational process. We believe that learning should last a lifetime.*

### Q / What kind of advice would you have for young culinarians?

*A: My recipe for success is to get the best education you can. Work for the best company, have the best attitude. Throw in a little luck and hard work, and you are going to be successful. Also, set your goals and aim high. You can change them along the way, but you have to have a map to get started. Seek advice from your elders. I do that all the time. Go and listen to them. My other advice is to work hard when you are 20 or 30 because you just can't do it in your forties. You've got kids, you've got family, you've got other commitments. Sometimes, you don't have the energy. Finally, have integrity and honesty, and manage from your heart. You treat people the way you want to be treated. I always say integrity takes years to build and days to ruin. No one can take that away from you but yourself.*

### DANIEL V. W. COUDREAUT, Director of Culinary Product Development, Metromedia Restaurant Group

### Q / What types of restaurants does Metro Media Restaurants run?

*A: Metro Media Restaurant Group has a number of chain restaurants: Bennigan's, Steak and Ale, Ponderosa, and Bonanza.*

### Q / Who do you report to?

*A: I report to the Senior Vice President of Marketing.*

### Q / What is a typical day of work?

*A: Essentially, I will come in and I'll hit my e-mails and voicemails. We have lots of market directors, franchisees, who will be communicating with me on a daily basis if needed. I act as a resource to these folks. Then, a large part of my day is spent in meetings trying to find out what the strategies are as far as where they are going with the direction of food.*

*Once I'm given that strategy in these meetings, then I will go out and start putting together the menu items and concepts that they want to have presented. They'll say, for example, that we really need to focus on chicken thigh meat because we have a contract for it. They will need x number of recipes to use chicken thigh meat in a food bar. So at that point, I develop specific recipes for that product. I spend a lot of time finding out what is the strategy, trying to coordinate all the different disciplines, whether it's purchasing, marketing, or operations. Then I go into the kitchen, develop the food, come out, take pictures, write up the recipes, work on the training materials with our training department, and move forward on a promotion. About 70% of my time is spent in our test kitchen, and a good amount of time is spent writing the recipes, costing out the recipes, and then there will be some time where I'm out actually in the field doing hands-on training shoulder-to-shoulder with our Market Directors, or what we call training the trainer.*

*I serve many masters, so to speak. I have to navigate the waters because Marketing wants the next wow product, Purchasing wants it to be cost-effective, and then Operations wants it to be very, very simple. So I have to find a product that nails each one.*

**Q / How many hours do you work a week?**

**A:** *I would say I work anywhere from 45 to 50. I determine most of my schedule. I don't punch a clock or anything like that. I work as long as it takes me to get the job done. Sometimes I'm traveling on the weekends. In order to really understand what the restaurant is going to be on a Saturday night, you have to be there on Saturday night.*

**Q / What was your career path to your current position?**

**A:** *I started as a dishwasher when I was about 14 years old. I worked in that restaurant as a breakfast and dinner cook, but mostly as a lunch cook. From there, I went into some catering at a conference center, and then worked as a line cook, sauté cook, grill cook, and shift leader at age 18 in a casual dining restaurant. Then I worked as a bartender and server. I was still doing catering on the side as a cook, so I was getting my feet wet in pretty much all of the aspects of restaurant management and operations. Then I moved to Manhattan and worked as a bartender/server. I moved over to a French bistro where I was the PM sous chef.*

*From that point, I enrolled in culinary school. I was going to get serious about this becoming my career, so I decided to go to school. For my externship, I opened up 64 Greenwich Avenue in Greenwich, Connecticut. When I graduated from school, I was recruited by Café Pacific, which is a highly rated restaurant in Highland Park Village, Dallas. I worked as their Executive Sous Chef, and then I went to the Four Seasons Hotel in Los Colinas as their Garde Manger Chef before being promoted to Chef de Cuisine of their fine dining restaurant. I was promoted again to Club Chef, and I oversaw four operations. I received most of my management experience when I was managing sous chefs and cooks, anywhere from about 30 to 40 people.*

*I was on the fast track within the Four Seasons family, but I stepped back a bit and wondered where my quality of life was, my family, and where I was going with this. I investigated the opportunity to work in Research and Development, my current position. Although I felt I had no experience in this area, it allowed me to have a better quality of life, with holidays and weekends off. That was more important, although the R&D profession has proven to be very exciting and rewarding. Essentially, the sky's the limit with where I am now. There is a lot of opportunity to improve restaurant products, and that is why they need culinary professionals that have education and experience in all culinary fields.*

**Q / What do you like most about your job?**

*A: I like the people. Restaurant people are great people. They are hardworking and dedicated. I like being an educational resource. I like being able to supply answers that are based on a foundation of knowledge. And if I don't have the answer, I know where it can be found. So it helps them to make their job easier.*

**Q / What do you like least about your job?**

*A: The pace of decision making. Coming from restaurant operations, if something doesn't work today, you don't continue to put it on the menu. You basically change it immediately. When you are in a chain restaurant operation, it takes longer for those decisions to happen. I don't like indecision or bureaucracy. It's just the nature of what it is. I think that is probably what tends to frustrate me more. On a daily basis I'd like things to move quicker.*

**Q / Give some advice to new graduates who want to go into a career of your nature.**

*A: Don't be afraid to work. Have some patience. If you want to work in research and development, you need a diverse knowledge and experience base. You have to be able to draw upon a lot to get through a day. We are running restaurants. If you think chain restaurants are any different than independent restaurants, there are very subtle differences, but we are still putting food on a plate for a guest to eat. And it happens 365 days a year, so you have got to be able to have a base of knowledge that allows you to do that. I think hard work and different experiences will help anybody trying to get into this side of the industry. I don't think you should be closed to any experience. When I first went into the hotels, did I want to go into Garde Manger? Not really. But it was probably the best experience I ever had because it was the busiest kitchen in the restaurant. I learned how to hustle, organize my time, how to read banquet event orders. I learned all of that as well as was able to move up the ladder. So you have got to be open to getting your hands dirty and taking on the hard jobs sometimes.*

## Organization and Job Descriptions

---

### JOB DESCRIPTION

**Executive Chef**

*Position Type: Full-Time/Permanent*

*Pay Basis: Salary*

*Reports to: General Manager*

**RESPONSIBILITIES:**

*Menu Planning*

> Develops dining room menus and revises twice annually.
>
> Develops catering menu and customizes menu as necessary.

*Purchasing and Inventory*

> Sets appropriate purchasing specifications.
>
> Purchases all food and supplies and ensures items are available on a timely basis.
>
> Obtains fair pricing by using the bidding process.
>
> Completes all purchasing and inventory records within guidelines.
>
> Maintains inventory controls.

*Food Production and Presentation*

> Accurately forecasts food production needs.
>
> Completes all food production records within guidelines.
>
> Coordinates activities of cooks and other kitchen personnel engaged in preparing and cooking foods.
>
> Prepares food as needed.
>
> Develops and standardizes production recipes to ensure consistent quality.
>
> Develops quality standards for food quality and presentation.
>
> Checks food quality, presentation, and portion control before meals.
>
> Ensures maintenance of cooking equipment.

*Sanitation*

> Maintains a sanitary and safe kitchen/storage environment for all employees.
>
> Implements HACCP plan.

*Financial*

> Manages/controls food costs.
>
> Prepares budgets in consultation with General Manager.

*Supervision/Human Resources*

Supervises and motivates cooking staff.

Interviews and hires cooking staff.

Trains cooking staff.

Evaluates employee performance.

*Leadership*

Serves as a role model for employees through teamwork support, attention to detail, quality control, and interactions with guests and team members.

Ensures guest satisfaction.

Ensures that employees receive fair and consistent treatment and opportunities to advance.

## QUALIFICATIONS

Bachelor's or related culinary degree is required.

7+ years experience as a Chef or Sous Chef required.

ACF-Certified Executive Chef certification preferred.

Foodservice handlers certification and knowledge of HACCP standards required.

Skilled in menu development within budgeted food costs.

Knowledgeable about plating and presentation.

Experienced in supervising and training staff.

Knowledge of Word and Excel.

Strong oral and written communication skills.

# JOB DESCRIPTION

## Pastry Chef

**SUMMARY**:

Under limited supervision, oversees food and pastry production for all food outlets and banquet functions. Develops menus, food purchase specifications, and recipes. Directly supervises all production and pastry staff. Maintains highest professional food quality and sanitation standards.

**DUTIES AND RESPONSIBILITIES**:

1. Hires, trains, and supervises the work of food and pastry production staff.
2. Plans menus for all foodservice locations, considering customer base, popularity of dishes, holidays, costs, and a wide variety of other factors.
3. Schedules and coordinates the work of chefs, cooks, and other kitchen employees to ensure that food preparation is economical and technically correct.
4. Conducts regular physical inventories of food supplies and assesses projected needs; orders all food and supplies for catering and cash operations.
5. Ensures that high standards of sanitation and cleanliness are maintained throughout the kitchen areas at all times.
6. Establishes controls to minimize food and supply waste and theft.
7. Safeguards all food preparation employees by implementing training to increase their knowledge of safety, sanitation, and accident prevention principles.
8. Develops and tests recipes and techniques for food preparation and presentation that help ensure consistent high quality and minimize food costs; exercises portion control over all items served and assists in establishing menu selling prices.
9. Prepares necessary data for the budget in area of responsibility; projects annual food and labor costs and monitors actual financial results; takes corrective action where necessary to help ensure that financial goals are met.
10. Consults with catering staff about food production aspects of special events being planned.
11. Cooks or directly supervises the cooking of items that require skillful preparation.
12. Evaluates food products to ensure that quality standards are consistently attained.
13. Performs miscellaneous job-related duties as assigned.

**MINIMUM JOB REQUIREMENTS**:

*Certificate or college degree from accredited culinary school.*

*State foodservice safe food handler's certificate.*

*American Culinary Federation certification preferred.*

*5 to 8 years of experience with some supervisory experience.*

| The Feel | A Day in the Life | The Reality | Earnings and Outlook | Professional Organizations | Interviews | **Organization and Job Descriptions** |

**35**

**CONDITIONS OF EMPLOYMENT**:

Successful candidate must submit to postoffer, preemployment physical examination/medical history check.

**WORKING CONDITIONS AND PHYSICAL EFFORT:**

Work involves moderate exposure to unusual elements such as extreme temperatures, dirt, dust, fumes, smoke, unpleasant odors, and loud noises.

Moderate physical activity. Requires handling of average-weight objects up to 25 pounds or standing and/or walking for more than four (4) hours per day.

Work environment involves some exposure to hazards or physical risks, which requires following basic safety precautions.

**Figure 1-1**
Sample Multi-Unit Dining
Organizational Chart

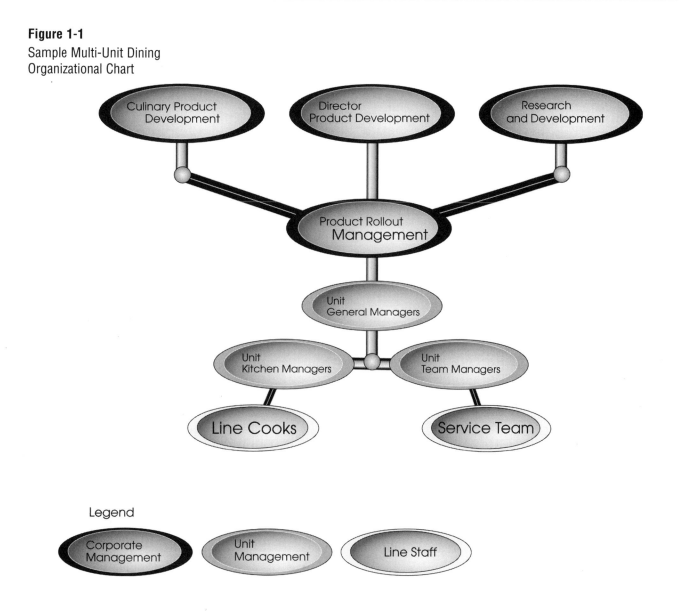

**Figure 1-2**
Sample Cutting-Edge
Dining Organizational
Chart

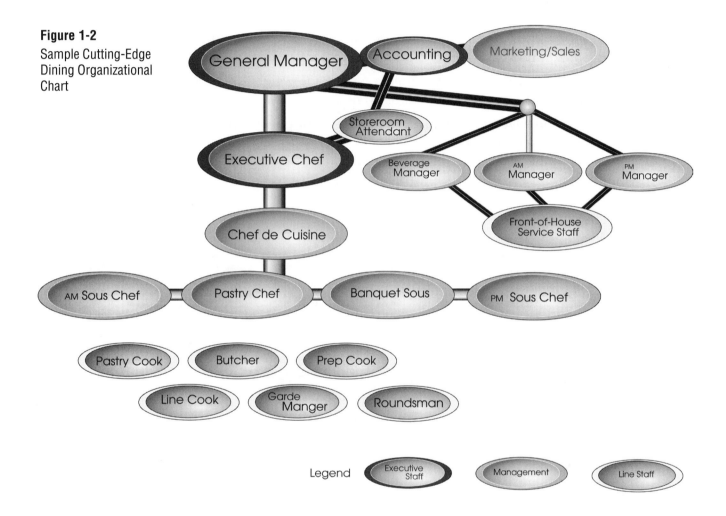

# EXERCISES

1. At the website of the National Restaurant Association (www.restaurant.org), click on "Careers and Education," then on "Get Advice." Click on "Job Possibilities" to see a listing of all sorts of restaurant positions. Which position(s) are you most interested in once you complete culinary school?

2. Using the Career Path Guide (page 2) and Education Path Guides (pages 5–6), map out a possible career and education path for a chef in a cutting-edge restaurant or a corporate chef in a multi-unit restaurant, including:
   - number of months in each job
   - each place of employment
   - how and when you would complete your formal education
   - how and when you would complete skills and competencies

3. Find two articles in recent restaurant-oriented publications about culinary trends. Summarize each article in one paragraph.

4. How would you rate the job outlook for culinary jobs in restaurants: excellent, good, fair, or poor? Explain briefly.

5. Would you consider working in restaurants? Why or why not? If yes, what type of restaurant would you like to work in, and where?

6. Look at the job listings at the National Restaurant Association website (www.restaurant.org). Click on "Careers and Education," then on "Job Bank —Find a Job." Print out two job listings that sound exciting to you and you could interview for while you are in culinary school or once you graduate. Look closely at the Required Skills and Experience section to see if you meet, or will meet, these requirements.

7. Working in restaurants is stressful, the hours are long, and you have to work many nights and weekends when most of the world is off. What would you do to ensure enough personal time to prevent burnout?

8. Go to the library and find five periodicals that relate to restaurants. List the names of each one, and compare and contrast them.

# Culinary Careers in Hotels

## Introduction

**HOTELS ARE AS VARIED AS THE MANY** business travelers and families they accommodate. In 2002, almost 30% of lodging customers were people traveling on business. About 25% were attending conferences and meetings, and another 24% were having fun on vacation (American Hotel & Lodging Association, 2003). The four basic types of hotels are commercial, resort, residential, and extended-stay.

Most hotels are commercial properties that cater mainly to businesspeople, tourists, and other travelers who need accommodations for a brief stay. Commercial hotels usually are located in cities or suburban areas and operate year-round. Larger properties offer their guests a variety of services, including coffee shops, restaurants, and bars.

Many large hotels have banquet rooms, exhibit halls, and spacious ballrooms to accommodate conventions, business meetings, wedding receptions, and other social gatherings. Conventions and business meetings are major sources of revenue for these hotels.

Resort hotels offer luxurious surroundings and a range of recreational facilities and restaurants. Resorts are located primarily in vacation destinations near mountains, the seashore, or other attractions. Some resort hotels also offer convention and conference facilities to encourage customers to combine business with pleasure.

Residential hotels provide living quarters for permanent and semipermanent residents. They combine the comfort of apartment living with the convenience of hotel services. Many have dining rooms and restaurants that are also open to the general public.

Extended-stay hotels combine features of resorts and residential hotels. These rooms in these facilities usually have fully equipped kitchens.

# The Feel

## MY FIRST DAY

As I stepped into the beautiful oasis covered with tall ficus plants, draping palms, budding begonias, and lavish lawns, I felt at home with a sense of accomplishment. The echo of running water filled the enormous glass-roofed atrium. Multicolored fish swam happily through man-made bridges and tunnels just as if we were nature. I stopped to absorb my new surroundings, which I would call "work" for the next six years.

Hotels come in boutique size, elegant and tasteful, or grand, with huge ballrooms and underground roadways traveled by electric golf carts. They are located in every country, in every state, and in every city. Many are tucked in fabulous out-of-the-way places, typically in a comfortable climate. The opportunity to travel and work in these magnificent landmarks is exceptional and well worth the journey.

## RESTAURANTS

Hotels typically have one restaurant serving three meals: breakfast, lunch, and dinner. A portion of the menu is directed by corporate guidelines, and the rest is created by the culinary staff. This arrangement is especially common in the large national brand chains that offer repeat-guest and frequent-traveler programs. Private or preferred hotels generally allow the Chef total individual creativity. Yet other hotels house multiple specialty restaurants that feature a wide variety of cuisines—traditional steakhouse or American festival, classic French, Asian flair, bistro style, or other concepts representative of the region.

## ROOM SERVICE

All hotels offer in-suite dining, or room service, with simple to extravagant choices that can be ordered around the clock. In the past decade, the emphasis on in-suite dining and an aggressive approach to customer satisfaction have made this department a particularly rewarding employment experience. Daily amenities (baskets of food gifts) have become another great revenue source for most hotels; these are ordered for guests or event clients and are carefully designed by the food and beverage department to reflect the hotel's ambiance or the area's attractions.

## CAFETERIA

Every hotel has a Team Member (employee) Café that feeds hundreds and even thousands per day, breakfast, hot lunch, and dinner. The employee cafeteria may contain full-selection salad bars, cold cut and sandwich stations, assorted beverages, grills with burgers and chicken, or hot dog wagons. The details depend on staff size, budget parameters, and corporate philosophy.

## CATERING

Catering is a hotel operation's biggest revenue source and its backbone. Catering encompasses small VIP all-day business meetings, politically focused lunches, elegant cocktail receptions, and five-course private fund-raisers with rock star entertainment. This is the heartbeat, the income that supports most of the other outlets in food and beverage (F&B). Banquet catering operations are calculated and intense, and they require constant attention. Generally speaking, catering revenues account for about 50 to70% of gross F&B income.

## BACKGROUND

The hotel atmosphere is corporate, with a team of executive decision makers at each location as well as a corporate office or owners. The commander-in-chief is the General Manager (GM), who is poised and confident and exudes years of experience mixed with expertise. The GM is supported by an executive committee that translates the direction, style, and philosophy to line-level staff with the goal of creating 100% guest satisfaction, exceeding all expectations. The culinary staff has a certain status and is perceived as talented, mysterious, dedicated, and professional. With tremendous support from other departments, the kitchen is the hotel's energy source, its heart, and the soul that dictates the rhythm of this gigantic machine.

# A Day in the Life

*"It is not the destination...but the journey that is the reward."*
Now, let's tour this amazing place, a hotel, and discover the opportunities and challenges faced every day by the Hotel Executive Chef. The work is endless, the dedication unparalleled. Skills, experience, and abilities are tested every day. It's 7 AM, and already the hum of the POS (point-of-sale equipment) is spitting tickets in two locations for the three-meal café. The waitstaff flows from the line to the exit like skaters gliding a well-rehearsed routine. The breakfast cooks move with planned steps that show their experience and flawless delivery. The Chef checks the restaurant and leaves the staff to continue their assignments.

| The Feel | A Day in the Life | The Reality | Earnings and Outlook | Professional Organizations | Interview | Organization and Job Descriptions |

**41**

## MENUS

The menus for the day are reviewed along with the prep lists and the food ordering sheets. I read each daily event to ensure that not a single request is overlooked. The Chef is the core of an operation that touches every department, from the bell service staff that greets the guests and parks their cars to the laundry staff that starches the uniforms to the concierge who recommends the cutting-edge restaurant.

## ROUTINE

As a Chef, the consistency of my routine is paramount, as I must keep track of all the outlets and operations every day of the year. My rounds of the kitchens set the tone for the day. First, a good-morning chat with the banquet cooks and a tour of the coolers and food storage to get a feel for production levels and supplies. A stint in the Garde Manger station to check quality, taste, and timing is next. Then I review the pastry department to check all parties and future events and to address any concerns. The day's events are getting under way, and the timing for the lunch rush is well under control. I stop at Stewarding, a thankless department that supports the kitchen's every move. All supplies are shined, clean, and ready for the day's functions. An older man stationed in the dish pit waves shyly and says, "Good day, Chef." I stop to acknowledge his kind words and ask how his two children are doing in school. Respect for all people in all departments is essential to great leadership. Now I'm off to the restaurants to check their setups, daily specials presentations, and quality. A quick cup of coffee in the office, some clearing of paperwork with my secretary, and it will be time to orchestrate the lunch rush. A review of yesterday's sales, payroll, and food expenses is essential to a successful operation, and it's my job to monitor and control them.

## AFTERNOON

Lunch is jamming, yet a well-trained staff can execute flawlessly in the eyes of the customers. The last meal is set before a guest, and my talented staff finishes service while I move on to the afternoon's proceedings. Wednesday is executive meeting day. The top eight people meet to go over financials, concerns, future changes, corporate specs, and important issues that need attention. The executive committee usually consists of the GM; the Directors of Human Resources, Engineering, Sales, and Operations; the Hotel Controller; the Rooms Executive; the Executive Chef; and the F&B Manager. Together we are the decision makers involved in the day-to-day operation as well as future assessments.

Thursdays I meet with the F&B Manager, and every day at 2 PM is the banquet menu reading, where I review all functions with catering and operations. Of course I must guard the financials of the operation, and one of the places where money is spent fastest is the storeroom. The purchasing department is responsible for price bidding, quality control, receiving, and tracking purchase orders. As the Chef, I oversee this process because the potential for excess expenditure is high. Now it's time for a quick meeting with my Sous Chefs to review evening business, staffing, and preparation already under way for the next day.

## THE CLOSE OF ANOTHER DAY

Almost before I can blink, the night's events are in progress, and my well-prepared staff is poetry in motion. We are rewarded daily by excellent customer feedback, countless thank-yous, and low employee turnover. I stop to look at my watch; the time has vanished! It's 8 PM, another 13-hour day has gone by, and I still have last-minute adjustments to attend to. At the peak of the season, the days blend together, and most of my executive-level colleagues work six-day weeks, as I do. Sous Chefs are scheduled five days per week, but some weeks that is impossible. In some hotels, a compensation day (an extra day off) may be granted in those situations.

At last I have got the crew settled, the meals prepared, the ordering for the next few days done, the prep lists established for tomorrow's big events, staff scheduled, and payroll submitted. The management teams are all fully informed of their responsibilities. My tall toque comes off, and I rub my face with both exhaustion and triumph. "Good night, Chef," one of my cooks says on my way out. I shake his hand and tell him, "Good job, James. See you tomorrow." As I walk down the long hallway to the exit, my mind wonders. "That's right!" I think with a rush of warmth. "I *am* the executive Chef in this magical place!"

## The Reality

You, as Executive Chef, are responsible for the activities of a great number of people and for production 24 hours a day, 7 days a week, 365 days a year. No, that is not your schedule, but it is important that you thoroughly understand how to get your employees to function well whether or not you are present. As every good executive knows, you merely drive the boat and decide the course and speed, as well as how many people it needs to operate efficiently. Management of a large culinary facility is full of pitfalls and variables, and the executive Chef needs top-notch supervisory, organizational, and people skills to make it run smoothly. Once you have accomplished this, the rewards are both personally and professionally satisfying.

Did you notice we haven't mentioned cooking yet? As you venture into upper management, the rule of thumb is to never work until everyone around you is working. Although your days of running your own station or working the line are

The
Feel     A Day
         in the Life     The
                         Reality     Earnings
                                     and Outlook     Professional
                                                     Organizations     Interview     Organization and
                                                                                     Job Descriptions     **43**

over, the challenges of nurturing and mentoring your staff, helping them become competent cooks, are endless.

## THE GRIND

The days are filled with corporate meetings, and you must be comfortable and confident taking others' opinions into consideration while intelligently voicing your own thoughts. All corporate ventures require teamwork, and you must be a team player to be successful in the culinary field. Your biggest asset is your ability to gain respect through performance, not intimidation. As you train, you will have the opportunity to work alongside many seasoned specialists; this team will challenge, support, and mentor you, and in return you will do the same for your colleagues. The knowledge you gain in this atmosphere in a single year is priceless and can only be learned on the job.

## PHYSICAL ISSUES

My feet hurt. No, my knees hurt. No, I feel it in my back. Good thing I don't have a headache! These are typical Chef complaints. The hurt comes with the territory. The pain grows with tenure, dedication, and achievement. The complaining is silent, as Chefs understand the job is physically strenuous. The injuries come because you are constantly exposed to fire, sharp objects, dangerous machinery, blazing-hot metal and steam. You will be wounded, burned, and cut countless times — never too seriously, we hope! — but all Chefs have stories that will quickly help you understand why they develop a grand respect for their tools and surroundings. You, as Chef, will be the chief protector of your employees, not only ensuring that they are in safe surroundings but that their training is thorough enough to protect them. Nothing is more heartbreaking than for one of your cooks to be injured. After all, your staff is your family, even when you are at the top of the ranks.

## THE BENEFITS FOR YOUR CAREER

You will absorb the experience of this job and soon become an integral part of the operation. You will see your effect on the grand scheme of things, and you will continue to grow every day. Your accumulated knowledge of sanitation, systematic production, feeding a well-orchestrated meal of 500 all at once, the artful creation of the perfect dish, and the ability to work with a challenging team will be put to the test daily. You must be the same talented professional whether you are tired, mad, sad, or stressed; after all, you are the coach, the general, the leader of the many cooks, dishwashers, stewards, and F&B staff who notice your every move and action. Realize that everything you do is watched and evaluated. Your demeanor, your image, is part of setting the attitude of the workplace.

### YOUR LIFE OUTSIDE THE JOB

Life outside the job is critical to survival inside the job. Have you ever heard of burnout? Too much of one thing can make a Chef a boring person. Whether it is raising a family, having a significant other, or indulging in a hobby, you must do something beyond your kitchen. Stability is key to the success of every great leader. Your time outside this type of profession is limited and somewhat dictated by business demands, which makes being a good planner even more imperative. Not to say you will have no external quality of life, but you must work very hard to strike the right balance. You must be proud of who you are on your time off. Take advantage of the perks of traveling and visiting new places. You have earned them. Downtime revitalization is essential for job and your health in this domain of unwavering pressures, stresses, multitasking, and deadlines. The trick to this part of your life is *balance*.

## Earnings and Outlook

Earnings in hotels from two sources are given here.

| 2003 Earnings in Hotels/Motels | WAGE ESTIMATES | |
| --- | --- | --- |
| | Mean (average) Hourly | Mean (Average) Annual |
| Chefs and Head Cooks | $19.04 | $39,610 |

**SOURCE:** 2003 OES *National Industry-Specific Occupational Employment and Wage Estimates,* Bureau of Labor Statistics, 2003

| 2000 Compensation in Hotels | WAGE ESTIMATES | |
| --- | --- | --- |
| | Median (Middle) Base salary | Median (Middle) Annual Bonus |
| Executive Chef | $47,500 | $4,000 |
| Chef | $36,000 | $2,750 |
| Sous Chef | $30,000 | $1,500 |
| Pastry Chef | $32,000 | insufficient data |
| Food and Beverage Director | $50,000 | $4,000 |

**SOURCE:** *National Restaurant Association. Compensation for Salaried Personnel in Restaurants, 2000,* National Restaurant Association in cooperation with Hay Group, Inc., 2001. Reprinted with permission.

The wages in these charts reflect the entire United States. Wages can rise sharply in busy urban and metropolitan areas and in highly prestigious properties.

| The Feel | A Day in the Life | The Reality | Earnings and Outlook | Professional Organizations | Interview | Organization and Job Descriptions |

**45**

Wage and salary employment in hotels and other accommodations is expected to increase by 17% by 2012, compared with the 16% growth projected for all industries combined. Over the long run, travel is expected to pick up as the economy improves and people feel more comfortable about traveling again. In addition, as more states legalize gambling, the hotel industry will increasingly invest in gaming, further fueling job growth.

Many openings will arise in full-service hotels and resorts and spas simply because they employ the most workers. Because all-suite properties and extended-stay and budget hotels and motels do not have restaurants, dining rooms, lounges, or kitchens, these limited-service establishments don't offer job opportunities for Chefs.

# Professional Organizations

General culinary associations, such as the American Culinary Federation, are discussed in Appendix A. This section discusses associations for culinary and foodservice professionals working in hotels.

## AMERICAN HOTEL & LODGING ASSOCIATION (AHLA) AND EDUCATIONAL INSTITUTE OF THE AMERICAN HOTEL & LODGING ASSOCIATION (EI)

### *Who Are They?*

The American Hotel & Lodging Association (AHLA) is a 93-year-old federation of state hotel lodging associations. The AHLA represents about 11,000 property members worldwide and assists its members with operations, education, and communications. The organization also lobbies in Washington, DC, for a climate in which hotels prosper.

The Educational Institute of the AHLA (EI), the leading source of quality hospitality education, training, and professional certification, serves the needs of hospitality schools and colleges and industries around the world.

### *Where Are They?*

American Hotel & Lodging Association
1201 New York Avenue NW #600
Washington, DC 20005
202-289-3100
**www.ahma.com**

American Hotel & Lodging Association Educational Institute
800 North Magnolia Avenue, Suite 1800
Orlando, FL 32803
800-752-4567

American Hotel & Lodging Association Educational Institute
2113 North High Street
Lansing, MI 48906
517-372-8800
**www.ei-ahla.org**

## *Publication*

Monthly publication: *Lodging*

## *Certification*

The Educational Institute offers a number of professional certifications, including Certified Food and Beverage Executive. The F&B manager must be an expert at providing quality service and have excellent leadership and organizational skills, technical proficiency, and a commitment to high standards.

Certification lasts five years, during which time requirements for recertification must be fulfilled. Certificate holders must maintain a qualifying position within the industry and earn points by carrying out various activities. The categories in which points are awarded include:

◆ professional experience
◆ professional development activities/seminars
◆ industry involvement
◆ educational services

EI provides a portfolio to help track professional development activities.

## Interview

**JOHN DOHERTY, Executive Chef, Waldorf Astoria Hotel**

**Q / What volume of business do you handle?**

*A: The Waldorf Astoria is a $55 million food sales operations hotel. About 70% of that is banquet driven. Furthermore, we support two restaurants. Oscars is a three-meal casual American brasserie that serves anywhere from 800 to 1,200 people a day. It has its own kitchen, Chef, and culinary brigade. The Bull and Bear is a steakhouse that has its own self-contained kitchen as well. Another area of food and beverage is room service that has its own à la carte kitchen for 1,400 guest rooms. We have 130 culinary staff, 7 of which are Chefs, 13 Sous Chefs, and then the remainder are Chef de Partie, Commis, Tourant, and about 10 or 12 utility personnel.*

**Q / Is a Chef a member of the hotel's executive committee?**

*A: Yes, in this company. That is very important because an Executive Chef today is first and foremost a businessperson who represents the culinary department. You need to learn how to conduct yourself as a professional busi-*

nessperson. In the eyes of the Board of Directors of a company and the corporate office, or if it's a small company the owner or General Manager, your role is to keep that business financially sound. So, controlling your expenses, payroll, food cost, and providing food that is going to make the business grow, as well as provide a profit, is your first and foremost objective. If you can do that, then you will have the freedom with your staff to buy fun china, display pieces, and interesting venues for your buffet. Without those resources, you can't improve your product. So, you have really got to be focused on the bottom line. The old philosophy of "I'm here to be a Chef and I take care of the food and the Food and Beverage Director takes care of everything else" is long gone. That is a clear road for disaster for any Chef.

**Q / When you first decided that you wanted to do this, what road did you take to get to your present position?**

*A:* The story started when I was 14 or 15, pumping gas on these long gas lines in the 1970s. It was a freezing cold winter, and I was miserable. My friend said there was a busboy opening where he worked, and I said, "If it's warm, I'm there." So I went and talked to the boss and was hired on the spot. First day at work, they sent me to the kitchen to wash dishes. I didn't care; it was warm. That lasted about two weeks, and then they kept pulling me to help them with the food prep, which I didn't mind. It was fun. Back then it didn't matter what I did if I was making money. Next I got to do cooking. It was a real mom-and-pop restaurant. They did everything from scratch. Only later on in my career did I really appreciate what they did, the pride they had. I noticed right away the pride and the care that they took in preparing food for their customers. It was inspiring, and I really loved it. The owner saw the interest that I took in their business. I noticed right away the pleasure I got in preparing food, taking the care, making people happy, and getting the positive feedback right away. I found the passion in this part of the business.

Next I got a job in another restaurant that paid more money. It was a very busy restaurant. This is where I really learned to work hard and meet the challenge of a fast business, yet still take the pride and care in everything that I did. My next-door neighbor was a professor at Nassau Community College and told me there was a school that you could attend to actually do this for a living. So I looked into it and I learned about the Culinary Institute of America (CIA), and that was it. I was a junior in high school, and that was when my decision was made.

At that point I started going to a technical high school in the afternoon. At the tech school, the Chef enrolled me in statewide competitions and doing the hotel show circuit. I really had a good jump start in my career. By the time I got to the CIA, I had already been cooking for probably close to three years and had done food shows and competitions and even won some awards. I was well prepared, which made getting through school easier. I don't remember why or how, but somebody gave me a book about the French Pavilion. That inspired me to only want to work at the greatest places.

*One of the Chefs at school in my first semester asked me what I wanted to do, and I said someday when I am 30 or 40 years old I hope to be good enough to work at the Waldorf. He said you get down there this weekend and you tell the Chef you will do anything. Tell them you will peel onions or sweep the floors, and he'll give you a job. So, that is what I did. I came down to the city, bewildered as a 19-year-old kid can be, and looked up at the Waldorf and held my breath and went to find the Chef. I told the Executive Chef, Arno Schmidt, that I had this externship program and I only want to do it here, and I'll do anything, whatever it takes. Believe it or not, he gave me a chance at it. So I came here at 19 years old between my first and second years at school and worked in all the different kitchens. It was a crazy, crazy place. It was a monster. The amount of covers that we did, we did parties up to 5,000 people, it was a crazy, exciting place, but I loved the energy. I loved the potential that it had. Every day the Chef would come out and do a special party for somebody in The Towers. It was very special food, and that is what I wanted to be part of. I did my externship and went back to school. I had it all set up that when I graduated I would come back to work here. Then I graduated on a Friday and came to work here that Monday. That was 26 years ago.*

A longer interview with John Doherty can be found in Chapter 13.

# Organization and Job Descriptions

## Hotel Executive Chef

**DIRECT REPORT**

Food and Beverage Director or General Manager

**MISSION**

To achieve food and beverage revenue, profit, and customer satisfaction goals by developing and implementing a menu that creates a luxury dining experience.

**POINTS OF INTERACTION**

*Guests*
*Employees*
*Hotel owners*
*Management team*
*The local community*
*Vendors*

**SUCCESS FACTORS:**

*Customer loyalty and satisfaction.*
*Meet or exceed food and beverage revenue, profit, and cost goals.*
*Maintain employee satisfaction and retention goals.*
*Assure a constant adherence to all quality standards and operational procedures.*

**RESPONSIBILITIES:**

### Financial

Develops and implements strategies for the kitchen that support achievement of goals.

Develops and implements business strategies for hotel that are aligned with overall mission, vision, values, and strategies.

Monitors status regularly and adjusts strategies as appropriate.

Develops the annual budget and participates in forecasting in conjunction with the controller and Food and Beverage Director.

Creates an environment for employees that is aligned with the hotel culture through constant communication and reinforcement.

### Guest/Employee experience

Communicates the service vision to and reinforces it for managers and employees.

Creates an environment designed to stimulate all senses through personal services, amenities, and experiences provided by employees.

Ensures that the hotel delivers the desired experience by reviewing hotel operations from the customer's perspective as well as from a business perspective.

Keeps current on pulse of the guests, constantly seeking opportunities to follow up on their experience.

Provides employees with the tools and environment they need to meet desired job performance goals.

Develops and implements strategies to support employee enrichment through creative learning programs and mentoring.

### Leadership

Researches customer preferences and develops a menu that is marketable to the customer base.

Creates the environment outlined by the hotel's business plan.

Incorporates regional flavors and methodologies that reflect the local area.

Instigates the culture that drives the kitchen philosophy and motivates employees.

Serves as a leadership role model for every employee and manager, fostering teamwork, attention to detail, quality control, and care for all facets of departmental operation.

Exhibits above-average personal standards in all areas related to the position.

Identifies, develops, and implements business projects through careful planning and detailed guidelines.

Demonstrates the ability to relate to, communicate with, and motivate team members to sustain acceptable performance.

### Culinary Operation

Coordinates service with restaurant and banquet operations.

Maintains product consistency by inspecting seasonings, portioning, and appearance of food.

Seeks out sources for quality product.

Monitors all perishables and meat for freshness.

Develops menu pricing structure and adherence to usage guidelines.

Oversees the inventory, purchasing, and disbursement of all supplies.

Ensures that proper sanitation practices are followed.

Recruits new staff and participates in staff selection.

Directs staff scheduling based on forecasted volumes.

Assures all controls and procedures are in compliance with policy.

**QUALIFICATIONS::**

*3+ years senior-level Executive Sous Chef/Executive Chef experience in high-quality volume-oriented environment.*

*AOS Culinary School Graduate required, Bachelors degree preferred.*

*Ability to handle multiple tasks and work well in environment with time constraints.*

*Excellent communication skills, both oral and written.*

*Ability to develop, implement, and maintain systems.*

*Ability to generate new ideas and implement them.*

*Effective leadership skills.*

*Must have state food handler's certificate.*

*Must be detail-oriented and able to perform work accurately and on time.*

*Ability to meet deadlines and work independently.*

*Ability to work flexible work schedule.*

| The Feel | A Day in the Life | The Reality | Earnings and Outlook | Professional Organizations | Interview | Organization and Job Descriptions |

51

## Pastry Chef

### Role Description

**Reports To:** Executive Sous Chef or Executive Chef

**Purpose:** To achieve food and beverage revenue, profit, and customer satisfaction goals by managing the pastry department for a five-star dining experience.

### Constituents:

*Guests*

*Employees*

*Management team*

*The local community*

*Vendors*

*INDICATORS OF SUCCESS*:

Achievement of customer satisfaction and loyalty goals
Quality of food product
Achievement of employee satisfaction and retention goals
Achievement of the hotel's financial goals

*AREAS OF RESPONSIBILITY*:

Participates in the development and implementation of business strategies for the kitchen that are aligned with the hotel's overall mission, vision, values, and strategies.
Participates in the development of the kitchen's business strategies.
Creates menu for the pastry department.
Creates an environment at the hotel designed to stimulate all senses through personal service, amenities, and experiences provided by employees.
Ensures that the hotel delivers the intended experience by reviewing hotel operations from the customer's perspective as well as from a business perspective.
Keeps current on the pulse of the guests, constantly seeking opportunities to follow up on their experiences.
Provides employees with the tools and environment they need to perform.
Develops and implements strategies and practices that support employee engagement.
Creates luxury for the senses by managing the preparation of baked goods, pastries, and desserts for a fine hotel restaurant.
Develops menu for baked goods and desserts.
Creates signature dessert items.
Creates daily specials based on availability of ingredients.
Recruits, selects, and schedules staff.
Develops and implements training plan for employees that includes culinary skills, customer service skills, and kitchen safety.

Ensures adherence to standards of food quality.

Maintains product consistency by inspecting seasonings, portioning, and appearance of food.

Ensures that unused food is stored properly in order to minimize waste and maximize quality.

Ensures that proper sanitation practices are followed.

Is responsible for bread ordering (with business fluctuation).

Is responsible for inventory and ordering of pastry supplies.

Is responsible for all pastry production, including breakfast buffet, à la carte, banquet, and room service.

*ANALYZES BUSINESS VOLUME AND PRODUCT USAGE DAILY.*

Anticipates business volume by consulting with Executive Chef.

Manages inventory of food; provides Executive Chef with recommended stocking levels and purchasing requisitions.

Keeps current with local competition; continually seeks to improve skills and culinary management expertise.

Must be a teacher, able to communicate and show staff new and current ideas.

*SUCCESS FACTORS:*

**Attend to Detail**—Ensures that data are accurate, work is thorough, and products meet the highest standards.

**Focus on the Customer**—Seeks to understand the (internal/external) customer and meet the needs of both the customer and the company.

**Drive for Results**—Works to achieve high levels of personal and organizational performance in order to meet or exceed objectives.

**Continuous Improvement**—Constantly assesses and adapts current practices to perform a task better, faster, or more efficiently.

**Information Sharing**—Provides information so coworkers, customers, and suppliers understand and can take action.

**Teamwork**—Works well in a team environment and motivates teams to sustain exceptional levels of performance.

**Development of Self and Others**—Continually works to develop own capabilities and those of others.

**Professionalism; Product or Technical Expertise**—Demonstrates the ability to apply technical, professional, or product expertise to real-world situations.

**Creative Thinking**—Develops innovative approaches and imaginative solutions that meet real needs.

**Strong Relationships**—Fosters trust and cooperation among coworkers, customers, and suppliers; develops and sustains personal contact in order to provide mutual benefit.

**Delegation**—Assigns tasks using such techniques as needs analysis, individual skills assessment, objective setting, and communication.

**Organization**—Demonstrates ability to proactively prioritize needs, put first things first, and effectively manage resources.

**Supervision, Performance Management**—Demonstrates ability to relate to, communicate with, and motivate employees to sustained high performance and quality levels.

**Planning**—Shows skill in determining whether tasks should be attempted, identifies the most effective way to complete the task, and prepares for unexpected difficulties.

The
Feel

A Day
in the Life

The
Reality

Earnings
and Outlook

Professional
Organizations

Interview

**Organization and
Job Descriptions**

**53**

**Figure 2-1**

Sample Hotel
Organizational Chart

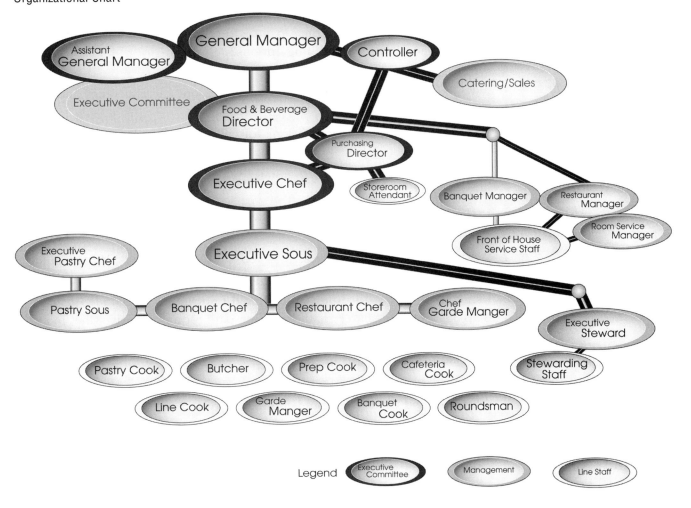

## EXERCISES

1. Describe some of the types of restaurants within hotels and give examples from properties you have visited. What types of hotels seem to offer more restaurant choices? Which offer fewer or none?

2. At a Disney property, you might find a snack bar, a fine dining restaurant, and a buffet with Disney characters. What challenges would the Chef face in each of these dining options? What is similar in each venue? Why is it important to offer several dining options?

3. To see articles from *Lodging,* go to the following website: www.lodgingmagazine.com. Read an article about current trends and issues in food and beverage.

4. How would you rate the job outlook for culinary jobs in hotels: excellent, good, fair, or poor?

5. Would you consider working in hotels? Why or why not? If you would, what type of hotel would you like to work in and where?

6. Look at the job listings at the AHLA website (www.ahma.com). Print one out that sounds exciting to you.

7. Working in hotels can be stressful—you always have guests! What would you do to make sure you got enough personal time to prevent burnout?

8. Using the Career Path Guide (page 2) and Education Path Guide (page 7), map out a possible career and education path for an executive Chef in a hotel, including:

   - number of months in each job
   - each place of employment
   - how and when you would complete your formal education
   - how and when you would complete skills and competencies

9. Go to the library and find three periodicals that relate to hotels. List the names of each one, and compare and contrast them.

# CHAPTER 3

# Culinary Careers in Cruise Lines

## Introduction

**CRUISING IS ONE OF THE FASTEST-GROWING SEGMENTS** of the travel industry. Because cruising appeals to a wide range of tastes, lifestyles, pocketbooks, and interests, cruise ships have become a popular vacation destination. The typical cruise ship carries 2,000 passengers and 900 crew members. Popular places to cruise include the Caribbean islands and the Mediterranean Sea. Many ships today offer vacationers luxurious, full-service spas and health club facilities; specialty restaurants and dining options to satisfy any taste; professionally staffed children's facilities; sports activities from golf to rock climbing; Internet centers; and many other services designed to cater to vacationers of all ages. Working as a Chef on a cruise ship is an unforgettable experience. Read further to learn about culinary careers in this distinctive work setting.

## The Feel

**Surmage**—"Listen up!"—Always said before calling the order in to the Station Chefs)

**Arboyer**—Usually the sous Chef or Chef tournant who calls the orders to the Chefs of the different stations

**Gloryhole**—The Steward who looks after the crew quarters

**Bulkhead**—A separating wall, such as a cabin wall, within a ship

## NOTES OF A YOUNG APPRENTICE

The first day of work was completely overwhelming while at the same time electric and exciting. The throb of huge engines pulsed through the massive vessel, and I heard the unusual clicking and clacking of pipes as the bulkheads began to move to the rhythm of the ship. The tugs took up the strain on the mooring lines, pulling the vessel into the main water of the dock area where we have been berthed, lining us up to go down the channel and into the open sea. The mind-awakening sound of the ship's horn penetrated deep into my skin, a wonderfully deep bellow that acknowledged the small helpful tugs. I felt a surge of excitement and the exhilaration of adventure.

It was a new beginning, being away from my homeland, meeting new people, and starting a career I had yearned for since visiting the finest hotels in England and applying for a three-year apprenticeship. My thoughts were brought back to reality as the strict arboyer began yelling out in deep, confident tones, "Surmage!" We stood at attention at our stations, ready to accommodate his every wish, hoping our presentations would pass his inspection as well as that of the Maître Chef de Cuisine. The variety of mise en place we needed to prepare was quite overwhelming to a newly hired beginner. The secrets: Don't speak unless spoken to, don't stop working unless relieved, and don't make any mistakes.

The night before we sailed had been a weary one. All the apprentices climbed into their bunks in the cabins assigned by the Gloryhole Steward. Each cabin held four men, two bunkbeds, two closets, and two drawers in which to store four months' worth of goods. That was going to take a bit of organizing for four young inexperienced mates who worked day and night and who would rather play cards in their spare time then clean or do laundry. At 2 AM we were up on the deck, cold and tired, for a mandatory safety drill emphasizing evacuation in case of an emergency. Sleep deprived, we slumbered until the crazy Gloryhole Steward started his banging, hooting, and hollering rampage at some ungodly hour, rallying the troops. The moans of my coworkers over how early it was and how tired they were could barely be heard over the banging of our illustrious Steward. But off we went to the head (bathroom) to get ourselves together, knowing full well the Chef was in the kitchen, ready to ride us for yet another day.

First and most important, we were inspected by the Maître Chef de Cuisine, then the Executive Sous Chef, then the Chef de Parties, and so on down to the Sous Chefs and the third Commis for cleaned, pressed uniforms, clean fingernails, a smooth shave, and mirror-shiny shoes. I was just a kid, 18 years old—what did I know or care about being inspected in the morning? After all, I didn't join the military. I came here for a career in cooking. Not until years later did I appreciate the military-style apprenticeship I received, which helped mold me into a professional.

I remember the first day I climbed the steep gangway abroad the liner. She was a beauty, sparkling with nearly three acres of exterior glass and containing a collection of five-star hotel activities and entertainment. The grand plaza, referred to as the atrium, was a huge three-story open area that seemed even more magnificent as I studied its details. It's 2 AM, and I am assigned to the midnight buffet, which allows me to sleep in an extra hour if I can survive the deafening sounds of the wake-up call. I've decided to start a journal not only for my culinary notes and recipes but also to describe the exquisite ambiance, the people, and the profession to which I will be exposed on this grand sailing ship. I must take a minute and write home, for they will never believe in the small town I come from the wonders I am exposed to daily. I am privileged to be aboard this mighty vessel. The experience will surely be well worth the journey.

## THE ACTIVITIES AND AMBIANCE

The impression of luxury begins when guests board the ships. Curved hallways invite exploration and curiosity; art and antiques make guests feel comfortable, yet they surprise and delight. Fine art is an integral part to the cruise experience, giving a feel of the classics with a touch of contemporary style. Guests are greeted as they arrive by an eager, well-trained professional staff passing savory hors d'oeuvre and vintage champagne to mark the start of an unforgettable experience. Imagine a full-service day spa that offers acupuncture for pain and stress management. Picture playful options such as a rock-climbing wall, jogging track, basketball and volleyball courts, Ping-Pong tables, and a nine-hole miniature golf course. Night owls can cruise to the martini lounge of the full-blown casino. The kids are all tied up with their own activities—scavenger hunts, sports tournaments, and arts and science workshops. The teen hangout is a nightclub complete with soda bar, dance floor, DJ music, and karaoke rooms—or teens can visit a video arcade packed with every game they ever wanted to play. Professional productions for the whole family to enjoy are staged in three theaters—and don't forget the comedy club, showing live performances from well-known television comics.

The ship has three pools: an indoor pool with a skylight roof, perfect for a cool evening dip; a spa pool located next to the gym, which also houses a beauty shop, massage center, sun tanning, and other health treatments; and the main pool on the sun deck. This one is adjacent to two large dining rooms serving a marvelous, delectable, and well-monitored midnight buffet. Tantalizing food stations satisfy the whim of every guest, whether off to the late-night party places or turning in before a busy day. Besides the elegant à la carte dinner that is available every night, authentic Thai, Mediterranean, sushi, and tapas restaurants offer the same quality and attention to detail the other food outlets are expected to maintain.

I had always heard of the phenomenal Garde Manger work for which cruise lines are so well known, but to witness its grandeur in person is amazing. Grand buffets are the highlight of the cruise ship experience, not only to consume but to observe and experience. To have creative control of these extravagant food stations is a privilege. Although the art of Garde Manger lost the intricate style of the classic kitchens of great hotels and country clubs, the fresh approach taken on cruise lines has revitalized these classic skills. Customers now expect a certain Garde Manger skill level on cruise ships.

## THE KITCHENS

Thirty-nine hundred assorted muffins, 400 pounds of exotic fruits, 1,600 pounds of beef, 1,430 pounds of game and poultry, 1,170 pounds of potatoes, 960 pounds of bananas, 910 pounds of ice cream, 852 pounds of shellfish, 551 pounds of butter, 390 pounds of bacon, 550 pounds of vegetables, 306 pounds of veal, 650 pounds of fruits, and 104 pounds of salmon are just some of the provisions used *daily* on this cruise ship. The space of the main kitchen is tight yet accommodating as long as everyone does his part to stay clean, neat, and organized. Everything is slick stainless steel. All the cabinets have covers so nothing slides out when the weight of the vessel shifts. The cooking battery resembles that of any hotel-size kitchen and includes everything from a Swiss brazier to double-stacked convection ovens. This is a purely from-scratch kitchen, except for some of the meals executed for staff members. Deliveries are made by helicopter, so you must order well in advance, which leaves you little room for mistakes! You must use whatever food products arrive because returns aren't an option.

In addition to the main kitchen, satellite kitchens serve the other restaurant options available to guests. You as the Chef must be in constant touch with these satellites to ensure their quality meets the same standards as the main preparation area.

The full pastry kitchen presents the same space challenges as the rest of the food prep areas. Stunning afternoon teas feature elegant pastries, intricate tea sandwiches, and assorted scones as an impressive culinary interlude before dinner service. The cool afternoon breeze and the spectacular horizon are the backdrop for these popular afternoon events. Keep in mind that guests are on a revitalizing vacation with a multitude of activities, so the rich pastries usually passed up at home are a well-deserved treat on board the ship, and they must exemplify the standard of excellence the cruise demands.

# A Day in the Life

Dining aboard these magnificent vessels hasn't changed since their inception. It is an event that lasts from the time you wake until you collapse at night, not to mention the 24-hour first-deck pizzeria. Imagine unwinding at an after-hours place directly across from the casino with selections such as gourmet goat cheese, artichoke and veal sausage pizza with organic baby greens, roasted garlic bread, and signature calzones. There are so many styles and concepts of restaurants that you are sure to find a cruise line that meets your needs. From casual to elegant, the experience is meant to be unforgettable.

## THE ROUNDS

Let's tour these intriguing outlets as they reveal the legendary culinary rituals for which most cruise lines are famous. It's 4:30 AM, and I feel as though my head just hit the pillow; I worked a mere 17 hours yesterday. This really isn't too bad, considering I get three months on and a month off to recoup. Besides, I have little or no expense so I can save, save, save. It is Day 2 of a 14-day cruise to the Panama Canal and the Chilean fjords. I have been fortunate to visit many wonderful places since I joined this cruise line five years ago.

The first place on my rounds is the main kitchen, which is the core of everything that takes place. One constant is that for the most part your crew is reliable, and unless someone falls ill, everyone is present for work. Preparation for breakfast is in full swing, with the morning sous Chef calling out checklist and preparation details to the cooks and Commis. "Morning, Chef," they shout with respect. They know what it takes to attain this position. Working on a ship quickly shows what kind of manager you are in terms of organization, philosophy, cooking skills, and production schedules. You can't fake anything in this environment, and if you aren't right on with your planning, your crew will lose all respect as they run frantically around the kitchen like a disrupted nest of ants.

The breakfast Arboyer barks out orders as the competent waitstaff ensures that all items are being put together in a timely fashion. Five seasoned breakfast cooks move in sync as they flip eggs, layer short stacks, and fold generously stuffed omelets. All quiets down for the most part by 10 AM, at which time a limited crew remains available for stragglers as the kitchen scrub-down begins. Everything must be cleaned and organized in preparation for the next meal period. You hold a brief recap with the Sous Chefs as the crew continues the endless task of meticulously preparing the next feast. Your creative assistance is needed for several VVIP suite functions involving the captain and some of our most prominent guests. You must spend a patient, well-focused moment reviewing the final product orders, staffing issues, and menu delicacies for the evening.

The ship is at full capacity with six restaurants and smaller grab-and-go outlets on every deck; these are designed to satisfy the guest on the move with an overstuffed sandwich on a ciabatta, handcrafted sushi, or salads with an abundance of toppings. The crew comes back from a short power break to plow through the festivities of the evening. All restaurants are booked with a ship full of influential clients. There's a burst of energy from everyone, placing the finishing touches on the exquisite Garde Manger work or the final chocolate filigree on the Pastry Chef's petit fours. The more experienced Sous Chefs are mentors to the younger, less experienced cooks and Commis, who shine with pride as they note their accomplishments of the day. You look around and know this is why you are in this business: the rush, the enthusiasm, the passion, the commitment, and the yearning for creativity. As the guests arrive in their beaded gowns, black ties, and tails, you can hear classical music throughout the ship's corridors, creating a feeling of wealth and sophistication. The crew is rehearsed and ready, needing little or no instruction to deliver the superb dining experience one would expect on this vessel.

## THE END OF ANOTHER DAY

At last, I have a minute alone atop the first deck to reflect on the accomplishments of the day. The cool ocean breeze eases the kitchen's heat from my face as I gaze over the same horizon witnessed by great sailors many years before. I remember hearing when I was growing up that people are drawn to water because our bodies are made up of about 75% of this magical liquid. There is a calmness when one is completely surrounded by water. The challenge of this 975-foot, 115,000-ton mass of molded beauty is intense, considering the diversity of clientele, the level of quality, and the many dining options. The days are long, hard, and demanding, especially in the beginning, when the crew is making last-minute adjustments, but the benefits combined with the gratitude are priceless. These floating resorts with their impeccable service create an atmosphere, a memory, an experience guests will remember forever. This is the essential reason I spent my early days training to become an accomplished Chef. The hum of the midnight air whispering against the water is mesmerizing as my thoughts drift into tomorrow's activities. Rest is first, before I dive into another full-scale day of events. One last look around, and I say to myself, what a great day. Then I think, no, what a great life.

## The Reality

The cruise ship is a floating amusement park, resort hotel, and dining mecca, so as you may imagine, it is no small task to get it all to function as advertised. In your charge is room service, multiple food outlets, and the special event functions. Don't forget that labor relations, product procurement, and standard operating procedures are included in that mix. You'd better have your act together or it will not come together; this is not a fake-it-until-you-make-it kind of job. You are master of the galley and must perform flawlessly to keep your young charges in line and respectful of your orders. This is perhaps one of the toughest leadership positions in the culinary industry. It is important to keep in mind that you are going to work side by side with the same people for three months, night and day. You will eat, drink, and sleep in the same tight quarters. You must maintain order as well as morale. It is a tricky spot, but given your years of experience and well-polished abilities, you can do it. Oh, by the way, I hope you don't get seasick!

## THE GRIND

Attendance is recorded, and production has begun. The day's events are demanding your attention, as are the details of supply, maintenance, and special events, which must be discussed with the Chefs in charge. It is the start of another day on board, and we have a port of call at which we restock and reevaluate our needs, then perhaps get a few moments to visit land. There is a grand lobster bake tonight, and then a dance and midnight buffet. You have a great love affair with this operation—the excitement, the globe roving—although it is also a bit lonely even when you are surrounded by comrades and your trusty staff. The routine can be monotonous, but as you know, this is true to some degree in every foodservice career. However, the monotony can be magnified by the space restrictions, the extended never-ending workdays, and the familiar ports of call.

A great benefit of working the cruise lines is that while you are on board, you really incur almost no expenses. In addition, some good tax loopholes are available because your earnings are usually based at sea. All in all, the lifestyle is unusual but rewarding once you acclimate to it.

## PHYSICAL ISSUES

Your feet and legs tire as the sway of the ship tends to strain the muscles even more than the typically long working day. You must get used to the confinement of the ship. Even though it is large, it is limited; there is no town to go out on at night, and there is no country to take a drive in. You must also grow accustomed to the constant motion. This is easy for some, but not so easy for others!

## THE BENEFITS FOR YOUR CAREER

You can gain a plethora of knowledge all in one place on a cruise ship, given your exposure to dedicated professionals, a range of food styles, and elaborate buffets. You can learn from the best, and because you are semipermanently on board, you can study intensively with almost no distraction. As you progress through the ranks, the travel becomes more accessible and your opportunities to see more of the planet expand.

## YOUR LIFE OUTSIDE THE JOB

As noted, cruise work generally is scheduled for three months on and one month off, so this is how your life goes. Your personal life must be orchestrated around this unusual arrangement. It takes some getting used to, but those long months off do feel wonderful. It takes a special person to build a life within these parameters as well as a special family at home to get used to the schedule. The life is not for everybody, but it is for some. You should consider this venue carefully.

# Earnings and Outlook

According to the International Council of Cruise Lines, cruise capacity is expected to increase each year and offer excellent opportunities for employment.

| 2003 Earnings in Cruises | CHEFS AND HEAD COOKS | |
| --- | --- | --- |
| | Mean (Average) Hourly | Mean (Average) Annual |
| Cruise Lines | $17.97 | $37,390 |

SOURCE: *2003 OES National Industry-Specific Occupational Employment and Wage Estimates*, Bureau of Labor Statistics, 2003.

# Professional Organizations

General culinary associations, such as the American Culinary Federation, are discussed in Appendix A. Here we present an association for professionals working on cruise lines.

## INTERNATIONAL COUNCIL OF CRUISE LINES (ICCL)

### Who Are They?

International Council of Cruise Lines members include the largest passenger cruise lines that call on ports in the United States and elsewhere, as well as individuals who work in the industry. ICCL offers many services to its members, such as advocating industry positions to regulatory organizations, and is dedicated to ensuring a safe and caring shipboard environment for both passengers and crew. ICCL advocates its positions to lawmakers, industry partners, and regulatory organizations.

### Where Are They?

2111 Wilson Boulevard, 8th Floor
Arlington, VA 22201
800-595-9338
**www.iccl.org**

### Publications

Monthly email: *Fast Facts*
Quarterly newsletter: *Even Keel*

# Interview

**JOHN DALTON, Vice President of Hotel Operations, Clipper Cruise Line**

### Q / What is your current position and company?

*A: My current position is Vice President of Hotel Operations for a company called New World Ship Management. I have a partner, and together we run New World Ship Management. We own the ships and lease them back to Clipper Cruise Line. Clipper Cruise Line is a very small cruise line, very different than what most people think of when they think of cruise lines. Clipper has four small ships. We carry no more than 140 passengers. We travel to seven continents. We do not focus on being an amenity cruise ship with the 17 meals a day, gambling, shopping, entertainment, celebrities, and so on. We focus on a much different type of a cruise. Our cruises are mainly educational. They are either historical or they are natural in their content. For example, we do a 14-day historical cruise along the Intracoastal Waterway of the United States that traces the Civil War. On board the ship we have historians, all PhDs and great at communicating their knowledge, talking about the Civil War. The other type of trip we do, and we do more of these types of trips, are natural educational trips. We presently have a ship in Antarctica, one down in New Zealand, one in Belize, and one in Costa Rica. On board these ships are scientists, lecturers, naturalists. We always have a geologist and an ornithologist. And when people see things of natural beauty, these things are explained to them. So when people come off our ship, they have a better understanding of where they have just gone, what they have just seen, and what part it plays in the entire world's ecology. People have fun as well. They really come to Clipper to learn something or see something they can't see for themselves.*

### Q / How many people directly report to you? Whom do you report to?

*A: I have what I consider a four-tier management. We have a lot of people working on shifts that report to the hotel manager on the ship. The hotel manager on the ship reports to one of our corporate managers. Corporate managers report to me. And those are our tiers or levels. Out in the field we have about 350 people.*

### Q / What does your job entail?

*A: I have been here 19 years. So in 19 years it really has been a tremendous transition. My responsibilities technically are anything that you would find in a hotel that would be on a ship. I am responsible for food, drink, room service, housekeeping, decorating, refurbishments, maintenance, entertainment, management, marketing, sales. I also do crewing, staffing, hiring, firing, and promoting. So anything you would find in a hotel, that is what I would do. I don't really have anyone I report to directly, but of course my partner and I are ultimately responsible for the company.*

**Q / What is your educational background?**

*A: I was not terribly academic, and even at a young age my parents recognized that in me, and I recognized that as well. I don't perform well in regular academics. After I graduated high school, I worked in our family businesses while I waited for college, but no traditional school was interested in me. Then my parents read about the Culinary Institute of America. It was called the Harvard of cooking schools, and they thought it was the greatest idea in the world for me because I liked to cook. About six months later, I enrolled in the school. I never really had the desire to cook professionally. I wanted to be in the foodservice industry, and I wanted to understand the bigger picture of the foodservice industry. But I never had any great talent or discipline.*

*So I went to culinary school and saw a bigger industry. Toward the end of school, all my classmates were really intense about where they were going for their internship. It didn't really matter to me because I just knew I was going to be part of the industry. And I saw this little index card that said "cruise line looking for a cook." I called up, and seven or nine days later I was in St. Thomas working aboard a ship cooking. My first contract was for seven months, which is a very long time. And there are absolutely no days off. You never get a day off working on a ship. I was 20 years old at the time and young and dumb and willing to do that. I used to get up every morning at 5:30 and get off every night at 10. But coming from a family business, I fully understood the requirements, and I was vested in it.*

**Q / Do cooks still work those kinds of hours nowadays?**

*A: Well, our contracts are for three months. Our people on board work from 14 to 16 hours a day. That is very normal. But when you work aboard a ship you work three months on and one month off. In terms of a year, you have three months' vacation. So three months' vacation, if you figure that out and you work it out over a year, it's 90 days a year, almost two days a week. And your time comes in chunks, so you can do whatever you want to do, and I think that has an allure for a lot of people.*

*Living on board is also very cheap. You never need your money unless you go off the ship at night to have a beer or buy a souvenir. You have no bills. We pay for insurance from the first day you work on the ship. We pay for uniforms, too. So sometimes when we first tell people what they are paid on board a cruise ship, they get taken back a little bit, and they think for that many hours that is not a lot of money. But when they realize that they have absolutely no living expenses, our lowest level of employee on board the ship could easily save or just walk off a ship in a year with $20,000 in the bank. People making $100,000 in Manhattan can't put $20,000 in the bank in a year. So you really have to view it as a great investment.*

**Q / So what did you do after graduating from culinary school?**

*A: I spent four years as a hotel manager aboard, and those were the most fun years of my life. I worked on our domestic ships at the time.*

**Q / Were you prepared for all the management aspects of that job?**

*A: Probably one of my regrets in life is that I did not do a four-year degree. At the time I was very focused on career, and after I was graduated from culinary school, I just didn't want to spend two more years in school. In retrospect, that was probably a mistake.*

*I certainly suffered in management skills. I made a lot of mistakes that I may not have made had I gone for a four-year degree. I suffered financially. It took me a while to understand how to read financial reports. It took me a while to understand flow of goods. It took me a while to understand the psychology of the employee and employer. I think I would have gotten a little bit more of that had I taken another degree or gone on for an additional degree after graduating from culinary.*

**Q / Through the years, what has been your management style? How do you motivate staff?**

*A: Never scream. I have been very blessed with people that I have worked with. I work with some really great people. Whenever I work with somebody new or when I bring people up the ranks of the company, I tell them the first six months, we are going to be doing everything together. Every letter and memo they send out and every time they counsel somebody, I want to be a part of that. And then once I see that they have the ability to handle it, I give them some authority, and they learn how to use it. I don't believe in forceful authority. There are often people who put out a great product. People either want to be part of this or they don't want to be part of it. It's kind of interesting in my career. I've certainly had to terminate a lot of employees. I would say probably 95% of the people that I terminate send me a Christmas card, and they thank me because it was probably a bad match.*

*Talking about termination, it should never be a surprise. I've never had anybody come into my office when I've had to terminate them and tell me they were surprised. They will always say to me that they kind of knew this was coming. "I know you had expectations for me and they didn't really pan out and I tried but I guess that it just didn't work." And you have to separate that from the person. They are not bad people, just a bad situation, and it's not the right match. And you try to do it as gracefully as you possibly can.*

**Q / What do you love most about your job?**

*A: It's two things that always keep me going. The people keep me going because I work with a great bunch of young, dynamic people. Because of the nature of our industry, we don't get employees staying on our ships for a long time. They may stay two years or three years, and then they move on. That is understandable because they decide to have families and settle down. So I'm always working with new people. And that's fascinating to be working with young people especially. I like to teach people. I like to see them succeed. I don't mind when they make mistakes, and I tell everybody, you are going to make a lot of mistakes. The only thing you have to be careful about when you make mistakes is that you need to be honest with yourself. You have to own up to it.*

*I also really enjoy being on small ships because there are things we can do that would probably be impossible on a large ship. For example, we had a woman who was gluten intolerant, so she couldn't have any wheat flour. They had pan pizzas for lunch one day, and the server brought one to her, and she said to her server, "I can't have that." And they said, "Yes you can, because the pastry Chef made it with rice flour." The woman started crying. She hasn't had pizza in 30 years, and she just started crying. She was just so touched that one person would go out of their way to make something special for her. And that is the ability that I have on board these small ships.*

**Q / What is the thing you like least about where you are?**

*A: It's a 24-hour business, seven-days-a-week business. I have ships all over the world every minute of the day and there are so many things that are out of our control. The industry went through this huge problem with noroviruses—a group of viruses that cause gastroenteritis. It happens all over. It happens on land base, it happens on cruise ships. I actually sit on an advisory board for the Centers for Disease Control on this topic. For cruise ships, it's reportable. On land it's not reportable, so it became this huge issue with cruise ships. You take all the precautions and try to make it not happen. But it does happen occasionally. And when it does, you kind of scratch your head and say, how can we control this? I don't have any control over 30-foot seas, hurricanes, and the weather. I don't have control over a ship getting stuck in ice in Antarctica and having to pay someone $150,000 to send boats down to get us out of the ice.*

**Q / What kind of advice would you give culinary students starting out now?**

*A: I think it is important to find a school that is commensurate with your needs. Is it strictly cooking you are interested in, or perhaps you are equally interested in foodservice management? You have to know your needs, keeping in mind that this industry is built on serving people.*

*Once you have finished school, it's time to have a game plan, a flexible game plan, because things will always be changing. You have to be very careful about planning your career and make sure you make the best investments. And you can't do any of this without discipline. Discipline is essential.*

**Q / What do you see as a trend in the industry?**

*A: The number-one problem in the industry, and we all know it but nobody talks about it, is service. Service in the industry is overall fair. It's been inconsistent, it's untrained, it's undisciplined, it's hit-and-miss. I think the next place for us to go is to make service commensurate with the food that's going out. We go to great restaurants, and we have great food. I think culinary people have to understand that they are involved in a partnership. And the partnership is food and service. It's not just food. You can be the greatest Chef in the world and make the most beautiful food in the world, but if it sits in the window for 15 minutes before a person gets it, the guest loses. I don't think Americans are going to tolerate bad service for much longer.*

# Organization and Job Descriptions

## Executive Chef

**DIRECT REPORT:** HOTEL MANAGER
CORPORATE EXECUTIVE CHEF

**POSITIONS REPORTING TO:** CHEF DE PARTIES
GALLEY ASSISTANTS

**ROTATION:** TWO TO THREE MONTHS ON BOARD VESSEL; ONE MONTH VACATION

**RESPONSIBILITIES:** The Executive Chef is responsible for all aspects of the foodservice operation, including the administrative and technical duties required to provide the passengers and crew with high-quality cuisine. He/she is also responsible for supervising the duties of other kitchen personnel. In addition, the Chef ensures the galley consistently operates under the rules and regulations of the Centers for Disease Control. Communication in all aspects of the foodservice operation. The Chef consistently communicates with the cruise line corporate hotel office personnel.

**MAJOR RESPONSIBILITIES/SPECIFIC TASKS:**

- Keep the entire foodservice operation in an absolutely sanitary condition per the cruise line and the Centers for Disease Control standards.
- Maintain a clear understanding of the Company Safety and Environmental Policy and how to access it.
- Maintain fluency with accepted rules and regulations set by the Centers for Disease Control and practice and manage these routinely.
- Know the proper use of all kitchen equipment and the proper procedures and detergents used for cleaning and sanitizing all equipment.
- Responsible for daily food production and quality of all menu items served.
- Ability to prepare all standard menu items within production guidelines.
- Supervise the loading of all food provisions on turnaround day.
- Stay current on developments in the culinary field so updates can be added or changes made to our menus with the approval of the corporate Executive Chef.
- Maintain and evaluate the Galley Assistant's schedule and work lists in order to maintain sanitation standards.
- Provide high-quality crew meals using creativity in utilization of all food products.
- Communicate effectively with the Hotel Manager and discuss any matter concerning the operation of the onboard hotel department.

- Responsible for shutting down and securing all ranges and other cooking equipment.
- Responsible for securing and locking all refrigeration/freezer/storage areas.
- Responsible for the general appearance of the Chef de Parties and Galley Assistants.
- Maintain an overall awareness of all quality-control functions of the vessel and report deficiencies to department heads.
- Fill in as Chef de Partie whenever necessary.
- Secure galley and all galley storage areas in preparation for rough weather.
- Attend daily departmental meeting held by the Hotel Manager.

Administrative Duties:

- Process all invoices, control forms, and weekly menu copies on a timely basis and send them to the corporate office for processing.
- Accurately and proficiently follow, complete, receive, and store scheduled provisioning requirements while staying within budget guidelines for all items ordered.
- Refuse or accept incoming merchandise.
- Responsible for daily menu planning in line with the guidelines established by the corporate Executive Chef.
- Adhere to the concept of utilizing a block or preplanned menu setup. The block menu is utilized to manage menu flow and passenger expectation. Coordinate menu printing and distribution with the Assistant Hotel Manager.
- Responsible for ensuring that all foods are biologically 100% fit for human consumption; allowed to make menu revisions or substitutions, when necessary, to ensure that the food is maintained at our high standards subject to budget guidelines.
- Assist (when needed) other officers ordering supplies necessary for the ship to run effectively.
- Maintain a positive relationship with all food vendors.
- Maintain a positive relationship with all onboard officers.

The
Feel

A Day
in the Life

The
Reality

Earnings
and Outlook

Professional
Organizations

Interview

Organization and
Job Descriptions

69

## CHEF DE PARTIE — HOT AND COLD FOOD PREPARATION

DIRECT REPORT: **Executive Chef**
**Hotel Manager**

ROTATION: 12 to 16 weeks on board the vessel; paid on per diem basis in addition to a split of the crew tips

POSITION RESPONSIBILITIES: The Chef de Partie is responsible for the actual production of all the hot and/or cold foods assigned by the Executive Chef.

### MAJOR RESPONSIBILITIES/SPECIFIC TASKS:

- Familiar with menu guidelines set by the Executive Chef and able to prepare all items per standards previously set.
- Able to operate the galley whenever the Executive Chef is involved with administrative duties and carry on regular production.
- Responsible for the preparation of "raw as purchased" stage to final serving and garnishing stage on all food items.
- Familiar with accepted rules and regulations set by the Centers for Disease Control and able to practice these on a routine basis. Ensure that the galley remains in an absolutely sanitary condition at all times.
- Know the proper use of all kitchen equipment and the proper procedures and detergents used for cleaning and sanitizing all equipment.
- Understand that the use of water is closely regulated and used according to amounts prescribed by the Executive Chef.
- Work flexibly with the Executive Chef to develop orders for food and supplies, cleaning schedules, and freezer pulls at any given time.
- Fully understand that any concern about any crew member is reported to the Executive Chef, who, in turn, discusses the situation with the proper department head.
- Assist with loading of provisions as needed.
- Responsible for attending all safety drills assigned by the Marine Department.
- Responsible for maintaining personal appearance and all uniforms in a clean and presentable fashion at all times while on duty. Also responsible for the cleanliness and maintenance of living quarters.
- Responsible for attending a daily departmental meeting held by the Executive Chef and/or Hotel Manager. Also responsible for attending weekly crew photos (dress uniform: logo Chef coat, neckerchief, black pants, Chef hat, name tag, and polished black shoes) and weekly crew meeting held by the master.
- Responsible for participating in a regularly scheduled general cleaning of all galley areas.
- Recognize and inform the Executive Chef of concerns about product usage.
- Maintain daily the food storage holds and walk-in and reach-in refrigerators.
- Maintain CDC-sanctioned time and temperature logs relative to the production of food items as directed by the Executive Chef and/or the Hotel Manager.

**Figure 3-1**
Sample Cruise Lines
Organizational Chart

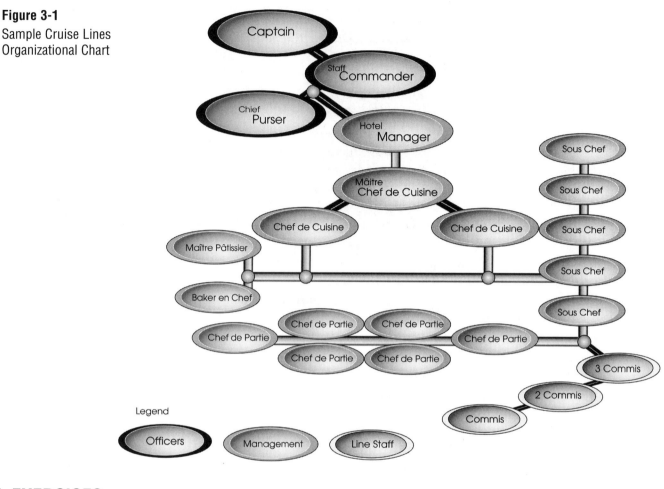

## EXERCISES

1. Compare and contrast the job duties of an Executive Chef on a cruise ship with those of an Executive Chef in a restaurant.

2. Would you like to work as a Chef on a ship? Which activities would you enjoy? What parts of the job might be difficult or not as much fun for you? Could you adjust to the hours?

3. Go the website of one of the major cruise lines (such as Carnival) and find out about the meals they offer.

4. Talk to someone who has gone on a cruise. Ask about the food and meals offered.

5. Using the Career Path Guide (page 2) and Education Path Guide (page 8), map out a possible career and education path for an Executive Chef on a cruise ship, including:

   - number of months in each job
   - each place of employment
   - how and when to complete one's formal education
   - how and when to complete skills and competencies

# Culinary Careers in Clubs

## Introduction

**IN ADDITION TO BEAUTIFUL COUNTRY CLUBS** complete with golf course and tennis courts, you can find city clubs, athletic clubs, faculty clubs, yacht clubs, town clubs, and military clubs all across the United States. Whichever type of club you consider, the kitchen and the food it produces are near and dear to the members, as witnessed by the number of places you can eat. As an example, one country club in Atlanta, Georgia, has no fewer than two formal dining rooms, two casual dining rooms, one men's grill, two members' grills, two bars/lounges, three meeting rooms, seven private function rooms, and two snack bars. That's over 20 places to eat! Needless to say, food and beverage revenue is a major part of the club budget, and the Chef is a highly visible presence.

## The Feel

### MY FIRST DAY

I drive to the entrance, which is guarded by two huge 10-foot Corinthian pillars connected by a trellis-style fence draped with pink baby roses. The plush green rolling lawns, the perfectly manicured flower beds, meet the eye through this magnificent opening. This is grace, elegance, and sophistication, like the grandeur of Tara in *Gone with the Wind*. Images of horse-drawn wagons being chauffeured through the entry flash through my mind. At the "club," men were men and women were ladies, and both were lavished with attention from the accommodating staff.

The circular drive ends at an elaborate porte cochère protecting the guests from inclement weather. The building is massive, with towering French doors that remain open in the summer months, welcoming the cool ocean breeze over the veranda. Balconies with seating line the second floor and overlook the exquisite grounds. The five-story weeping willow at the dead center of the rounded drive could tell tales of the diplomats, royalty, stars, and politicians who frequented these grounds in the past 100 years or so.

It's the first decade of the new millennium. The world has changed in many ways, but the rules and traditions of these private country escapes are mostly unaffected. These buildings have encouraged success, built fortunes, and created a model for the social order in America. Huge deals have been made with merely a handshake on the ninth hole of the golf course. Golf is dignified enough to accommodate all ages and both genders. It offers a fair playing field by employing a handicap, symbolizing a culture that has lasted for centuries.

## A Day in the Life

I arrive at the loading dock at 9 AM. A sign at the parking spot right in front has my name stamped on it in bold, unmistakable print. I make my way down the long corridor that mimics the main floor of the clubhouse. This secret lower floor is where the staff moves around the building to the many locations where members await service. The exciting part is that all these service areas connect in the underworld of these glamorous operations. No matter in what kind of foodservice operation you work, some form of chaos or sense of urgency pervades the back of the house. This is the world the guests will never see or hear; they just enjoy the result of the staff's dedicated work and effort.

On this Wednesday in June, two weddings are booked for the weekend and three corporate retreats for Friday. Thursday is King Crab Night, not to mention Men's Mixed Grill, Ladies Bridge Luncheon, Afternoon Tea, and regular à la carte service in the main dining room. My primary mission as Executive Chef is the creation and execution, along with the food committee, of exciting new activities for members to enjoy themselves, both as families and as couples. I am constantly challenged to devise a new outing or surpass one from the previous year. The assignment is to always be conjuring up a new event that excites the membership, maintains their interest in the food and beverage operation, and encourages their attendance.

### PURCHASING

First stop of the day is the purchasing office to check on all the orders placed. This is the backbone of all food operations, where you can commit expensive mistakes that are so difficult to identify if you are out of tune with your vendors and pricing. For instance, if an inferior or overpriced product comes into the establishment and no one catches it immediately, nine times out of ten it is used or prepped before the problem is discovered. I work to be proactive in purchasing early in the day so I am well prepared and in harmony with product pricing, specs, and availability. This is basic good practice.

Once I review the next five days of needs, it's off to the main kitchen, making a quick stop in the laundry room to pick up clean and starched whites, share a fast story with the workers about a movie or TV show I saw, and remind them today is Texas BBQ in the staff café. Never forget your support staff, whether they work directly for you or not. A few minutes of friendly conversation is appreciated and shows your respect and consideration for all.

## STAFF CAFÉ

Down the long hallway, freshly painted for the 200th time, I make my way to the next inspection point, "the staff café." This sounds like an unimportant checkpoint, but this foundation area can become a bigger thorn in the side if not set up with the same standard and professionalism as your kitchen. On a busy day in midseason, the staff café serves 60 to 70 people per meal period, usually including the General Manager, the golf and tennis pros, the office staff, the grounds crew, and all the other staff members necessary to run the facility. Some of these people eat here three meals a day, and it's imperative that they see you as a great and caring Chef. They rarely eat your food off the menu, so this is the only indication of what and how you cook. Club members often ask staff about the food and the Chefs, so doing a great job in your café is in your best interest.

## THE KITCHEN

Onward to the hub of the vessel, where production is well under way. Stocks simmer happily in 60-gallon tanks, and the Swiss braising kettle, humming with gas flames, is filled with a rich beef stew overflowing with colorful blanched vegetables. The smell of buttery cottage dill rolls, a recipe from my early years of culinary school, permeate the kitchen and hallway. Fifteen years after entering a professional kitchen for work, I still feel a warm rush of enthusiasm and a sense of thrilled expectation for the day ahead. How lucky am I to come to work every day, eat what I want, make what I like, and do what I love. Joe, my morning Sous Chef, gives me a tour of the hot and cold line, reviewing production levels, prep list, lunch setup, and, of course, the specials for the grill and main dining room. There are already 100 covers for lunch in the main dining room and 50 in the grill. As the Chef, I must have my nose in all the staff's business, ensuring that their work expresses my style and my philosophy. All the food they make must meet my standards of excellence.

## THE MIXED GRILL

Known in these social havens as the Men's Mixed Grill, this venue is masculine in design, with huge leather armchairs, giant coffee tables, and Baccarat crystal ashtrays. The bar, of handcrafted mahogany, curves into the matching colored walls like the bow of a grand sailing ship. The room is punctuated with private alcoves lined with thick benches and tobacco leather-covered couches. A row of imported and domestic brown liquors features the Maker's Mark Manhattan, the neat Basil Hayden bourbon, or the savory 30-year Dalmorie, a priceless single-malt scotch. The glassware is strong and sophisticated, as is the china and dinnerware.

The menu is that of a classic New York steakhouse — straightforward, no frills, large portions, and simple yet tasteful selections. Menu classics include colossal shrimp cocktail, oysters Rockefeller, clams casino, the daily selection of shucked clams and oysters, iceberg and marinated tomato salad, and, of course, a Chef's pâté du jour. T-bones, 20-ounce porterhouses, New York strip steak, and thick filet mignon are all served with our home-brewed steak sauce. In addition, we serve grilled 10-ounce salmon and swordfish, sautéed Dover sole, and the unforgettable 2½-pound lobster (market priced). Desserts range from the all-time favorite, crème caramel, to traditional New York-style cheesecake with a variety of fresh fruit toppings, to the sinful eight-layer chocolate cake. Done well, this kind of menu is still one of the most enjoyable and unpretentious, and it fits perfectly into this equation. Picture a packed room full of men, business deals on the table, and a no-nonsense, no-fuss, get-to-the-point dinner. Lives have been changed, decisions taken, disasters avoided, and industry developed over a warm snifter of Hines Rare and Delicate and a fine Cohiba cigar.

The grill is open for an intense business lunch or a midafternoon snack after a round of 18. Hearty sandwiches, inviting salads, and approachable entrées like veal piccata and calf's liver with bacon and onions are the choices in this bit of the Old World in the midst of the modern-day United States. Grills can be found everywhere, from the big-city alumni clubs to the faraway hidden oases tucked deep in the 'burbs.

## EVENTS

The plush ambiance, picture-perfect locations, and grand ballrooms make these venues extremely desirable for special occasions. Everything for social gatherings is already in place at clubs, which now produce affairs like a boutique hotel. Typically, a banquet manager or maître d' handles the booking of the party as well as the scheduling, planning, details, and execution. The back of the house employs a banquet event order like that used by all catering facilities, preparing large events as it supports the other food outlets; this takes careful timing and attention. Catering in all food and beverage venues is the gravy of the financial picture. In a club, producing catered events augments membership dues and helps support renovations and keeps the NOI (net operating income) out of the red.

The
Feel

A Day
in the Life

The
Reality

Earnings
and Outlook

Professional
Organizations

Interview

Organization and
Job Description

75

## HALFWAY HOUSE AND SNACK BAR

"2 dogs, 1 chick wrap, 2 Bobs, no make that 3 Bobs." This is in the language typically spoken by the staff at the summer concessions at most suburban clubs. Primitively laid out makeshift kitchens have been remodeled to replicate cool sidewalk cafés overlooking Olympic pools filled with hundreds of children during the summer months. Just sign the chit. That is the billing system used to charge the members as droves of kids feast daily on typical snack bar foods. Both poolside and halfway houses (located halfway through a round of golf) have upscale menus as well. Although the foot-long dog with all the toppings and the double bacon cheeseburger still rule, more creative and lighter alternatives have become standard items. A baby pizza oven cranks out individually designed pies. A panini grill adds a twist to the upscale sandwich selection. Sensational salads with a variety of sensible toppings make an attractive alternative on a sweltering hot summer day. The staff comprises mostly temporary summer employees, but many of them have worked through high school and well into their college years, which is a big help at such a seasonal facility. Many teachers and other professions that take the summer months off work as responsible and competent supervisors of these very busy outlets.

## THE END TO ANOTHER DAY

As you can see, being the Chef at a prestigious club is both a wonderful experience and a great career. These jobs are among the most desirable in the industry today. Club members care and are knowledgeable about high-quality cuisine, to which they have grown accustomed through their world travel. They crave and desire foods like those they eat at named metropolitan restaurants around the country. The transformation of menu items such as roasted duck with sauce à l'orange into trendy selections featuring flavored oils, light vegetable broths, and reduction sauces accompanied with delicate techniques of cooking, complete with exquisite taste and presentation, has led to new dining room standards.

Classics never change; they merely get updated with a new twist and feel. Being a Chef allows you to be more creative, better versed, and more informed than our clientele—because we have to.

It's a beautiful Sunday evening in June. The sun is setting over the 18th hole, shedding a gleaming golden blanket of light with a trail of fire-engine red. It reminds me of the glistening open-fire broiler I stood in front of for the past four hours. The tour of this culinary haven has come to an end, and it's time for this Chef to head for home. It was another crazy, busy day, but the satisfaction of coordinating all the events and executing superb food is an unexplainable feeling worth all the hard work and effort.

# The Reality

The Chef position at a club is perhaps one the busiest jobs at some points and one of the slowest at others. Nobody golfs with snow on the ground or in 102° heat, but you will be up to your ears in work when the season is on and members swarm the grounds. This is particularly true because in the slow season you lay off a lot of staff or lose them due to labor cuts. Yes, probably every year just prior to the busy season, you will need to hire a good deal of new staff, and then you will train them and lose them at the end of the season. This is an annual challenge; consequently, developing a solid training regimen, and clear-cut guidelines is essential. Did I mention that most of these transient staff members will have no intention of being cooks? Young people on summer break who need a job will support much of your culinary operation. This is not necessarily a bad thing; it just increases the depth of the challenge at hand and reinforces your need to be a great communicator, manager, and organizer.

Next, you will find that some clients consider themselves owners or stockholders, and they will not hesitate to communicate that attitude in many ways. This situation is manageable, but it requires a personality that is at the least politically sensitive. You need not have a subservient nature, but you do need to enjoy interaction with people and to be capable of dealing with demanding customers in a calm and confident fashion. These customers will think nothing of devising a menu item and telling you how it should be prepared for them each time they decide to dine with you. You are a host and the provider of a sublimely hospitable restaurant; particular requests are your specialty. Once mastered, this is a skill of which any Chef can be proud, as it indicates that you have an endless repertoire and thoroughly understand the true nature of our business. Further, once you have won over your membership they will support you in every endeavor, whether rebuilding the kitchen or buying a home. Remember, club members are often influential business people and society moguls who have the power and knowledge to get almost anything done.

## THE GRIND

At times you will wonder if you can see each outlet every day and will not be sure how you are going to monitor every activity. Most of the properties are vast, and you will need your golf cart to travel. Some clubs have outlets four or five miles away from the main property. Your job will keep you busy, and you will most certainly not be bored in season. However, once the season ends, at some point you will find yourself standing almost alone in the cavernous back of the house, which was only weeks ago filled with the buzz of your fully charged brigade dealing with the frenzy of business. Your staff is now almost all laid off or otherwise put on hold. Many will return next year, but a reasonable proportion will have moved on or no longer need the job.

| The Feel | A Day in the Life | The Reality | Earnings and Outlook | Professional Organizations | Interview | Organization and Job Description |

77

With that said, the other side of the grind is the part of the club that grows old and does not change—that is, the old favorites on your menu. Some menu items are standards (they don't call it a club sandwich for nothing), and you should be understanding enough to embrace these things as if they were your own creations, for your first rite of passage will be to execute them as the past 20 Chefs have. Do not under any circumstances try to change these unless the food committee puts it in writing and sends a letter to the membership, or you will feel their wrath! On the other hand, you can be creative with all the functions and special menu nights as well as the seasonal changes for your dining rooms. Don't worry—there is almost always room to express your culinarian self!

In addition, as we mentioned above, many private clubs are now opening their doors wider than ever as catering venues to nonmembers, acknowledging that this is a great way to strengthen the club financially without assessing the membership. So, as you may have guessed, this will help keep the place busy in the off-season and mean you can retain some ancillary staff. The catering aspect of club work is a rising trend and a sensible one that is expected to continue.

## PHYSICAL ISSUES

The pitfalls of the culinary world exist in clubs too: the cuts, the burns, the bruises, the long hours that make the feet burn, and the ever-present danger of lower back pain. However, as pointed out already, clubs offer a relatively generous respite as well as a good amount of slow time in which to recoup, regroup, and heal. The expression "no pain, no gain" is appropriate in the club world as well.

Don't forget to watch out for your employees! You will have a lot of inexperienced staff moving around. Preserve their safety through good training and careful supervision.

## THE BENEFITS FOR YOUR CAREER

Once you have established yourself at the club, you will enjoy considerable access to successful business experts. The people who generally frequent country clubs are leaders and learned individuals who will support your endeavors and, in many instances, help you to a better quality of life. In short, never underestimate what they can do for you. As well, you will most likely become involved in the associations that govern the management and Chefs of America's clubs. These associations are filled with some of the best in our industry, and you can gain a great deal of knowledge from their networks.

### YOUR LIFE OUTSIDE THE JOB

You will probably get at least a month off during the slow season—the payoff for the grueling pace you endure in the high season. Club life does offer enough downtime to allow you a reasonable quality of life with your family.

# Earnings and Outlook

The salary for an executive Chef working in a club during 2002 ranged from approximately $45,000 to $70,000 (National Restaurant Association). In addition, an executive Chef may receive a bonus based on performance such as food and labor costs, etc. The salary and bonus vary with the size of the country club, food and beverage sales, and location. Salaries are generally higher in large establishments and in affluent areas, and sometimes reach $100,000.

# Professional Organizations

General culinary associations, such as the American Culinary Federation, are discussed in Appendix A. This section mentions an association for professionals working in clubs.

### CLUB MANAGERS ASSOCIATION OF AMERICA (CMAA)

*Who Are They?*

The members of the Club Managers Association of America (CMAA) manage more than 3,000 country, city, athletic, faculty, yacht, town, and military clubs. CMAA provides its members with professional development programs, networking opportunitites, publications, and certification programs.

*Where Are They?*

1733 King Street
Alexandria, VA 22314-2720
703-739-9500
**www.cmaa.org**

*Publication(s)*

Monthly magazine: *Club Management*

The Feel | A Day in the Life | The Reality | Earnings and Outlook | Professional Organizations | **Interview** | Organization and Job Description

**79**

*Certifications*

Certified Club Manager (CCM)
Master Club Manager (MCM)

# Interview

**FRITZ GITSCHNER, CMC, AAC, Executive Chef, Houston Country Club**

**Q / As the Executive Chef of the Houston Country Club, to whom do you report? How many people do you supervise?**

*A: I report directly to the General Manager, and I have about 65 people who report to me. The people I supervise include employees from purchasing, stewarding, and the culinary operation.*

**Q / What are some of the challenges you face in your job?**

*A: The biggest challenge I have faced for many, many years now is staffing. Some people work for me for a year or two, then somebody else hires them off, pays them more, and off they go. And that is ongoing. I had to come to terms with that, because it takes a long time to train a person to finally do well. There is an investment of time and sweat. There are also many people involved in getting the person trained to operate at the level where you want them. Then six months or a year later, they are gone. I look at it this way, if they come in here and put 120% into what they do and want to be here, then I'll be happy with that. As long as they are being professional and showing up on time and doing what they are supposed to do, I am glad to have them for the time they are here.*

*In order to better prepare a potential employee, we ask that they spend time in our kitchen working with us because it gives them the opportunity to see our operation before we hire them. Some people say "Forget it, it's too crazy!" Not everybody is cut out for it. But the people we do have in our kitchen are very good people and I am glad to say many of my Sous Chefs move on to become Chefs and Executive Chefs.*

*Another challenge is meeting the needs of the members who see the club as an extension of their home. I can't compare it to a hotel or resort where I have previously worked. In a hotel or resort environment, people do not notice if the menu is the same for years due to the constant rotation of different clients through the establishment. In a club environment, where the clientele are mostly the same, almost everyone notices the smallest changes. If you do well in a hotel or restaurant, maybe once in a while, you will get a letter of thanks. But if you do not do well in a club setting, you will definitely hear about it. In our business, it is not what is fact based, it is what is perceived. You have to build that perception. There has to be a consistent delivery of high-quality products and service; it all has to do with being consistent. It has to do with being visible to members, talking to them, doing interesting things with them like when we bring them into the kitchen to eat a meal.*

**Q / For someone looking to work at a country club, how many hours do they need to put in?**

**A:** *I was one of those people early on in my life with the belief that the more hours you worked, the more productive you are. I disagreed with that later on in my life. My belief today is to try to give my Sous Chefs two days off. There are days when we work 12 to 14 hours, but I am very adamant that they are taking their two days off. They all have families and things to do, and you are just burning people out if you overwork them. I'm not necessarily applying the rules to myself. But then I always have other endeavors in which I am involved, like the Bocuse d'Or, the Culinary Olympics, and my own business-es. You just cannot do that in five days.*

*If you really want to succeed in the profession, you can't perfect your skills and the craft in a 40-hour week. I always say that you have to ask yourself how much you are willing to put into it, and that is what you are going to get out of it.*

*I could run the club, which makes about $5 million a year, in about 45 to 50 hours a week, and that includes running our own pastry, bakery, and butcher shop. I could probably run on fewer hours per week, but it depends if I have the right people in place and if they have good organizational skills.*

**Q / What are some of the things a good club Chef focuses on?**

**A:** *You need to have consistency in a club because if you don't, you have a lot of headaches. In hotels, you can get away with less consistency because you have a constant change of clientele. In a club, you have members here some-times four or five days a week. They know exactly how many olives you have in that salad, and some of them will be down there counting, believe me. Or, let's say you served a chicken piccata five years ago and you put it back on the menu, it had better look exactly the same and taste exactly the same as it did the first time. But knowing this just makes you a better Chef, because you may put systems in place to keep those consistencies there.*

**Q / How did you get started in the culinary field?**

**A:** *I did my apprenticeship in Vienna in a restaurant with classic French cuisine. For the first year as an apprentice you cleaned potatoes and the stove, and I was really questioning if I was in the right profession. It was tough, a little like the military, but it was good training. I got good, solid culinary skills.*

*I took my apprentice exam and moved to Holland to become Chef Garde Manger. I lied on the application because the requirements were that you had to speak English, and I didn't speak any English. I arrived in Holland and, of course, they found out very quickly, so they sent me to a language school. They kept me. I had not necessarily fine-tuned my skills, but I was full of ideas. I was very fortunate to have a Chef who would let me try things to see if they worked.*

| The Feel | A Day in the Life | The Reality | Earnings and Outlook | Professional Organizations | Interview | Organization and Job Description |

**81**

*Then I moved to Saudi Arabia, where I had a multinational culinary group of people to learn from. It was an interesting job from a cultural perspective. It was also interesting to see how people live in the Middle East and to go to the markets. I worked there for two years and then moved to the Princess Hotel in Bermuda, which was beautiful.*

*Next, I returned to Austria to open a restaurant for the government as Executive Sous Chef. I never worked in a kitchen since then or before that with a stainless-steel ceiling. It was a beautiful kitchen. Of course, the people in government had great parties and great dinners and lunches until the press got a hold of it and the whole thing stopped. Then I moved back to Saudi Arabia to work as the Chef at the Riyadh Marriott Hotel. I stayed there for two years, and then worked six months at a Scandinavian hotel in Kuwait. I came to the United States next and opened the Harbor Beach Marriott Hotel as the Executive Sous Chef in Fort Lauderdale, Florida. Later I became the Chef. Then I received an offer from Trust House Forte Hotel to be the Executive Chef of their property in the Bahamas. I accepted the position and the Bahamas were absolutely gorgeous. We had a house right close to the beach. It was a great lifestyle, and I had a great job. I stayed there for five years. I was very involved with the Bahamas Culinary Association there. We did expos and culinary competitions; it was a lot of fun. We fine-tuned the apprenticeship program. At one point we had 120 apprentices in our program.*

*One day while working in the Bahamas, I was asked by someone I had worked with in Holland if I was interested in the Chef's position at a hotel in Aruba. At that time, I was very interested in preparing to take the Certified Master Chef exam, so I said "yes" if they would support me in my quest to become a CMC. They agreed, so that is how I ended up in Aruba. I flew back and forth from Aruba to the Culinary Institute of America about four times to get the education I needed. I took the test while I was still employed in Aruba.*

*While in the States, I was interviewed to take over the St. Regis Hotel in New York, and at the same time the Houston Country Club. I didn't have much interest in the Houston Country Club right off the bat, but I took the interview anyway because I had never been in Texas and wanted to see it. I walked into the kitchen here and my first impression was "There is no way that I'd work in this kitchen." It was like being in the 18th century. During my interview, the General Manager asked me to "cook something." At that time, the banquet Chef was having trouble with a salmon dish because the juice for the salmon made the pastry wet. I took the salmon, wrapped it in freezer wrap, put the pastry around it, baked it, and pulled the freezer wrap out. It came out perfectly! After that I was the boy who could walk on water. They gave me a budget to build a new kitchen and then build some interest in the restaurants of the club. I took the job and had to work in less than tolerable conditions for about a year. Later on they did a serious update to the kitchen, which made it everything that I wanted.*

**Q / Was there any particular thing that made you choose your path to get where you are now, or was it just a series of events that you kind of followed?**

*A: I think it was more a series of events that I followed along. I never started off in the culinary field saying I wanted to become a Master Chef. It was not until I went to the Bahamas really that I learned about the American Culinary Federation (ACF), and the ACF opened up a whole new chapter, culinarily speaking, for me. I learned about culinary competitions, the Culinary Olympics, and professional camaraderie.*

**Q / Is there something that you might have done differently if you look back?**

*A: I think the way I developed as a Chef worked. There were mistakes that I made. Would I make those mistakes again? Not necessarily. I think the biggest challenge was to balance my personal life with my professional life. It was not until much later in my life that I realized it. I had a family, and I really didn't see my kids growing up. That is something I would do differently. But I was also maybe a little bit too young. I was 24 when I got married and 25 when my daughter was born. Maybe it would have been better if I waited until I was 30 or 32. I would have been more established in my career, and would have had more time to enjoy my personal life. I think the younger Chefs today realize this already much earlier than I did.*

**Q / What do you think was the most valuable part of your education?**

*A: The most valuable part of my education has to do with my youth and where I grew up. I really didn't know a supermarket existed until I was 15 or 16 years old. We grew up in the country. We grew our vegetables. We made our own hams. We slaughtered the pigs. We had chickens. We had to fetch the eggs. We made our own bread. These were not tasks we did to prove our skills in the kitchen; we did these things so we had food on the table. It was a necessity of life.*

*You cannot learn everything about food and cooking from a book or television program. You've got to experience it, such as how to make sausages or how a strawberry tastes as it is picked, how potatoes taste coming directly out of the ground, or how mushrooms picked from the woods taste. If you have never experienced these things, then you really have no gauge to compare. We never had items on the menu that were not grown in the area. For me, saltwater fish was something I was exposed to much later in life. All we had was river fish from fresh water. We also had seasons: asparagus seasons, bean seasons, etc. This was reflected in our menus.*

**Q / What do you like the most about what you are doing?**

*A: People. Definitely people. If you understand how to work with people and motivate them, you can bring out something they didn't even know they had. Only then can they produce great food.*

The
Feel | A Day
in the Life | The
Reality | Earnings
and Outlook | Professional
Organizations | Interview | Organization and
Job Description | **83**

**Q / What do you like least?**

*A: Employees not coming to work. Sometimes you have a better chance of winning the lottery than getting all your employees to show up. That is the most frustrating thing you have to deal with as a Chef. You can deal with things like being burned or ingredients that were not prepped, but if you have a person who doesn't have the motivation or the drive and just doesn't care, it throws a wrench in the system. It's like having a case of apples with one rotten apple. If you keep that apple, it will affect all of those around it.*

**Q / What advice do you have for young graduates deciding to pick a career similar to yours?**

*A: I think that the opportunities for young culinarians are more and much greater than when I grew up. The field still requires a certain individual who can survive under stress to produce really great food. So my advice would be to test those various areas you enjoy and find out what shoe fits you. Then focus in on being the best that you can be.*

**Q / Where do you think the industry is going?**

*A: I think what you find today is a consumer who is much more educated than ever before, and a lot of it has to do with the media. Because of what they are learning, their expectations of eating in your dining room, be it a club or restaurant or hotel, are going to be greater than ever before. Consistency, quality, and preparation techniques have to be much better executed than in the past when Chefs could get away with certain shortcuts. The demand on the new Chef is going to be greater, and cooking alone just doesn't cut it anymore. You have to have a well-rounded education. That well-rounded education means that your culinary skills have to be good, your administrative skills have to be good, and you have to be a good people person. If you are lacking in any of the areas, you will notice.*

# Organization and Job Description

## JOB DESCRIPTION
### Fox Chase Country Club

*Job Title:* Chef

*Position Type:* Full Time/Permanent

*Pay Basis:* Salary + Bonus

**GENERAL DUTIES:**

Working Chef for private country club with exclusive clientele. Responsible for all food production including meals for restaurants, banquets, and other outlets. Develop menus, food purchase specifications, and recipes. Supervise production. Develop and monitor food and labor budget for the department. Maintain highest food quality and sanitation standards.

**SPECIFIC DUTIES:**

1. Plan menus for formal dining room, terrace, grill, snack bars, banquet room, and other outlets as requested, considering guests, popularity of various dishes, holidays, costs, and many other factors.

2. Hire, train, and supervise the work of staff in the kitchen and grill.

3. Schedule and coordinate the work of Sous Chefs, cooks, and other kitchen employees so food preparation and presentation is technically correct and cost-effective.

4. Develop recipes and plate presentation for high quality, keeping in mind food costs and budget.

5. Approve the requisition of food, sanitation supplies, and smallwares.

6. Supervise portion control over all items and establish menu selling prices with Clubhouse Manager.

7. Ensure that high standards of sanitation and cleanliness are maintained throughout the areas at all times.

8. Train all cooking personnel on sanitation, food safety, and accident prevention principles.

9. Prepare all necessary data for the budget, project food and labor costs, and monitor actual financial results. Take corrective action where necessary and help assure that financial goals are met.

10. Attend food and beverage meetings.

11. Consult with the event coordinator about food production for special events.

12. Personally cook or directly supervise the cooking of items that require skillful preparation.

13. Evaluate food products and interact with guests to assure food quality and guest satisfaction.

14. Plan and manage the employee meal program.

**QUALIFICATIONS:**

Self-motivated, organized, well-spoken individual with excellent interpersonal skills. Extensive prior supervisory experience as a Chef required. Prior experience in large private country club preferred. Chef certification preferred.

**Figure 4-1**
Sample Club Organizational Chart

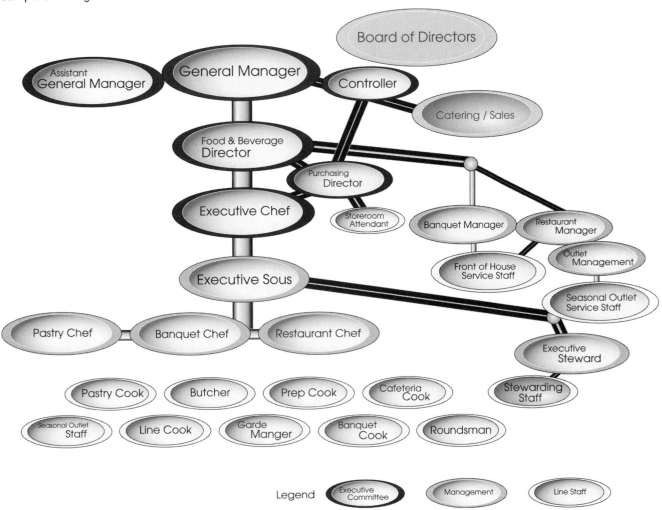

## EXERCISES

1. Investigate two clubs in your local area. What types of clubs are they? What types of clientele do they serve? Clubs that offer on-premises catering often have websites. What kinds of catering does each club do?

2. Compare and contrast the job duties of an Executive Chef in a restaurant with an Executive Chef in a country club.

3. Why do you need to have a lot of experience to be an Executive Chef in a club?

4. Would you like to work as a Chef in a club? Which activities would you enjoy? What parts of the job might be difficult or not as much fun for you?

5. Visit the website of Club Managers Association of America (CMAA) at **www.cmaa.org.** Click on Membership and then click on Club Management Magazine. On the homepage for their magazine, click on Dining Room to obtain a list of articles. Read an article from the list and summarize it in one paragraph.

6. Using CMAA's website, list ten benefits of association membership.

7. Using the Career Path Guide (page 2) and Education Path Guide (page 9), map out a possible career and education path for an executive Chef in a club, including:

   - number of months in each job
   - each place of employment
   - how and when you would complete your formal education
   - how and when you would complete skills and competencies

# Culinary Careers in Catering

## Introduction

**CATERING TAKES MANY FORMS,** from fancy box lunches at conferences to weddings in the country and corporate affairs in the city. Many catering events are in locations you might never see were you not in this business: country estates, corporate headquarters, galleries, openings of huge department stores and malls, institutions like Ellis Island and Radio City Music Hall. You can cater to people in the media, fashion, business, and the arts, so you can say that many of your events are star-studded. There are many reasons why catering is such an exciting and rewarding career. Every event has its own identity, from the location design to the brilliant tabletop composition to the menus. This chapter is full of details about working in this wide-ranging segment.

## The Feel

### THE JOURNEY OF A CULINARIAN

It was 1985, but I remember my first day at this remarkable catering company like I remember watching the rising of my first soufflé. The excitement went through my bones like a jolt of lightning, yet I had trepidation and apprehension. This was my first full-time position out of culinary school. Weeks before my first day, I thought about what my job would entail, who was on the culinary team, what kind of food they cooked. My restlessness and enthusiasm sent me to my new home several times late in the evening just to make sure I knew the most efficient route. Can you believe they say Chefs are obsessive!

It was finally the night before. I anxiously laid out my freshly pressed whites, razor-sharp knives, polished black clogs, a tiny notepad, and a new Sharpie marker. My plan was to get a great night's sleep before the big day. The only problem was that my eyes would not close and my brain would not shut off. I stayed focused on the clock, which flipped numbers like laps at a racetrack. Of course, the second I managed to get to sleep, the buzz of the alarm had me jumping out of bed. I dashed to the shower, got dressed, and off I went to my new adventure.

The only problem was it was 6:15 AM and I wasn't scheduled until 7:30 AM. I was too nervous to eat, but I knew about a local diner ahead because of my dry runs weeks before, so I pulled in for a cup of coffee. I sat staring at the wall and drinking one cup of coffee after another, looking like a kid on the first day of school. After my third cup, the gentleman sitting next to me leaned over and said in a fatherly tone, "First day of work?" We both laughed as I commented how noticeable I must look. We made small talk for about 20 minutes, which calmed my nerves and helped pass the time. The gentleman excused himself, as it was time for him to go to work, thanked me for a charming morning, gave me his card, and said to give him a call if I ever needed any paternal advice. I thanked him, and he was out the door. I glanced down at the card—and my knees nearly gave way. The gentleman was the owner of the catering company I was going to go work for! "What are the chances of that?" I thought.

When I arrived at the facility, the receptionist asked me to take a seat. I tried to stay focused and not appear too nervous when someone tapped on my left shoulder and nearly sent me through the roof; I turned to find the man I had just shared coffee with. We both laughed again as he took my hand and said, "Let me take you around and introduce you to everyone." I knew from that day forward, this was to be the best job of my life.

## THE KITCHEN

I had heard the word *commissary* before. Examples include military venues that produce large quantities for multiple units or the massive central kitchens that manufacture products for large chain restaurants. My first days at this catering facility started in the commissary—which, although an interesting locale, is not a glamorous one. The facility was spacious compared to others where I had worked, and the accommodations were impressive. The kitchen was designed expressly for off-premises catering. It held 15 workstations, each 8 feet long, that could accommodate 30+ Chefs and cooks prepping. It had separate walk-ins for raw products and huge walk-ins for storing prepped party menus ready for transport. It featured a cooking battery of convection ovens, a Swiss brazier, steam kettles, and chargrills. It was impeccably organized, professionally ruled, and quality driven. On the main wall at the entrance to the kitchen were huge corkboards covered with in-depth prep lists, event locations, and staff schedules. On the opposite side was a row of clipboards, one for each day of the week, containing the event orders for the parties. These were the bible for each event, carefully detailed and precise in every aspect: time, date, location, directions, staffing, equipment, pack sheets, menus, etc. This system ensured that nothing was forgotten, because in catering, once you leave the kitchen, you either do without or run to the local store, which gets old rather quickly.

# A Day in the Life

I vividly remember my first off-premises event, just several days after I started working. It took place at the Metropolitan Museum of Art in New York. I had visited this museum several times to view the exquisite exhibits, but never did I imagine the experience of producing a catered event in such an interesting venue. I was in awe every moment — from unloading the truck in the museum's loading dock to moving our equipment through the gallows of the back hallways to heavily guarded rooms. We set up makeshift kitchens in hallways adjacent to the Great Hall, the Temple of Dendur, and the American Wing, whose walls were hung with priceless artwork. We were forbidden to go from one area to another without an armed museum guard as an escort. I knew, even with my limited experience, that catering a party of 800 is challenging no matter the location; executing this job seemed like it would require a miracle. But our work appeared to be flawless, and I was incredibly proud of my job and the company I worked for. I remember how impressed I was with the camaraderie of the staff members and how they made sure I felt comfortable in my assignments.

I sensed a standard of excellence the second I walked in the door of the commissary. The kitchen staff was dressed in standard checks with hats and aprons, starched and professional, just like at a culinary school. The shelves were stacked with books of standardized recipes, food specifications, and photos of all the menu presentations, past and present. My first-day experience, although exhilarating, was overwhelming. "How will I keep up with all these people?" I thought to myself. "They are all so knowledgeable, confident, and focused." My fears stayed a secret, and I remembered the words of my first culinary teacher: "Fake it till you make it; never let them see you sweat; be busy; be polite and be quiet, and you'll get it."

I remember the sales and administrative offices on the first floor and the kitchen on the second, which created a cohesive work team that is imperative to the success of any operation. Every day for lunch the staff had a family meal. A designated cook, usually a first-level cook or supervisor, prepared the meal of the day for about 45 people, depending how busy we were. The cooks presented their goods with pride and accomplishment while the owner of the company said encouraging words about the fabulous job all were participating in. Then we broke bread together, like a family at Sunday-night dinner. It was a great way to pull a team together and create a powerful workforce committed to giving 110%.

The days seemed to fly by as my responsibilities became more challenging. It was apparent that I was passing many staff members who had been there long before me. I was a hard worker and left no one to fall. When I finished my tasks, I helped others finish. I voiced my opinions without sounding conceited or obnoxious. If something needed fixing, I made sure the right people knew about it.

## THE EVENT

Many catered events are in locations you might never see were you not in this business: country estates, corporate headquarters, galleries, openings of huge department stores and malls, institutions like Ellis Island and Radio City Music Hall. You cater to people in the media, fashion, business, and the arts, so you can say that many of your events are star-studded. There are many reasons why catering is such an exciting and rewarding career. Every event has its own identity, from the location design to the brilliant tabletop composition to the menus.

It's a Thursday morning in midseason. The commissary is in full swing, preparing the 18 events scheduled over the next two days. Your workday as a manager depends on the amount and kind of business booked. Typically, when business is strong, the Chefs work 60+ hours per week; in the slower seasons they may work 40 to 45 hours and perhaps have a four-day work schedule.

An event is a three-part process: prepping the raw materials and cooking as much as possible before the event; loading and travel; and final reheat and cooking at the event site. A lot can be prepped ahead without sacrificing quality: mixing composition salads (minus the greens, of course), blanching vegetables, creating the three-potato and apple flan to accompany the sliced filet. The pastry Chefs prepare their goods in advance, short of heating or scooping and, of course, plating. They might have created delicate friandises for each table and dessert consisting of a rich warm chocolate torte with white chocolate sorbet and macadamia tuile. The main courses are usually roasted and cooked at the event site for the best quality possible. You as the Chef need to plan carefully so your event is technically flawless without sacrificing quality.

The third part of the event is the final execution of the menu on site. All the cooking methods must be coordinated so everything is ready at the same time.

We are a high-volume caterer and take enormous pride in making everything from scratch. Our quality is the direct result of our ingredients, prep work, production, transport, and final event execution. Our production is crucial and watched carefully by the Executive Chef and all the Sous Chefs. In busy seasons we may have as many as 30 people prepping every day in each of our two commissaries. Tens of thousands of hors d'oeuvres are made weekly, hundreds of gallons of stocks and sauces are prepared, 350+ filets of beef are cut, and hundreds of cases of vegetables are carved, chopped, and blanched. The accuracy, the speed, and the organization of the production are critical to meeting the strict timelines necessary to complete the events. Each area of production has its own banquet Chef with a team of two or three Cooks. Production usually runs from 7:30 AM to 5 PM; however, during crunch times everyone stays until all the work is done.

## OFF-PREMISES EXECUTION

When an off-premises Chef is scheduled to do an event, he or she usually shows up at the shop three hours before the truck departure. Along with a commissary Sous Chef, the off-premises Chef checks off the event order form line by line to make absolutely sure every item specified is ready. The Chefs ensure the quality of the production items in every respect, from transport containers to quantity. After checking and rechecking supplies, the off-premises Chef is briefed by the Executive Chef about last-minute changes in details, cooking instructions, plate presentation, and so on. After that, the truck is loaded and we're off to the site, where the fun is about to start.

At the scene, the seasoned caterer experiences the controlled chaos as the juice that jazzes the senses. Trucks unload rental tables, gold-leaf china, crystal glassware, and elegant sterling-silver utensils. The florist hovers over the head table to demonstrate the elaborate table design, as the captain for the evening places an exquisite table setting as an example for the floor staff. The place is buzzing with streams of people moving in hundreds of direction, doing hundreds of jobs. Can we possibly be ready on time? Once again, between experience, patience, and the gods of catering, the event falls into place.

Off-premises kitchens are usually makeshift facilities that bring a pack list of comforts from back at the ranch. You become a master of sterno cooking, using hot boxes as your oven to perfect a sit-down meal for 1,000 guests. The off-premises Chef—referred to as the back-of-the-house (BOH) manager—works in tandem with the front-of-the-house (FOH) managers to orchestrate timing, handle logistical issues, and avoid typical obstacles.

The Chef is responsible for leading the staff through the execution of the menu through the entire affair. He or she must coordinate the meal timing while supervising the entire back of the house and communicating with the front of house. The captain relays the information to the Chef in charge for firing (getting ready) the various courses, yet the Chef must be extremely detail-oriented—for example, he or she must know where the kosher meals are and how many fish or vegetarian entrées are needed—while remaining composed at all times.

The Chef and crew usually work the event for about 9 hours, making a 13-hour day when all is said and done. A separate crew unloads and cleans up at the commissary so the Chef can head for home. The client usually makes a point of finding the Chefs to thank them for a superb performance; this of course is relayed immediately to the support staff. Another rewarding day is done—but tomorrow is just around the corner.

## MY JOURNEY TODAY

My journey from school goes back some 20 years. When I first started with this company, I never dreamed that someday I would be a main player, a partner in a long-established family-owned business. Once I started working here and was exposed to the company's culture, values, and mission, I knew it was something I wanted to be part of for a very long time. I didn't feel it necessary to move around to different jobs for the sake of moving. I found myself in a position that was both challenging and enjoyable. My responsibilities grew, as did my financial reward. I was clear from the beginning that I required continual growth to feel motivated and creative. Today, although my first love and focus is the food and beverage side of the business, I have responsibilities in marketing, public relations, strategic planning, human resources, and new business development. I love what I do and feel fortunate for the opportunities that have been in my path. With hard work, dedication, and stability, anything is possible.

## The Reality

They don't call it schlep for nothing! The reality is that it takes a great deal of planning, precision, checking, and rechecking to get a banquet from one place to another without glitches. That said, we are talking about places that provide at least what is known as on-premises and off-premises catering. Most caterers really only do off-premises, but if they made the investment to secure a permanent facility they probably do both. Clever entrepreneurs take the time to arrange interesting venues such as boats, museums, private beaches, islands, and even wilderness locations, while some stick to tamer settings like a roof with a great view or a beautiful park. Whatever the venue, if the food isn't prepared there, that means transporting the entire event—usually to a place with limited or even nonexistent facilities. To a great extent, excellence in catering involves maintaining a list of interesting venues that are at your disposal and sometimes even securing an exclusive deal with them.

Logistics, timing, organization, and mise en place are the words that drive the day. There is always a tight time frame for setting up the event, producing it, and breaking it down. The process requires a thorough understanding of what occurs at an event, which will serve as a guideline for your staff, who need to check onsite deliveries and make sure all is in place prior to the event beginning. Then your army descends on the function site like bees to honey, each soldier with specific directions and responsibilities. Some of your forces are veterans, others are in training, and many are seasonal regulars. Each must know his or her job and quickly get ready, as we have guests in two hours. It helps if you know the site and all its idiosyncrasies, or at least have conducted a thorough site visit, making notes and anticipating possible pitfalls. Some caterers invest in specially outfitted trucks or elaborate portable work areas and equipment to aid the efficiency of the process, but such an investment is so big it requires a pretty substantial company to support it. Most caterers still rely on rental equipment that is often not in the greatest condition. Whatever the equipment, you can be sure that it is experience and skill that makes the magic of turning an ordinary place into a beautiful event.

| The Feel | A Day in the Life | The Reality | Earnings and Outlook | Professional Organizations | Interview | Organization and Job Description |
|---|---|---|---|---|---|---|

**93**

## THE GRIND

The truth of the matter is that most catering companies do the majority of their business inside of two four-month periods: March through June, and September through December. This fact alone drives the real grind behind catering. For most caterers, this type of readiness and shut-down relies on a huge network of part-timers who can be available on short notice, or even a rent-a-waiter staffing service, although full-time jobs do exist. You, as a caterer, will probably be able to secure enough intermittent business to support yourself and a small permanent staff. Sometimes it is prudent to attach a small café, deli, or gourmet store business to help support your core staff. All of these choices can provide a great base income and pay the bills for the much-needed production facility all year round. In all cases you must find a creative way to sustain a nucleus that can be rapidly expanded on short notice, or you will have trouble maintaining consistency of product. This would be the downfall of your business, for one of your most marketable attributes is trustworthiness—the reputation for producing what you say you can. This is especially true because almost all large events—weddings, museum openings, society events, special birthdays—are very important to the hosts. You can be sure that no bride wants a moment of doubt about your ability to pull off her wedding. The idea here is to promise what you can do and do what you promise.

## PHYSICAL ISSUES

The lower back of caterers, especially those in small businesses, are at the greatest physical risk. As we have discussed, catering calls for moving a great deal of stuff, some light, some incredibly heavy, bulky, messy, and hot. Be careful, and teach your staff to lift correctly. The investment in providing them with the proper work tools will pay off in the long run.

## THE BENEFITS FOR YOUR CAREER

As a trusted caterer, you become an important piece of the community. You serve government officials, hospital fund-raisers, local moguls, and many other important people at important events. You must give assurance that family special events are also flawlessly handled and will exceed their best expectations. The fact that you are the food guru will help sustain your business too. Market this position, guard this position, and be the Chef they all believe you are. Becoming joined to the creative end of local culture confers many benefits and connections that will help you move through life successfully.

## YOUR LIFE OUTSIDE THE JOB

As you probably have guessed, because most of the big business is planned far in advance, a caterer's life is pretty predictable, so you can enjoy the fruits of your labors. This can even mean time to get away, especially when you work a light schedule part of the year. Much of your job occurs outside the office; you get to move around and are not tied to the stove all day, especially if you are the Chef-owner.

# Earnings and Outlook

Salaries range widely with the sales volume of the catering company and its location. Catering jobs in big cities such as New York pay much more than jobs in less urban areas. Here are examples of salary ranges for some culinary positions in a major metropolitan area as of 2004.

| | |
|---|---|
| Executive Chef | $90,000 – $125,000 plus bonus |
| Executive Sous Chef | $60,000 – $75,000 |
| Sous Chef | $40,000 – $58,000 |
| Lead Cooks | $12 – $15/hour |
| Freelance Chefs/Cooks (off-premise) | $18 – $32/hour |

# Professional Organizations

General culinary associations, such as the American Culinary Federation, are discussed in Appendix A. This section mentions only associations for culinary professionals working in catering.

## INTERNATIONAL ASSOCIATION OF CULINARY PROFESSIONALS (IACP)

### Who Are They?

The International Association of Culinary Professionals (IACP) is a group of approximately 4,000 food professionals from over 35 countries. IACP provides continuing education, networking, and information exchange for its members who work in culinary education, communication, or in the preparation of food and drink. Its mission is to "help its members achieve career success ethically, responsibly, and professionally." Many of its members are cooking school instructors, food writers, cookbook authors, Chefs, and food stylists.

*Where Are They?*

304 West Liberty Street, Suite 201
Louisville, KY 40202
502-581-9786
**www.iacp.com**

*Publication*

Quarterly: *Food Forum*

*Certifications*

Certified Culinary Professional (CCP)

## INTERNATIONAL CATERERS ASSOCIATION (ICA)

*Who Are They?*

The International Caterers Association (ICA) includes both off-premises and on-premises caterers from around the world. ICA provides education, mentoring, and other services for professional caterers. It also promotes the profession of catering to the public, vendors, and others.

*Where Are They?*

1200 17th Street NW
Washington, DC 20036
888-604-5844
**www.icacater.org**

*Publications*

Bimonthly newsletter: *CommuniCater*
Membership includes complimentary subscriptions to *Catering Magazine*, *Event Solutions*, and *Special Events Magazine*.

## Interview

**ALISON AWERBUCH, Partner and Chief Culinary Officer, Abigail Kirsch Catering**

**Q / Tell us a little bit about your company and position.**

*A: Abigail Kirsch is an upscale, high-volume on and off premises catering company headquartered in New York. We operate five in house catering venues as well as have a thriving off premises business throughout the tri-state* *area. Our annual revenue is approximately $40 million and we employee over 250 full-time and 500 part-time and on call associates. My title is partner and chief culinary officer.*

*I have varied responsibilities and have an overview and general accountability for all aspects of our businesses. More specifically, on a day-to-day basis, about 60 % of my responsibilities entail overseeing all the culinary and back-of-house operational aspects of the operations. My direct reports include three executive chefs, our corporate pastry chef, our corporate director of purchasing and our operations manager. The other 40% of my responsibility overlaps with all other aspects of our business including finance, sales, marketing, human resources, front-of-house operations and new business development. Although I am not directly responsible for these areas. I do work closely with our key managers in these areas.*

**Q / Who do you report to?**

**A:** *I don't report directly to anyone. I have three partners in the business, so I have a lot of accountability to them, especially for the aspects of the business that I'm responsible for. I also have accountability to everybody that works for us for providing the necessary tools to create the work environment in which they can thrive.*

*And I definitely have accountability to all of our customers and clients in that we're meeting their expectations and giving them the services and the products that they're expecting.*

**Q / Do you meet with an executive committee on a regular basis?**

**A:** *Yes. Our executive team consists of myself, my partner who is the chief executive officer, the two founders of the business, our corporate director of human resources, our corporate director of finance and our corporate director of operations. We try to meet quarterly. In addition, the same group meets with our general manager from each property on a quarterly basis. Our senior managers have individual departmentalized meetings within their properties.*

**Q / How many venues does your company operate?**

**A:** *The overview of the business is we've got five venues that we operate full time and, for the most part, each venue has a General Manager. Our venues include the Light House and Pier Sixty at Chelsea Piers, Tappan Hill in Tarrytown, the New York Botanical Garden, and the Abigail K Yacht. We are also affiliated with numerous historic venues and unique sites in New York, New Jersey, and Connecticut, that can be used for off-premises catering. These include Ellis Island and the Metropolitan Museum of Art.*

**Q / How did you first get into the culinary profession?**

**A:** *At a young and tender age. I always wanted to go into the culinary field, but I didn't think of it as a profession then. So, after high school, I started out pre-med at the University of Michigan and, after two years of undergraduate school, I didn't like the classes I was taking. I worked in doctors' offices and* *hospitals to get some experience, and I realized that the medical field would not be anything that I'd be interested in and I had no passion for. Growing up, my family was very into food—all kinds of food, including ethnic and gourmet foods. I always loved to cook as a child, and I followed my mom around. Wherever we traveled, wherever we went, things were geared around food. On holidays, time would be spent in the kitchen preparing food. It was just part of my background, and I loved it. I thought that it was something that I really wanted to do. But I didn't have any practical experience in it, so while I was in college I started working in an entry-level job in a pantry at an upscale restaurant preparing cold appetizers and salads. I loved it, so I stayed there for two years. Then, after I graduated from the University of Michigan, I traveled to Europe for several months and worked in catering before starting at the Culinary Institute of America.*

**Q / What degree did you graduate with?**

*A: I ended up graduating with a bachelor's degree in business. I switched my major after two years and selected classes that might practically help me if I went into the foodservice industry. I thought it would help me in the long run and, for some reason, if the culinary and foodservice track wasn't right, I'd have something to fall back on. I always felt that having a broad education in business, no matter what, would be helpful. So after I graduated, I continued to work full time for a couple years before I applied and went to the Culinary Institute of America. I worked in a hotel and with a small catering company, because in the back of my mind I thought it might be something I'd enjoy. When I started working for the catering company, it was something I really enjoyed and felt this is what I ultimately wanted to go into.*

**Q / As opposed to restaurants?**

*A: At the time, with my limited experience of three or four years, but catering was something I felt very passionate toward. I liked the creativity, the quality of the food, and the ability to develop new ideas all the time. In the hotel and restaurants, things were much more repetitive. At the time, the food just wasn't as exciting, the service wasn't exciting, the presentations weren't as exciting. And it was just a lot more commercial and a lot more corporate as opposed to catering. The company I worked for was independent and entrepreneurial, which intrigued me.*

**Q / Where did you do your externship during culinary school?**

*A: I did my externship at a catering company in New York called Glorious Food, which has a tremendous reputation. It was difficult to get into this externship because they had never taken an extern before, and they had no women working in the kitchen.*

*It took numerous interviews and hounding the Chef-owner to allow me to do the externship. But I did the externship and loved it. I ended up deferring going back to school so that I could stay at Glorious Food about a year and a half. Every time I thought I was going to be ready to go to school, we were hit-*

*ting a busy season and the Chef didn't want me to leave. They kept giving me more and more responsibility, so that I was really doing all the tasks and had a lot of the responsibilities that some of the Sous Chefs there had in running events on my own. I found it challenging and exciting. After a year and a half I went back to college and got my associate's degree, and then decided to look for a job in catering in the New York area.*

**Q / Did you want to go back to Glorious Food?**

**A:** *At that point, I wanted to gain as much varied experience as I could. I thought my long-term aspirations were to open my own business. My philosophy, which is a lot of people's philosophy, is to learn from other people, be able to then pick the best from everything you've learned, and make mistakes when you're working for other people. And my thought was to work for three to five years after I graduated and then start to gear toward opening my own business, and my thoughts were to work for maybe three different companies within that period of time. So I started to research different catering companies. I interviewed with Abigail, the founder of Abigail Kirsch, and really enjoyed meeting her, in part because it seemed we had a lot of the same characteristics and qualities as far as food philosophy and people philosophy and our backgrounds. My trial with them, all I had to do was make a crudité basket, and they thought it was the most beautiful one they ever saw. So I think that's why I was hired. I was hired as an entry-level catering chef.*

*Over the first two years, there were three different Executive Chefs who had started, and none worked out. I learned their business quite quickly, although I didn't have nearly enough experience, I think, to be an Executive Chef at the time. Every time one of the Executive Chefs they had hired didn't work out, I was the one that really ran the kitchen, and I would be the one that trained the new Executive Chef. So, after the third Executive Chef didn't work out, I was ready to move on and learn something new. At that point in time, they offered me the opportunity to be Executive Chef, and that was reason to stay because it provided me with a great opportunity and great challenges and a lot more managerial responsibilities. And what transpired after that was that the business continued to grow fairly rapidly and, as the business continued to grow, my job became more and more challenging. So that, every several years when I thought I had somewhat mastered my present responsibilities, something new came on the horizon, and it kept me very challenged and busy.*

*Only five years after I started with this company, the husband and wife who were the founders and one of their sons offered me the opportunity to become a partner in the company. That was something I had to think about long and hard because I knew, if I took that opportunity—and it was a great opportunity—then I would be, at least for a period of time, giving up my aspirations of leaving, my aspirations of having my own independent company, my aspirations of working with other people. But it was just an opportunity too good to pass up, because it gave me opportunities for growth and responsibilities. So I became a partner in the business in 1990, and then that's also when the*

| The Feel | A Day in the Life | The Reality | Earnings and Outlook | Professional Organizations | Interview | Organization and Job Description |
|---|---|---|---|---|---|---|

99

*business was making a real transition and we wanted to start to really grow the business from just off-premises catering to owning and operating on-premises catering facilities. In 1990, we opened our first catering facility and, over the next ten years, we opened four other catering facilities in addition to the off-premises catering.*

*The business had seen tremendous growth in the 1990s, and the last several years we decided we wanted to go in another pattern of growth and are really gearing ourselves up for that now.*

*The business of catering has evolved tremendously over the years. When we started out in the early 1980s, we were one of the first companies that focused on quality food — restaurant-style food and quality service where we trained service staff just like in the finest restaurants. We planned everything from the tenting to the electrical to the special lighting to the floral designs and music. And, when we first started out, that was very unique. Very few offered these services and really were able to execute at the highest level. Nowadays there are a lot of catering companies that offer these services. But we feel what we offer is unique because of our expertise in all aspects of logistics, planning, experience, and overall meeting and exceeding our guests' expectations. Also, our company's culture is that the customer comes first and our associates come first and that no matter what we have to do, we're going to do the right thing to make the events great. We have unique, outstanding presentations, and we stay on the cutting edge of what's going on in the culinary world with trends, yet not being too trendy. I think a lot of people don't understand catering, especially on-premises catering, and they think of it as a banquet hall, the rubber chicken. They often don't correlate it to fine dining like restaurant dining. We are constantly changing and improving where there's reason to change and improve, not only with our menus and our presentations but in all aspects of our business.*

### Q / How do you remain cutting edge?

*A: First of all, I don't think you need to remain cutting-edge by job-hopping and working all over. Even though I didn't work with a lot of other Chefs previously, I think I stay cutting-edge by knowing a lot of Chefs in the industry, networking with them, keeping up with professional and other periodicals, dining out a lot, traveling, staying involved in a lot of different organizations that are related to the hospitality industry, and, at times, taking continuing education classes. So that I'm able to observe a lot, see a lot, network a lot, travel, and so on. That's how I stay stimulated and motivated.*

### Q / You hit a certain level in your career where it needs to be self-motivating.

*A: Self motivation comes when you love what you are doing and truly enjoy creating and making changes.*

### Q / Can you talk a little bit about your management philosophy?

*A: First of all, I'm a people person, and I think that's really important because you have to enjoy working with people, teaching people, listening to people, learning from people, having a good time with people. I think that one of the key factors to having a good team around you is being able to really enjoy who you're working with and having them enjoy where they're working, having them feel like they're learning and being challenged. Your team needs to feel that they're being appreciated, motivated, and given opportunities. A lot of times you can keep them challenged simply by asking their opinion and by communicating to them a lot. So that's a big key to the philosophy.*

*My goal is to be able to promote somebody from within or have them be able to move on and get a job that's of a much higher level because of what they've learned from us. Culture is really important. A good work environment involves treating everybody with respect, having a lot of different programs in place to make people feel good about where they're working, giving them incentives, providing fun things for them to do outside of work. The basic core philosophy is the same, but obviously with different levels between a line employee and a manager and a senior manager, the way of motivating varies a little bit. But that's the basic philosophy.*

**Q / What are the hours of employment like for people who work in catering?**

*A: Some people think catering is a real part-time type of job. But there are definitely full-time opportunities and, especially if you're in a metropolitan area, there are a lot of large-volume catering companies that offer great jobs with great benefits. Compared to the à la carte industry, what catering allows is definitely a much more flexible schedule. It's not for everybody, but typically each workweek varies quite a bit, because your shift and your hours are dependent on events, so that's different from a restaurant where you have set days and hours. So, for some people, they don't like the uncertainty. But, for a lot of people, they like the variety of it.*

*Catering is definitely seasonal, and there's not a region in the United States where there isn't seasonality in catering, and the seasonality varies based on locale. In the New York area, the slowest times in catering are the dead of the winter—January and February—and then in the summer it slows down a little bit—in July and August. Then the rest of the year, business is pretty steady. There's steady work through March to December, and as far as the workweek is concerned, it varies company by company. But for full-time associates, it is typically a five-day workweek. You'll have six-day weeks in the busy times, five days on the most of the typical workweeks, and then maybe a three- to five-day workweek in winter and in the summer when it is a little slower.*

*The hours vary based on level of responsibility, but typically we shoot for an eight-hour day. For managers and Sous Chefs, it's more likely to be a ten-hour day. I always tell anybody we're interviewing and hiring for a Chef position that, on an annualized basis, it's about a 50-hour workweek. There may be some weeks they're working 55 or 60 hours in our peak times and, in our slower times can work 30 to 35 hours. Basically 80% of the year our Sous Chefs work a five-day workweek. We really try to keep it to that except for the real extreme times. I've always had the philosophy that if you don't burn out*

| The Feel | A Day in the Life | The Reality | Earnings and Outlook | Professional Organizations | Interview | Organization and Job Description |

**101**

*your associates with excessive on-going work, then they will perform at a higher level and they will stay working with you longer. This philosophy has proven true in our organization.*

**Q / What is your biggest challenge?**

**A:** *The seasonality of it makes the job extremely challenging. The only positive of seasonality is that it gives everybody a little bit of a break. But, from a revenue perspective, from a planning perspective, from a staffing perspective, from a buying perspective, the peaks and valleys make it really difficult to plan and manage a business. Everything you think you've mapped out can change on a dime, so that for running a profitable business, it's extremely challenging to keep quality and consistency. So we always say if we could take our same revenue and have the same volume week in and week out or comparable volumes, we'd be running a very different business. So that, to me, is always the biggest challenge and the biggest frustration at times.*

*The other difficult thing I'm finding in the last three to five years, and it's more industry-wide, not just catering, is really finding and hiring the right staff. Because whether it's front of the house or back or the house, whether it's management or line, it's just difficult to find professionals who are really committed and interested in working in our industry and learning and growing in the industry. So that all I really want, all I really look for, is somebody who is excited and passionate, dedicated and motivated, and wants to learn.*

**Q / What do you love most about your job?**

**A:** *Everything else I love. I love the fact that catering is always changing. No two events are ever alike. The clients vary; their needs vary. The type of event varies. It could be a social event, a corporate event, a benefit event. The menus need to vary all the time, as does the style of presentation. You're dealing with a variety of people, a variety of ideas, a variety of talents, and a variety of ways that they envision their event to be. It's very different when you have a restaurant and you create your one style or philosophy and people come to you for that.*

*Most caterers have to be like chameleons because, if we only have one style or philosophy, we're going to have a much smaller pie of people that want to use us. But that's what I love about it, because if you're creative, that's what keeps me interested and challenged. It also keeps our staff really interested and challenged because if they had to produce and repeat the same food and the same events day in and day out, week in and week out, then it would be like a factory. So that's one thing I love.*

*I love the fact that we run our business like a tightly run corporate business, but with a lot of entrepreneurial spirit. And I love the fact that I think we're the best of both worlds because, if we have corporate structure, but with the entrepreneurial spirit behind it, we're passionate about committing to change and to perfection and trying to always look at what we can do better. I think that, in a true corporate environment, that's not always the case. And we're able to make the change very quickly. So that, if we meet on a Monday and we realize there is an issue with an event over the weekend and that we need to change our system of doing something, we're able to get the right team in place to brainstorm it, to think it completely through, and then within a day or two, make a change companywide. It's a great thing.*

**Q /** **What kind of advice would you give a new graduate who was looking to go into catering?**

*A: I think anybody, whether they're interested in catering or another field that they're interested in, would say to first do some research to find the top five, eight, ten people in that field, because I think it's really important to associate yourself with top people in the field. That doesn't mean it's the people that do the highest volume or the highest revenue but operations where you think you're going to have the opportunity to learn the most from a quality, operational, and business standpoint.*

*Then, I think it's a great idea to try to set up appointments to see if you can meet with them just to spend an hour or two to learn and understand what their business is like, and ask the questions that are most critical to you. Especially if you're starting out in your career, you may want to know what the workweek is like, what type of room there is for advancement, what your career path will be like once you gain the experience at this establishment.*

*The next step I think would be, if you're really uncertain whether you want to go into catering or another field, to see if you can trail someone in the kitchen for a week or two so that you can really get a feel for it. Because I don't think there's anything worse than, especially in one of your first jobs, starting a job and finding out two weeks, four weeks, six weeks later, that that field isn't for you.*

*For some people, especially those without much experience, I feel it's better if in catering they start to work for a very small company where an owner or head Chef is very hands-on, where you'd be working side-by-side with them, so you'd be able to not only learn the culinary aspects but really start to pick their brain about other parts of the business. In a small company, you're also going to have the opportunity to do a lot different things. From a culinary perspective, typically there are a few people in the kitchen and you have to do everything from making hors d'oeuvres to making stock to helping with dessert.*

*In a larger company like ours, it's more departmentalized, so that you don't get quite as involved in making a variety of items day in and day out. You rotate, so it's over an extended period you have the opportunity to do that. But the upside of a larger company is that you get great experience in volume production and volume execution of events. You know, once you learn that and you learn the right ways of doing it and you learn great organizational methods, then that applies to a variety of industries. So a lot of people who leave catering and enter other segments of the industry feel they bring wonderful organization and discipline to their stations.*

| The Feel | A Day in the Life | The Reality | Earnings and Outlook | Professional Organizations | Interview | Organization and Job Description |

**103**

**Q / In catering, there are many opportunities. Can you describe some catering careers?**

*A: Catering is similar, if you look at an organizational chart, to almost any operation in the hospitality industry. Most catering organizations have an owner/CEO or a General Manager at the top of the organization chart. Then typically the umbrella under the General Manager would be managers in finance, human resources, sales, culinary, purchasing, operations, and maybe marketing. To go the next tier down of management, there are the Sales Staff, Event Managers or Maitre d's, Beverage Managers, Executive Stewards, Sous Chefs, Facilities Managers. The next tier of associates includes Sales, Administrative Support, Captains and Waiters, Cooks, Stewarding and Warewashing Staff, Receiving Clerks, Housemen, etc.*

*In the back of the house, there is an Executive Chef, Executive Pastry Chef, Executive Sous Chefs, and Catering Chefs. In our company, we have a corporate director of purchasing, and the different facilities each have a purchasing manager. I'm very strong on promoting from within and promoting Chefs to become purchasing managers. We actually have an opening right now and we're just discussing it today and to me what is more important in that position is somebody who has a culinary and a food background more than a traditional purchasing background. It's more and more becoming the norm. Some of the bigger companies now have Master Chefs that, in fact, are in charge of purchasing—for huge corporations. Oftentimes in the past a traditional purchaser knew nothing about food and, even if they were good at negotiating bids, really didn't know if the quality was coming in right and if there was more waste in the long run, and the Chefs were able to work with food.*

*The one position I forgot about is food and beverage management. And that's a position in our industry now that, in a slight way, is being phased out in a lot of operations because the Executive Chef becomes the food and beverage manager. In catering, there are still companies that have food and beverage managers, depending on the size of the company, so that's another opportunity.*

*One other aspect of catering that's interesting is that there are a lot of opportunities to work part time and freelance to pick up extra money.*

**Q / What trends do you see in the industry?**

*A: A trend that is definitely happening is that to be a Chef now, you can't just be a great culinarian. You have to be a very well-rounded person with great business acumen. You have to be a good people person, a good leader, and a good motivator. You have to have human resource skills, as well as be an administrator. You have to wear a lot of different hats. I think the days of somebody being a great Chef but not being able to manage a kitchen, motivate employees, and not able to understand business are on their way out.*

# Organization and Job Description

## JOB DESCRIPTION

**BANQUET CHEF**

POSITION TYPE: *Full-Time/Permanent*

PAY BASIS: *Salary*

REPORTS TO: *Executive Chef*

### RESPONSIBILITIES:

Maintain the highest quality of food production and operational standards in order to exceed guest expectations and maximize profits.

Supervise all phases of food production and service related to banquet functions.

Control the quality of food production.

Control the quantity of food production so as to avoid excessive leftovers or waste.

Supervise banquet kitchen and staff.

Communicate with others to plan and coordinate efforts for scheduled functions.

Maintain an open and positive working relationship with union members in agreement with the collective bargaining agreement.

Maintain an effective training program for all employees.

Do yearly performance evaluations of all employees.

Recommend hiring decisions.

Maintain sanitation standards.

Maintain equipment.

Assist in monthly and weekly forecasts.

Provide input for annual expense budget and capital budget.

### REQUIREMENTS

Two-year associate's degree or four-year bachelor's degree in culinary.

Minimum two years' supervisory experience with proven team building, problem solving, and training skills in a large operation.

Excellent communication and interpersonal skills.

Proven ability to motivate, guide, and coach diverse employees.

Strong organizational and time management skills.

Ability to work under stress and in a fast-paced environment.

Flexible with hours, days off, assignments, and additional duties.

**Figure 5-1**

Sample On- and Off-premises
Catering Organizational Chart

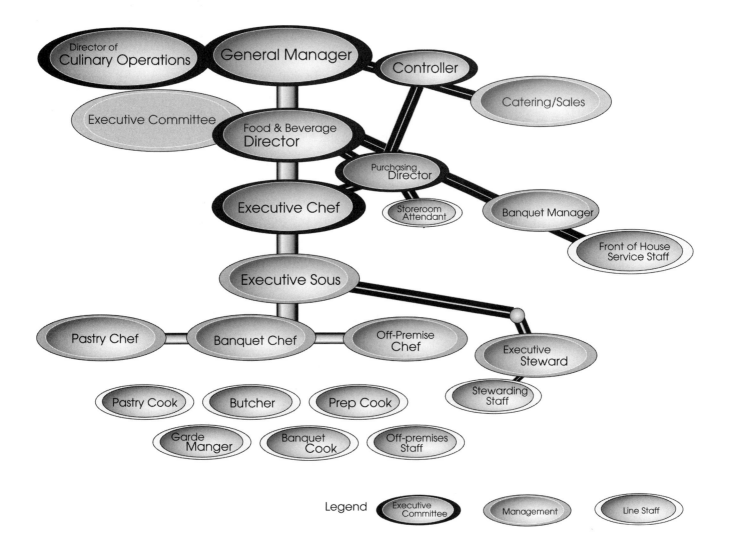

## EXERCISES

1. Go to the website of the International Caterers Association. Click on "Newsletter" and read two articles that are of interest to you. Summarize each article in one paragraph.

2. Investigate the names and types of two caterers in your local area. Do they do off-premises or on-premises catering? Are they small businesses, or are they large-volume caterers? Where do they prepare their food? What types of clientele do they serve?

3. Go to the website of *Special Events Magazine* (**www.specialevents.com**). On the left you will notice Articles by Subject. Pick a subject you're interested in and read two articles about it. Summarize each article in one paragraph.

4. Compare and contrast the job duties of an Executive Chef in a restaurant with an Executive Chef in a catering business.

5. Why do you need to have a lot of experience to do a job such as Alison Awerbuch describes in the interview?

6. Would you like to work as a Chef in a catering business? Which activities would you enjoy? What parts of the job might be difficult?

7. To see what a large-volume catering business can do, go to the website of Abigail Kirsch Caterers (**www.abigailkirsch.com**).

8. Using the Career Path Guide (page 2) and Education Path Guide (page 10), map out a possible career and education path for an Executive Chef in a catering business, including:
   - number of months in each job
   - each place of employment
   - how and when you would complete your formal education
   - how and when you would complete skills and competencies

# Culinary Careers in Supermarkets

## Introduction

**SUPERMARKETS ARE GRAND PLACES** that serve the shopping needs of modern America in one huge area. They stock everything from pastry and breads to authentic Chinese ingredients and house themed store cafés. From the delicacies of the culinary department, turn left and enjoy the convenience of a day care center, pharmacy, optometrist, dry cleaning, even a video rental outlet. The conveniences differ from store to store, but the goal is the same: to make a weekly event an experience rather than a chore. With supermarkets offering more and more prepared foods, it's only natural that Chefs gravitate here and take the worry out of "What's for dinner?" and "How do I prepare it?"

## The Feel

### THE BUTCHER, THE BAKER—IT'S 1968

Ring, ring, ring, ring—the sound of groceries being rung up on our family's weekly shopping visit. The colorful foods were wrapped with brand names that stayed printed in my head for many years to come. We were special, growing up in a second-generation Italian family that considered TV dinners a once-a-year treat. Grocery stores were a necessity, with specialty stores still the norm: The butcher, the baker, and the fishmonger all figured in the weekly ceremony of filling the house with food. Mealtime was a ritual with all members present, creating an atmosphere of family bonding, establishing strong and lasting relationships. In the 1960s, 1970s, and even the 1980s, chain grocery markets or superstores didn't provide us with the ethnic or sophisticated foods needed to satisfy our palates. Deli departments were mediocre at best, with industrial salads and unappealing displays of convenience food. Breads were marshmallow in texture, bland in taste, with little or no selection. The cakes were simple; inexpensive ingredients such as shortening replaced butter, and canned fillings were added as a cheap afterthought in a crisis situation.

## THE SUPERSTORE MAKER

In a blink, the focus moves forward to 1997. Am I crazy? I think to myself. Am I making the biggest mistake of my career? I am moving my family to a new home so I can work at a supermarket. We all know the trend is changing and cooking is becoming America's number-one obsession, but are consumers really ready to be educated? Do they want to know the difference between a snap and a snow pea, between summer and winter truffles? Do they want to eat exotic baby greens? Do they care how to make a béchamel sauce for macaroni and cheese? My heart throbs with apprehensive excitement. The two gentlemen who recruited me had convinced a staunch hotel Chef that the American food market was the next explosion in our industry. But was I ready for a uniform of half suit and tie, half Chef whites?

I would be the liaison between field and office, preparing culinary programs, developing recipes, doing customer relations, photo shoots, designing a magazine, and implementing the vision for growth shared by the top executives in the company. I was up for the challenge and believed this career change was the right move. It took about a year to get my feet planted and our philosophy appreciated. We were all on the same page, from the owner to the store attendant, striving to be cutting-edge superheroes. In one year, we became the envy of the industry, and I was the Operations Director of Food Preparation. We supplied a package to our clientele with all the accoutrements, from cut mise en place, marinated and stuffed proteins, cooking instructions, secret sauces, the right cookware, wine to serve, desserts best suited, breads, and even after-dinner chocolates. We believed the more we educated the consumer, the better we would look —a philosophy that has been successful.

## A Day in the Life

Picture a lavish, well-lit store with aisle after aisle of farmers' market produce, extraordinary fresh fish and meat counters, a spectacular selection of domestic and imported deli meats and cheeses, four varieties of our own brand of extra-virgin olive oil culled from different Italian regions, and spoils such as Russian beluga and fresh black truffles. These treasures lie alongside the mundane necessities of everyday life such as milk, bread, paper towels, and toothpaste. Now

| The Feel | A Day in the Life | The Reality | Earnings and Outlook | Professional Organizations | Interview | Organization and Job Description |

**109**

take a tour of our abundant salad bar, Chinese buffet, bakery/pastry shop, and sushi station; admire our restaurant-quality open kitchens. Seven years ago, prepared foods were just home replacement meals—unglamorous prepared complete dinners craved by the working yuppies of the mid- to late 1990s. Part of the reason for the new mega-markets was the rebirth of the professional culinary regimen I was hired to implement and orchestrate to a level of excellence. The attraction for me in coming to work for this supermarket company was its unmistakable commitment to hiring restaurant-quality Chefs to create restaurant-quality food.

## FIRST ON TODAY'S AGENDA

Conference call with all regional Chefs. It is important when you multi-task to be in contact with your core staff who are in different locations. The only way to success is constantly educating your people and encouraging a trickle-down effect of information and attitude mixed with periodic touring of each location. You hire professionals who share your passion for quality and perfection. The conference call lasts about 45 minutes, recapping newly launched programs and working out potential obstacles to success.

Today I am in the Chef whites part of my uniform. For the next three days we will be shooting (photographing) finished dishes for our monthly magazine; this gives me the opportunity to get my hands dirty, which I love and miss the most. Preaching philosophy and instilling passion in my staff is a full-time job, both rewarding and satisfying. "Lead by example" are the words that have echoed through my head for most of my career. If I want my staff members to wear a hat, then I must wear a hat. If I expect an organized workplace, then I must be organized myself. The Chef is at the helm of this vessel; he or she is the mentor to all staff members, and they watch every move. Many of them strive to become Chefs themselves, which is a compliment to one's leadership as well as a responsibility.

## THE MARKET CAFÉ

The culinary brigades in food markets have been instrumental in reinventing retail food cases. Concepts reflect those of the Klienmarket in Germany; the downtown street markets in Tokyo, Japan; the Peck Market in Milan, Italy; the Mercato in Valencia; and the quaintness of an Italian trattoria. We as supermarket Chefs take pride in maintaining the integrity and the authenticity of the foods we prepare. Other stores offer cheap imitations; however, we are committed to educating our customers. Our reputation is built on trust and loyalty, on delivering a quality product few can match.

Food prep is the buzzword, the core of the Executive Chef's responsibilities. Anything related to a food product that requires preparation is supervised by our culinary department. The Perishables Manager, who mainly handles meat, fish, poultry, and produce, relies on the Executive Chef's expertise to provide prepared oven- and grill-ready foods that enhance sales from their elaborate display cases. Such delicacies include fillet of sole stuffed with Jonah crab; macadamia and horseradish–crusted salmon; spinach, dried tomato, and fontina steak roulades; mustard and rosemary marinated pork tenderloin; and a delicate lemon sole rolled around a scallop, shrimp, and dill mousse.

Every store has a repertoire of established recipes that are logical to implement, possess a refined appearance, offer superior taste, are affordable, and supply balanced nutrition. The hot and cold stations of the café are an integral part of the daily operation. Thick-piled sandwiches are made on a variety of breads with a vast selection of creative toppings, cheeses, and spreads. Sushi hand-rolled to order may be your desire of the day. Select from tuna, fluke, smoked eel, giant clam, pulpa (octopus), sea urchin, or the American tradition, California roll. Next in line is the authentic Chinese buffet serving our famous Chinese sesame chicken, pork with scallions and oranges, sirloin strips with Chinese long beans and oyster sauce, quick-fried vegetables with edamame (fresh soybeans), tofu and jasmine rice, or the picture-perfect jumbo shrimp with snow peas, peppers, and napa cabbage all packed in the customary Chinese quart and pint containers. The local Tong Hoy restaurant had better get ready to rumble.

## PASTRY/BAKERY

The bakery/pastry shop completes the shopper's experience. World-class hand-crafted pastries and cakes are designed with masterly skill and the best of ingredients. The selection of baked goods has something for everybody. Notice the freshly baked artisan breads, authentic buttery breakfast Danish, scrumptious oversize muffins, and assortment of designer cookies that look like they're straight out of European bakery cases. Luscious caramelized banana and white chocolate brioche pudding, sinful dobos torte, gorgeous raspberry chocolate mousse cake, mouthwatering chocolate caramel hazelnut tart, delectable chocolate coffee cupcakes, and a brilliant linzer torte in three flavors are just a few of the sweet endings you can get here for the finale of your fabulous meal. And then the miniatures! Hand-filled candies, ten kinds of fudge, tiny pastries, and many more classics are executed with the same skill and devotion as all the food preparation. I was taught a long time ago to savor a little of a fabulous treasure, then to consume a lot of something mediocre, which sums up how to handle the temptations of this department.

## RETAIL LINE

Next stop is the retail line, a bountiful presentation of quality selections to replace a home-cooked meal; note particularly the comfort-food standards like

| The Feel | A Day in the Life | The Reality | Earnings and Outlook | Professional Organizations | Interview | Organization and Job Description |

**111**

meatloaf, mushroom gravy, pot roast with root vegetables, and fried chicken with Yukon gold whipped potatoes. The next level supplies a more elegant evening and features the Chef's more adventurous choices. Move down the line to the pumpkin ravioli with citrus-sage butter, wild mushroom and three-cheese risotto, grilled lamb tenderloin à la Grecque (olives, tomato, cucumber, and mint), or grilled pork loin with roasted baby vegetables and apple chutney. Then hit the spread of accompaniments—salads and side dishes that are as delicious as they are enticing. Fabulous platters of these appetizing selections give shoppers an abundance of choices such as broccoli rabe with roasted elephant garlic, eggplant rollatini with fresh mozzarella and ricotta salata, or the pear provolone salad with walnut vinaigrette.

## THE CLOSE OF ANOTHER DAY

The creative opportunities for a supermarket Chef are endless, as is the tremendous customer respect for your many years of cooking experience. It is a great feeling to know you have a new class of students (the customers) interested in your advice, knowledge, and abilities. As a supermarket Chef, you can endlessly improve your skills by self-motivating continuing-education classes; you can spend a few weeks scattered throughout the year to "stadge" (apprentice) at respected restaurants that share your passion for exchanging ideas and information. These are some of the great advantages in being a Chef in our changing world of food.

Today passed like a twister in Kansas, fast and furious. The energy is so positive and the team is so motivated that I feel honored to be around this dedicated group. It is a wonderful feeling to know I am at the peak of my productivity, the prime of my career, and to see I am able to make a difference.

## The Reality

Working as a Chef in a supermarket is not for the inexperienced. On the contrary, it is for those who have many varied experiences and a thorough understanding of the culinary profession. Supermarket Chefs need an unlimited repertoire of dishes and cuisines to support their production. They must focus on sanitation training and HACCP standards to prevent mishaps; they must distinguish between hot foods to be held hot and home meal replacements, which are designed to be reheated. In short, this job requires a top-notch pro with quality skills and training.

Note that this is a corporate atmosphere, and chances are the Executive Chef will be called on to participate in much more than just food production. The Chef must be savvy in the areas of public relations, marketing, business planning, and goals, and must be able to present his or her ideas. Just look at the organizational chart in Figure 6-1, and you will get an idea of all the people the supermarket Chef interfaces with. As the Executive Chef, you will be involved with many types of professionals and expected to have the skills required to interact with sharp corporate business people. You must have strong management skills as well, because your staff will be large and made up of various types of people, including salaried and hourly employees. That is not a negative; it merely indicates that the motivators for your staff will be varied and that you should expect to use many methods to achieve the goals you have set for your business.

## THE GRIND

To a great extent this job has so many facets that there literally is no grind — it is more like a run! You face different issues from day to day and have a wide variety of challenges to conquer. The opportunities are endless and the creativity motivating as you devise new menus, items for home meal replacements (HMR), and store demos. You may even do some teaching in your on-site show kitchen. Your life is busy and interesting at every turn. Of course, you must regularly deal with the financial aspect of your enterprise and must monitor the processes you have implemented to make sure all is well and profit is generated. You must keep communication flowing to your department heads and motivate them to succeed.

## PHYSICAL ISSUES

Mental stamina is the primary issue in the supermarket; the real physical pain was felt prior to this job while you built your cooking experiences. Of course, this is an on-your-feet job, so your feet and legs will speak up once in a while, but that comes with the territory.

## THE BENEFITS FOR YOUR CAREER

The broader market offers discount groceries galore, and you know how much that means if you have a family. But a career like supermarket Chef, which offers so many challenges and opportunities, is always giving you a new light to follow. You can take advantage of extensive networking and get to know a completely different sector of vendors, manufacturers, and grocery specialists. There is enormous support for you to take continuing education or professional development courses. Mega-markets are committed and dedicated to developing well-educated, inspired professionals. The trek is enlightening and built on the sound culinary base you developed in the early years.

### YOUR LIFE OUTSIDE THE JOB

You should be able succeed as a supermarket Chef and at the same time enjoy a high quality of life with your family. Of course, work demands rise at busy holiday times, the occasional ramp-up for a new store, and the promotional rollout of new products. You must travel to headquarters for meetings and planning sessions, and make the occasional trip to check out competition.

## Earnings and Outlook

| 2003 Earnings in Grocery Stores | CHEFS AND HEAD COOKS | |
| --- | --- | --- |
| | Mean (Average) Hourly | Mean (Average) Annual |
| Grocery Stores | $18.50 | $38,490 |

SOURCE: *2003 OES National Industry-Specific Occupational Employment and Wage Estimates,* Bureau of Labor Statistics, 2003.

Employment of Chefs, bakers, and other culinary professionals will grow as the population increases and as more grocery stores offer prepared foods, bakery items, and catering services.

## Professional Organizations

Supermarket Chefs may be members of the National Restaurant Association, American Culinary Federation, and the Research Chefs Association. Information on these associations can be found in Appendix A.

## Interview

**JIM SCHAEFFER, Operations Director of Food Preparation, Wegmans Food Markets, Inc.**

**Q / As Director of Food Preparation for Wegmans, to whom do you report, and how many people report to you?**

*A: I report to Jim Berndt, the Senior Vice President of prepared foods for Wegmans. I have three direct reports in the office. One person does all the nutritionals, labeling, and costing for all of our products. I also have an*

*Executive Chef of Asian cuisine. And another training Executive Chef also works for me. We work directly with all of our Executive Chefs in the stores, doing the openings, remodels, and training, but we also set the standards and specifications for our programs like packaged sandwiches, packaged salads, seasonal soup cycles, different food festivals like Mardi Gras, and Oktoberfest. It is very exciting.*

*We do a lot of different things. For example, we have done a dinner at the James Beard House. I spearheaded that event along with our Chefs from some of our different stores. Also, last week we did an event in our restaurant, Tastings, with David Bouley for two nights. Tastings is our only restaurant, and it is attached to our Pittsford store in Rochester. It has an open kitchen and serves lunch and dinner.*

**Q / What is *Wegmans Menu Magazine*?**

*A: It's a publication that is put out four times a year. It's designed for our customers and gives them menu ideas, recipes, and cooking techniques such as pan-searing, braising, or roasting vegetables. We're in our fifth year of publication. It's been a great success.*

**Q / Can you give me a little bit of a typical day, or if that's not enough, a typical week of what you would accomplish?**

*A: The beauty about it is that it is very, very different all the time. Today I did everything from working on a design for an outside cooking kiosk that had a fryer, a grill, a beverage component, and cash register, to meeting with a soup vendor, to meeting with people with Consumer Affairs about proper seafood temperatures that will be communicated to customers. Tomorrow I have employee evaluations, a financial accounting meeting, which is a couple of hours, and a brainstorming meeting about an upcoming store managers' event where we cater for 120 people. The final meeting of the day is about a central production facility we are planning to open in 2006. We currently have a cook/chill facility where we do our own sauces, soups, and purees. There is a fine balance between what we do outsource centrally and what we do in store. I also of course spend time in stores, cooking, fine-tuning programs, and speaking with our employees and customers.*

**Q / Do you go into core recipes?**

*A: Exactly, and I'll do all the photo shoots for that and then we will do product rollouts, division by division, to make sure everyone gets it, tastes it, feels it, and then they can go back and be the cheerleader for their store.*

**Q / How many stores are there?**

*A: There are 67 stores, and we grow two to three stores a year. This year we have two stores that we are going to open, and next year we have three on the boards. The stores are split into regions, and we have Senior Vice Presidents who are responsible for anywhere from 5 to 12 stores.*

## Q / Where did you go to school, and how did your career unfold?

*A:* I got into this business by accident. I went to a restaurant, the best restaurant in my hometown in Wisconsin, to work as a busboy with a friend of mine. From day one they put me into the kitchen, and I've been working in the kitchen now for over 30 years. All my friends were playing football and I was cooking, and that was not an attractive profession 30 years ago. But I needed to have a job, so I did it, and then once I got out of high school, I wasn't old enough to get a job in a factory so I was kind of forced to stay in food-service, and luckily for me I began to enjoy it.

Growing up, I was a poor high school student because I didn't apply myself. I always thought that going to college was definitely out of reach; my family couldn't afford it. But it was not until I worked a couple of full-time jobs and I started accumulating this money that I figured, the only way I am really going to excel at this is if I get a formal education. I used to think it's all going to be experience, but it's really those classroom settings that are equally important. I wouldn't be where I am if I hadn't had that. I'm sure I would have done well but not as well without the formal education.

I had cooked for five years before I went to culinary school, which I think for me was a big advantage because I was a little more mature and I was paying for the school by myself. So when I got to school I was really focused. I spent a lot of time in the library and the audiovisual area because it was my money. I really didn't have time to mess around. It wasn't my first time away from home, because I had lived in Chicago for three years by myself. Sometimes students go there right out of high school, and their focus is not necessarily there yet.

I went through my two years at culinary school, and I did my externship at the Fairmont Hotel in Philadelphia, where I worked with many talented people. Although a short period of time, I really grasped a lot of knowledge and made a lot of good contacts. After graduation I went to the Playboy Hotel and Casino in Atlantic City, and I opened the hotel as the Banquet Chef in March 1981. The hotel had 18 Sous Chefs, 220 cooks, 190 stewards. Back then, the hotel had $36 million in revenue, which was significant. There were nine restaurants plus banquets. Because of the volume, I got a lot of exposure and a lot of experience that paid off later on, in my years with Hyatt. In 1983 I became Executive Sous Chef at the Playboy.

In 1986 I decided to learn more in another job. I was fortunate that Hyatt Hotels gave me the opportunity to become an Executive Chef at 27 years old. I went from a hotel where expense was not a problem to a 400-room union property that had difficulty paying the debt service every month. From a staff recruitment standpoint, I was competing against hotels and restaurants that could pay their employees more money than I could, so I had to not only provide a paycheck but also had to provide mentoring, upward mobility, and a great work environment. I was really fortunate to develop a great team of people and promote a lot of people.

*In 1989, I was promoted to regional Executive Chef at the Hyatt Regency in New Orleans, a 1,200-room convention hotel. I was in that position for eight years. Then in 1997, I moved on to Wegmans Food Markets, Inc.*

**Q / What is your management style?**

*A: That has always been one of my strengths. I lead by example. I work hard because I want the people to work hard. I wear a hat because I want them to wear a hat. When the stuff is hitting the fan in the restaurant, in the banquet, wherever, my staff knows that I'll be there. They can always count on me and, therefore, I can always count on them. That's huge. No matter what you are doing, you lead by example. The bottom line is that it is going to be up to them whether or not they do their best.*

**Q / What do you find was the most valuable part about your education?**

*A: No question about it, it was skill development. If you don't have a really rock-solid foundation, I think you'll have problems. There's a lot to learn in this field, which makes it a very humbling career.*

**Q / Why do you feel that you excelled in your career? What are some of your best attributes?**

*A: I've always prided myself on being a complete package. There's always going to be people who can cook better than you, but I think the ability to manage a team and mentor people is very important. I'm a very consistent person; I'm methodical. You have to be a professional every day. I've also always been pretty humble and modest. I'm usually one of the first people here and one of the last to leave. There are days that are very, very demanding, and I meet those challenges.*

**Q / What are the things you like the most about where you are now?**

*A: I like it because we pioneer a lot of things. I like it because of who we are. We are always on the cutting edge on whatever it is. We are the envy of the industry. That's pretty exciting to be part of that team. Also, I've been blessed with the ability to explore a lot of things that I only read about in the past. I've traveled to Europe, Japan, Thailand, and Hawaii to check out food and food trends.*

*It's a great company. It's the ninth-best company in America to work for, and we're striving to become number one. Chances are pretty good we will accomplish that. Everyone just operates at a very high level of integrity. We want to deliver the best product to the customer, the top of the line, a great product at a great value.*

**Q / What advice would you want to give young aspiring Chefs?**

*A: Stay hungry, work hard, stay focused, be a good person and a role model. Lead by example. Eventually cream rises to the top. It's always going to be that way, always.*

# Organization and Job Description

## JOB DESCRIPTION

### Supermarket Cook

*Reports To:* **Sous Chef**
*Position Type:* **Full-Time/Permanent**
*Pay Basis:* **Salary**

JOB DUTIES:

1. Prepare meals, including hot and cold foods.

2. Service customers; meet and go beyond their expectations.

3. Display and make products available to customers.

4. Educate customers on meal solutions.

5. Adhere to food safety and sanitation guidelines.

6. Act as a team player by supporting and encouraging coworkers.

7. Maintain two-way communication with management and team members.

QUALIFICATIONS:

Minimum two years' cooking experience

Excellent knife skills

High school graduate, some culinary school preferred

## EXERCISES

1. Visit the Wegmans website (**www.wegmans.com**). List six features on the site that could be useful to home cooks.

2. Compare and contrast the job duties of an Executive Chef in a restaurant with those of an Executive Chef in a supermarket.

3. Why do you need to have a lot of experience to do a job such as Jim Schaeffer describes in the interview?

4. Would you like to work as a Chef in a supermarket? Which activities would you enjoy? What parts of the job might be difficult?

5. Visit several local supermarkets and compare the types of prepared foods they offer. What food-borne illness concerns are possible with supermarket foods?

**Figure 6-1**

Sample Supermarket Organizational Chart

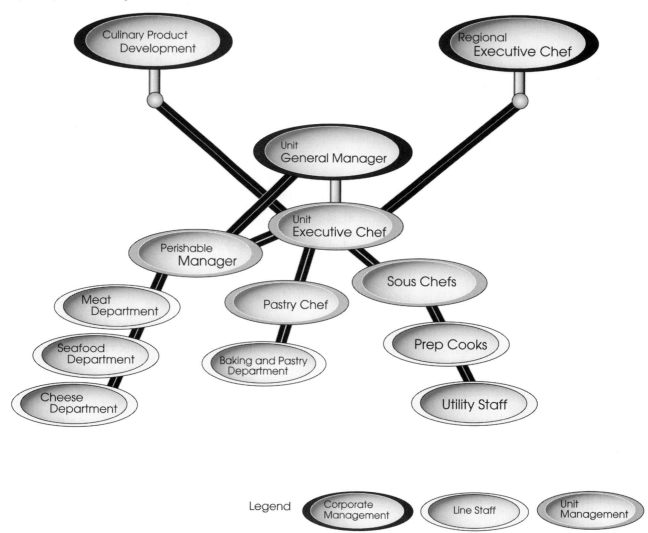

6. Visit the website of Whole Foods Market (**www.wholefoodsmarket.com**). What distinguishes this chain of supermarkets from most others? On their home page, click on "Products," then on "Catering." Print out a catering menu for the Whole Foods Market closest to where you live. What clientele are they trying to attract with their catering menu?

7. Using the Career Path Guide (page 2) and Education Path Guide (page 11), map out a possible career and education path for an Executive Chef in a supermarket, including:

   • number of months in each job

   • each place of employment

   • how and when you would complete your formal education

   • how and when you would complete skills and competencies

# Part 2

# INTRODUCTION TO FEEDING THE MASSES

The chapters in this part of the book discuss on-site foodservice, which in the past was called noncommercial or institutional foodservice. Typical on-site foodservices include those in universities, schools, hospitals, places of work (such as in business and industry), and the armed forces. Defining on-site foodservice is not easy; usually, it involves food served in a location where foodservice is not the primary business. For example, a hospital cafeteria offers meals to hospital employees who come to the hospital to take care of patients. Likewise, a university dining hall feeds students who are at the university to get an education, and the U.S. Army feeds soldiers who work to protect the nation. On-site foodservices are a part of larger operations with a broader purpose—such as providing healthcare or education. The foodservice is there as a convenience or benefit to provide meals to the employees, students, patients, soldiers.

On-site foodservices are different from commercial foodservices in additional ways.

- As an elementary school student, you probably had two choices for lunch: Bring your lunch from home or eat in the school's cafeteria.

You and your classmates were basically what is called a "captive audience" because you had to eat the school lunch if you didn't bring your own lunch from home. You had no other choices. Perhaps as a high school student you were able to leave the building for lunch to go to a deli or quick-service restaurant nearby. You were not quite as captive as in elementary school, but you probably didn't go out to eat every day because you had a limited amount of time. The customers in on-site foodservices often have other foodservice choices, but they are not as convenient or are more expensive. You could say that on-site customers are captive to varying degrees.

- Because a given number of people attend a school, work at a bank headquarters, or are hospital patients, you can actually track the participation rate, or what percent of potential customers use the foodservice. Many on-site foodservices use marketing techniques like their commercial counterparts to raise customer counts and revenue.

- As a Chef in a commercial restaurant, you have a much smaller group of regular customers than the on-site Chef who sees many of the same people day in and day out. Therefore, one of the challenges of the on-site foodservice is to provide enough variety on a day-to-day basis to satisfy customers.

- Because the on-site Chef knows more about the number of customers who are likely to be coming in today for lunch, he or she can plan purchasing and production with more certainty than can commercial restaurants. For example, university foodservices sell board plans to their students. Based on the number of board plans they sell, they can have a good idea of their volume of sales, and therefore production volume, than many other foodservices. This is a plus for the Chef.

- Because of the greater predictability of production, on-site foodservices are usually not as frenzied as their commercial counterparts, which experience more fluctuation in customer counts and menu item popularity. Also, working hours tend to be shorter and more predictable in on-site foodservices.

As you can see, on-site foodservices have their own benefits and challenges.

On-site foodservices are either self-operated or independent, meaning that the Chef and other employees are employed directly by the institution or managed by an outside company known as a managed services company, foodservice contractor, contract company, or management company. A managed services company is a business that provides foodservices at institutional, governmental, commercial, and industrial locations for a specified period based on contractual arrangements. The contract may be based on profit and loss, management fee, or a mix of the two.

When a managed services company has a contract with a hospital, for example, to run its foodservice department, the hospital becomes the client of the managed services company. The managers for the hospital foodservice are provided by the managed services company, which may also pay the salaries of the hourly employees, depending on the contract.

Managed services companies provide management to approximately 80% of business and industry, 50% of college and university, 40% of hospital, and 10% of school foodservices. A business may decide to contract out its foodservices to a managed services company for various reasons: to bring in the contractor's employees and save money through reduced payroll, to let the contractor take care of hiring and paying employees, to reduce food costs due to national purchasing contracts, to allow the contractor to renovate or build new foodservice facilities that otherwise wouldn't be developed, and to provide management expertise the client normally wouldn't have. On the other hand, a hospital may decide to stay independent for other reasons: to retain revenue from cash operations, to eliminate management fees, to allow the foodservice managers to have more flexibility and ability to innovate, to retain control, and to avoid the divided allegiance of managers who serve two masters—the managed service company and the client.

The bottom line is that neither option, self-operated or managed services, is better; rather, each has its niche. As a Chef, keep in mind that managed services companies do generally offer more opportunities to work in varied venues, more possibilities for promotions, and excellent development and training programs. Examples of managed service companies include Aramark, Restaurant Associates, Sodexho, Canteen, and the Compass Group.

## FEEDING THE MASSES CAREER PATH GUIDE

Figure P2-1 shows the Feeding the Masses Career Path Guide. It is only a guide for you to get an idea of what type of work experience you need to move forward in your career. The Career Path Guide indicates the number of months in various positions that ideally prepare you for the Executive Chef and other positions noted at the bottom of the chart.

The Career Path Guide should help you understand the time commitment needed to reach top management positions in the foodservice industry. By no means do we claim that these paths are the only way or that the time frames are precise. Some people will excel much quicker than we indicate, and some may need more time. The length of your workweek, your dedication, and your commitment are all factors in the time frame needed to obtain your final goal. In addition, you may be able to hold two positions at the same time. For example, you could probably get experience as an expediter while working as a grill, sauté, or pantry cook. Similarly, you could simultaneously be a roundsman and a butcher or a garde manger.

**Figure P2-1**
Feeding the Masses Career Path

## On the Way

### Nonmanagement positions*

| Position | | | | | | | |
|---|---|---|---|---|---|---|---|
| Prep cook | 8 | 8 | 8 | 8 | 8 | 8 | 8 |
| Pantry cook | 6 | 6 | 6 | 6 | 6 | 6 | 6 |
| Grill cook | 12 | 12 | 12 | 12 | 12 | 12 | 12 |
| Sauté cook | 12 | 12 | 12 | 12 | 12 | 12 | 12 |
| Expediter | 6 | 12 | 6 | 6 | 6 | 6 | 6 |
| Breakfast cook | 6 | 4 | 6 | 6 | 6 | 6 | 6 |
| Banquet cook | 6 | 8 | 12 | 12 | 8 | 12 | 12 |
| Butcher | 4 | 6 | 6 | 6 | 6 | 6 | 6 |
| Roundsman | | 8 | 8 | 8 | 8 | 8 | 8 |
| Retail counter cook | 6 | | 6 | 6 | | | 6 |
| Garde manger | 8 | 8 | 8 | 8 | 8 | 8 | 8 |
| Pastry cook | 8 | 12 | 8 | 8 | 8 | 8 | 8 |
| F. H. service staff | 4 | 6 | 4 | 4 | 6 | 4 | 4 |

### Management positions**

| Position | | | | | | | |
|---|---|---|---|---|---|---|---|
| Sous chef | 12 | 12 | 12 | 12 | 12 | | 12 |
| Banquet chef | 12 | 12 | 12 | 12 | 12 | 12 | 12 |
| Restaurant chef | | 24 | 12 | 12 | 12 | 12 | 12 |
| Executive sous chef | | | 12 | 12 | | 12 | 12 |

**Finish as**

| B&I Exec. Chef Staff Restaurant | B&I Exec. Chef Exec. Dining | B&I Exec. Chef Catering | Regional Chef B&I | Executive Chef Assisted Living | Executive Chef Hospital | Executive Chef University & School |
|---|---|---|---|---|---|---|

*Number of months in the position at 40 hours a week.
**Number of months in the position at 55 hours a week.

# FEEDING THE MASSES EDUCATION PATH ADVICE

Figures P2-2 through P2-4 give education path advice for business and industry, universities, and hospital foodservices. The Education Path Guide displays the level of importance (through the relative height of the balls) of formal education, knowledge, and competencies needed for positions in different settings.

When you plan your educational path, the key is to remember that education is an investment in the future on which you will draw over your entire career. You may want to go to a two-year intensive cooking degree program, then work in the kitchen to perfect your skills, and return to school for another two years or so to get a bachelors degree. Make a plan to maximize your time and your education.

The foodservice, hospitality, and culinary professions are first and foremost hands-on professions, but you should understand that they present minds-on challenges as well. As your hands carry out the precision tasks of common kitchen activities, your mind must retain, compute, and comprehend these important skills you are perfecting. As you absorb the academic aspects of your profession—for example, financial reporting, computer techniques, menu costing, and menu development—the hands-on aspects make these skills come alive. The hands on, which is the passionate and creative part of your profession, complements and motivates all facets of the work. Did you ever notice that when you are committed to a topic, you find that learning about it is easy? You're not studying, you're absorbing; you're not being tested, you're remembering.

Don't be intimidated by the Education Path Advice. Keep in mind that any academic program you need to complete will be filled with examples and situations related to the hospitality field that is your first love. Take one of the subjects in the education path that may sound threatening. For instance, food and beverage (F&B) financials may be scary because of the word *financials*, which makes it sounds like the class is all math problems. This is true—except the math problems all represent real-life situations you will need to know how to handle as your career unfolds. Financials and math problems are not as scary when you encounter them in the business you are committed to and passionate about.

Continue to advance your abilities with classes that will be an asset to your career. You could learn Spanish, which is spoken every day in many kitchens. Or you could take a public speaking course to help you do cooking demonstrations or other group presentations. These investments in your future will definitely pay off.

The educational tools we have mapped out are a matrix of the many directions available. We hope you use them to inspire, plan, and evaluate a successful journey toward a rewarding culinary career.

**Figure P2-2**

Business and Industry
—Education Path Advice

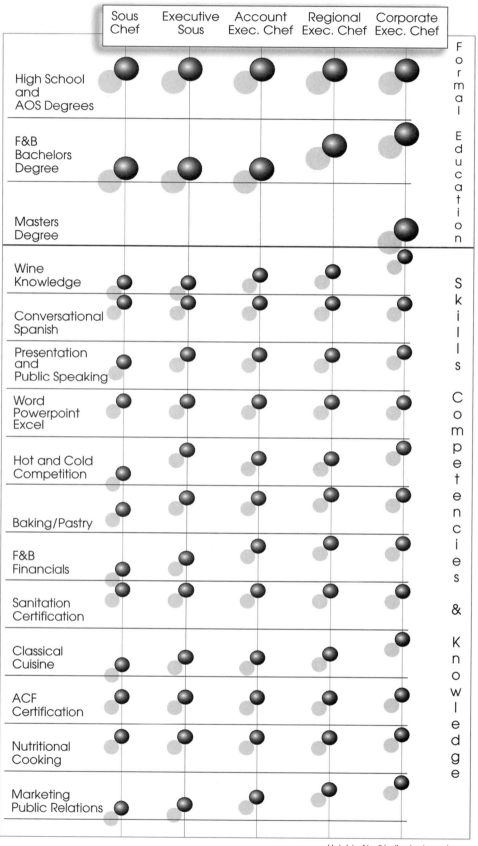

Height of ball indicates importance.

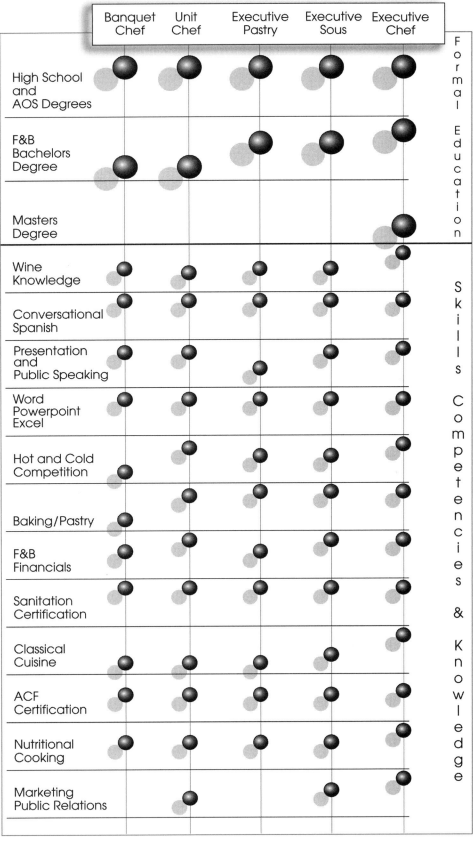

**Figure P2-3**
Schools and Universities
—Education Path Advice

Height of ball indicates importance.

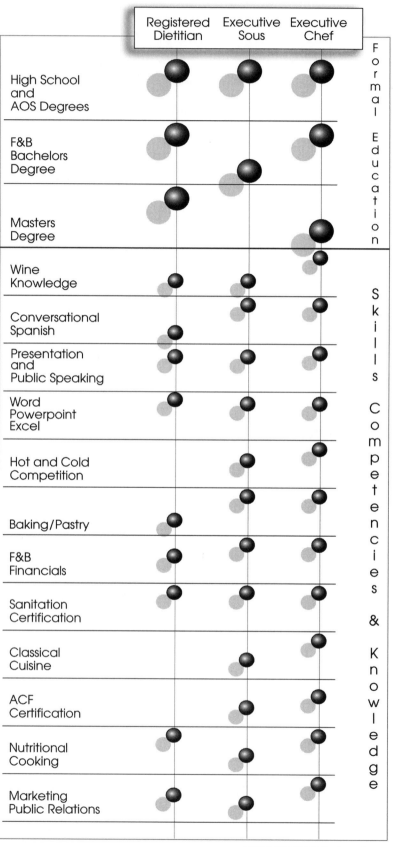

**Figure P2-4**

Healthcare—Education
Path Advice

Height of ball indicates importance.

## ADDITIONAL ON-SITE FOODSERVICES

Part 2 discusses on-site foodservices in business and industry, universities and schools, healthcare, and the armed forces. On-site also includes in-flight foodservice, recreational foodservices, and correctional foodservices, which are discussed here.

In-flight foodservice provides meals and snacks to people traveling on airlines. As you can imagine, in-flight foodservice is quite specialized because the food must be prepared, stored, transported, and then, in the case of hot foods, reheated on the plane and served. Tens of thousands of meals are prepared and delivered to planes for in-flight service. Scheduling is particularly important to avoid delaying flights.

Airlines often outsource foodservice to catering companies, such as LSG Sky Chefs, that specialize in preparing meals for flights. Caterers are awarded contracts, much like a managed service contract, and prepare the food in commissary kitchens located near airports. Serving freshly made food is a challenge when it may be hours before the food is served on the plane.

Although some airlines are cutting back or even eliminating in-flight food, there is no clear trend. Other airlines are putting more emphasis on good food and hiring world-class Chefs to plan menus and oversee food production. Chefs who work in this segment may become members of the International Inflight Foodservice Association (see further information on IFSA in Appendix A).

Recreational foodservices include food served at sports areas and stadiums, theme parks, racetracks, state and national parks, zoos and aquariums, and even the Olympic Games. Some recreational foodservices are run by managed services companies. For example, ARAMARK has much experience providing foodservices for the Olympics. Food choices offered in recreational venues have improved over the years. For example, baseball stadiums have always been known for their hot dogs and popcorn, but these days more ballparks are offering local dishes and upscale foods. For example, San Francisco's 3Com Park has vegetarian burritos and a 40-clove garlic chicken sandwich on the menu. In the Northeast, cheesesteaks are a best-seller in Philadelphia stadiums, and clam chowder is popular at Boston's Fenway Park. In some cases, a national brand such as McDonald's operates in a zoo or other recreational facility.

Correctional foodservices provide meals to people in prisons and jails. Correctional foodservice managers provide nutritious, cost-efficient meal service for confined populations. The American Correctional Food Service Association (ACFSA) is an international professional association serving the needs and interests of foodservice personnel in the correctional environment. ACFSA sponsors conferences that help its members sharpen management skills, promote professionalism, and keep up-to-date on developments in foodservice and corrections. ACFSA also has a quarterly journal, *Insider*, and two certification programs: Certified Correctional Food Service Professional and Certified Correctional Food Systems Manager.

# Culinary Careers in Business and Industry

## Introduction

**CHEFS IN BUSINESS AND INDUSTRY** prepare meals and snacks for people who are working. Business and industry foodservices (commonly referred to as B&I) provide meals, for example, in financial and insurance companies, manufacturing plants, service companies, and high-tech businesses. Sales in B&I foodservices come from the café (which used to be called the cafeteria, but is often quite upscale nowadays), executive dining rooms, catering, and vending.

B&I foodservices must be innovative to compete effectively with restaurants in the area. Companies also want a pleasant dining room and good food so employees stay on the premises and enjoy themselves during breaks and at mealtimes. A well-run foodservice has a positive effect on employee morale.

B&I foodservices often charge reasonable prices. In some cases, the company subsidizes the cost of the operation. This means that the revenue brought in by foodservices does not completely cover costs, and the company kicks in whatever money is needed to continue operating.

## The Feel

### THE WORLD OF B&I

During my tenure in B&I, I was invited to travel to other facilities to teach, assess, and elevate the culinary level of our company. I was fortunate to visit an array of dining locations. Some were as big as ball fields, others small yet accommodating. The script for B&I properties is huge, from the corporate giants to the small companies sharing centralized cafés. All have a common goal: to feed the business world of America a quality breakfast, lunch, and sometimes dinner. Some run seven-day schedules, but most operate a standard workweek that allows for a normal quality of life.

## THE LAW FIRM

I feel like I landed on the set of *Law and Order*. I'm here at one of the most prestigious law firms in Manhattan to evaluate the executive dining room and conference catering. The plush interior and elegant ambiance set the tone for how the company operates both within its walls and externally. My first and most important stop is a fresh latte at the coffee bar. The surroundings are swank and stylish for a cafeteria, better referred to as an employee restaurant.

Breakfast starts with an array of dense and delicious muffins, scrumptious scones, flaky croissants, gooey sticky buns, buttery cheese danish, and New York's finest bagels with all the trimmings. An abundance of fresh fruits in season complements an assortment of yogurt and granola parfaits. There are eggs any style, hickory-smoked bacon, thick maple-cured ham, chicken-apple sausage, Belgian waffles, buttermilk pancakes, corned beef hash, chipped beef on toast, and don't forget the Krispy Kreme doughnuts. Not every facility is this complex, but new standards in corporate cafeterias call for a significant range of choices. This trend will continue to mature in years to come.

Lunch is where it all started, like the first food court found at the Galleria. What is your whim today? Sushi, classic Italian, burgers, pizza or calzone, authentic Indian, or a trendy panini station. There's a salad bar with every topping imaginable and an entrée of the day presented with the care and planning of a home-cooked meal. The beverages, exotic coffees, cappuccinos, and lattes are as creative as at any upscale New York eatery.

## THE BRIGADE

A brigade of culinarians in starched whites lines the perimeter of these food stations, ready to prepare the many choices available to you. Lavish buffets lines with induction chafing dishes are cutting-edge in efficiency and presentation. The look, the style, the feel are important components of the execution of the meals. The pizza station is equipped with an imported wood-fired oven, allowing the smell of pizzeria-quality calzones to permeate the eatery; calzone is today's top choice. The parsley and sun-dried tomato sausage with roasted peppers and onions would give the San Gennaro Feast in Little Italy a run for its money. Every week, many places offer a theme cuisine such as A Taste of Tuscany, Foods of the Mediterranean, Western Hoedown, Texas BBQ, and the foods of many traditional holidays.

In an atmosphere similar to that of country clubs, where you feed mostly the same clientele every day, you need to be predictable even as you surprise your guests with tempting dishes, imaginatively displayed, and mouthwatering demonstrations done to order. There is a balance of well-prepared foods with an emphasis on healthy fat percentage, salt content, and cholesterol count, according to nutritional guidelines. The variety alone can accomodate almost any dietary need. Atkins and South Beach Diet are common terms when lunch is your primary focus.

## THE KITCHEN

The kitchens are contemporary, for the most part, updated and renovated along with the food and seating areas in a time frame acceptable in the industry. Equipment is purchased as trends direct the next culinary spotlight. Tandoori ovens, open-fire grills, and rôtisseries, to name a few items, give versatility to the menu design. The layout of most kitchen facilities compares to that of a hotel, country club, or catering company. The à la carte cooking batteries are professional in design and can satisfy almost any Chef's needs.

## CONFERENCE CATERING

The bread and butter, meat and potatoes of a B&I account is the daily cafeteria, now called a Food Court, Employee Restaurant, Café Dining Facility, or other company-dubbed name. Beyond that portion of the foodservice is a whole world of hotel-style catering, complete with linen, plated meals, banquet staff, and a catering manager. Larger facilities may employ a Banquet Chef dedicated, under the supervision of the Executive Chef, to run last-minute high-level events.

Picture many small conference rooms scattered around a law or accounting firm, investment bank, or other corporate headquarters. These are where client entertaining, board meetings, brainstorming sessions, and weekly staff conferences take place. These functions typically begin with a continental breakfast consisting of sliced and whole fruits, pastries, muffins, bagels with assorted jams, preserves, butter, cream cheese, and coffee service. The philosophy is that there's no need to go outside when you have the convenience of quality catering a phone button away. For companies, the savings is significant in many ways. Your staff remains on the premises and is therefore more productive, while at the same time your in-house costs are considerably lower than going out.

Catering continues with three-course plated business lunches, in-house meetings with extravagant hot buffets and impressive salads, all-day workshops with roll-in deli and wrap stations. Creative catering is the tune for these new sources of revenue. Giant office buildings have foyers the size of basketball courts, with atriums three or more stories tall. What a great space for a reception, a plated dinner, a corporate announcement, or the company holiday party. Why? Why not? These questions are being answered by aggressive foodservice companies thinking harder and smarter for their clients and themselves.

# A Day in the Life

## THE FIRST DAY

I wake up every hour on the hour the night before I start my new position. After 20-plus years in the business, why such anxiety? I'm confident, mature, and seasoned in my craft. Why do I keep dreaming of oversleeping? Could it be all the new things I'm going to do tomorrow that I've never done in my life? It's the first time I'm taking a 40-minute train ride to work, the first time employed in New York City, the first time operating a kitchen in a B&I corporate setting—and the first time in 20 years I won't be required to work weekends.

It's 5:30 AM. The alarm sounds and I jump to attention, like a soldier in the military. The address of the prestigious investment firm where I will be stationed for the next four years is in midtown Manhattan. The world of corporate dining has been a well-kept secret in the foodservice industry. Its reputation from years past was mediocre food, run-down kitchens, and untrained staff. This reputation was fought by the trendsetters in the field who should be recognized for their dedication to converting the way America thinks about corporate cafeterias. The philosophy of corporate dining over the past several decades has changed with the transformation of our eating habits, the demand for higher-quality food, and the need to retain conference catering in-house.

Its 6:45 AM, when I enter the 30-story glass-walled building, to be greeted by a platoon of security staff. "Morning, Chef," they call as I flash my company ID. "Morning, gentlemen, what's up in the good world?" I reply. "You, Chef, you," they quickly snap back. A thumbs up, and I hop into the private elevator that whisks me up to the VVIP executive dining and conference floor. It looks straight out of an old black-and-white movie: dim lighting, arched ceilings, plush carpets, hand-crafted woodwork, and priceless artwork, not to mention a mysterious corporate ruler who is feared and respected. It's an environment I am used to, given my years of training under demanding and talented Chefs.

There's a buzz on the floor of waitstaff zipping around getting the extravagant continental breakfast ready. The smell of freshly baked croissants, buttery fruit-topped breakfast pastries, and made-from-scratch low-fat muffins are the first scents I experience. The managing directors at this site, my audience, share my belief that breakfast should offer a plethora of choices that are satisfying but not the traditional fat-laden breakfast items. Seventeen of your favorite cereals, all in 16-ounce boxes, are stationed on a gorgeous imported marble credenza for self-service. Note the array of golden pineapples, two kinds of grapes, five whole-fruit selections, and several melons in season, cut into perfectly trimmed wedges. Depending on flavor and time of year, there are ruby red or pink grapefruit, fresh berries, and five kinds of dried fruits, from Turkish apricots to medjool dates. Decorative silver bowls are lined neatly on crushed ice, brimming with freshly cooked egg whites, nonfat yogurts, and low-fat cottage cheese. Ornamental stands line the tables stacked with New York bagels, country white loaves,

The
Feel   A Day
in the Life   The
Reality   Earnings
and Outlook   Professional
Organizations   Interview   Organization and
Job Description   **133**

Jewish rye bread, eight-grain health bread, accompanied by a variety of spreads, marmalades, imported and domestic preserves.

The finale and daily best-sellers are the cooked steel-cut oats and the Swiss-style birchermuesli complete with organic honey, Devonshire cream, raw brown sugar, golden raisins, and toasted nuts. This fabulous display looks as if you opened a *Saveur* or *Bon Appetit* magazine. Not only is the presentation perfect but the fruits are at their peak of ripeness and the quality extraordinary. Then, just before the directors get tired of the same assortment of items, you, the Chef, dig into your magic file of product tricks and introduce a new twist to your already magnificent breakfast. There is a great deal of room for creativity and freedom at these venues, and in return your clients appreciate your hard work. This is an art, a passion, an obsession that must be preserved and nurtured daily.

As breakfast service commences, the lunch prep is in full swing, with the culinary staff making circular paths around the kitchen like the Indy 500. Intensity, dedication, infatuation, perseverance, enthusiasm are expressions to describe the type of commitment this kitchen inspires.

## EXECUTIVE DINING

The air reeks of money. High-energy power lunches, client entertaining, political affairs, financial reviews, and recruiting receptions are all in the palms of your hands. Senior managing directors, superstars, athletes, political figures, and the upper crust are your audience. Menus showcase diversity: from a torchon of foie gras, red onion marmalade, tiny greens, and our own butter crackers, to nutritionally balanced selections, or a Coney Island frank with all the toppings. Let's not forget to mention the killer hamburger and house-cut fries with a fresh-baked dill roll, or appetizers of lobster head cheese, house-made spicy chicken and seafood sausages, jumbo lump crab cakes, and hand-crafted ravioli followed by a selection of fresh fish, certified Angus steaks and chops, lamb three ways, poached chicken breast with keeper squash hash, tiny mushrooms, natural jus, and other composed entrées wrapped up into about 100 covers per day. This is a rewarding experience focused on the pure essence of cooking, flavoring, and technique.

## THE END OF THE DAY

The rush is over, but the staff works diligently to get the place back in order, for tomorrow's another day. It's kitchen downtime, a quiet, peaceful moment when a Chef just sits, stares, and plays back the day's events. Mental notes are made to correct a flaw or to praise a fine job. Critiquing your own work keeps your job fresh and challenging. Tweaking is a word for the lazy; bold changes, risks, and perfection are where my drifting thoughts roam. Well, time to go. Remember! Chefs never sleep; they merely rest. Tomorrow is another day. It's great to never be finished.

# The Reality

In this segment of the industry, everyone's opinion, which includes your client, daily customers, and boss, drives your initiatives. The important thing to remember is that you provide an amenity, a service that is purchased by management and provided for an employee, manager, partner, guest. If you cease to produce a valuable service (based on the opinion of those you feed), your account will go out to bid to the competition. As every Chef in this business knows, you will need to transfer to another account or find another job. The good part is, if you consistently run your operation with a clear objective—to make sure your customers see the difference you make in their dining world—then you will become a fixture and, barring a corporate change, you will have a good job for years to come. Always remember, your function is to create the repast for a busy executive, a harried secretary, or a particularly demanding partner. You are the oasis in their corporate world, and they will look forward to visiting two or three times a day if you can keep the oasis green and pleasurable. Use this to your advantage and make their day.

## THE GRIND

To a great extent, you run an operation that requires daily attention and a real understanding of how to supply variety without exceeding cost parameters. Oh yes, you will be given cost guidelines; corporate foodservices often operate at a loss or on a subsidy basis, and you must fully understand and manage the costs as prescribed by the agreement between the foodservice company and its client. All agreements differ in the details. Fundamentally, however, the agreement is that the foodservice company provide the agreed-upon services at a predetermined level of cost in exchange for a specified fee. With that said, I am sure it is obvious that cost management is a skill in which you must excel.

The fun part is that you must change your menus quite often in order to keep the clientele interested, and you probably will have a reasonable food budget to play with. Of course, your freedom is delineated by context: the type of establishment you serve and the level of expenditure afforded by the employer. Regardless, your challenge is to provide daily offerings of acceptable quality and suitable variety in a timely fashion of perceived value to the employees.

The manner in which employees pay for their food varies. They may simply pay for it, or they pay but are reimbursed, or they receive an allowance. What does not vary is that the diners consider your service a benefit and will be critical of your performance. This is not as threatening as it sounds, for most of us in the industry are driven by passion and love for our cooking style. We take pride in what we do and are constantly challenging our skills and abilities. The main point to understand in this setting is that you feed the same core clientele every day, with a primary client as your liaison.

Although in many ways corporate foodservice is less stressful than a 24/7 operation, you nevertheless must be on your toes with your culinary bag of tricks close at your side. On the other hand, you may well be supported by the managed services staff, which supplies information for monthly promotions, a repertoire of fabulous recipes, culinary demos, and food shows with trendy new products and equipment.

## PHYSICAL ISSUES

As with most Chef positions, the legs get a good workout, but this life tends to be a bit gentler on the system. Burns, cuts, and bruises are facts of life for all Chefs; however, the activity level is a bit more placid than that in a hotel or restaurant. Your days are shorter, as are your workweeks, but the job is every bit as demanding in other ways. The corporate environment usually features an excellent support system, and your facility is probably in better condition than many restaurant kitchens.

## THE BENEFITS OF YOUR CAREER

Being the Chef in a corporate foodservice is highly rewarding, and both you and your staff will enjoy the workplace. You are the one the clients turn to for a recipe for home or help with a menu for a special occasion. You are the person to whom they turn for dietary advice, cooking tips, and maybe outside catering. With respect to your staff, you are the leader who takes the time daily to teach them, look out for their best interests, and make sure their employment needs are met. In short, you are a very important part of the work lives of many people every day, and that is very rewarding. As with many culinary jobs, there is a family feel to staff relations, and you are the older brother or sister.

## YOUR LIFE OUTSIDE THE JOB

Happily, corporate foodservice work leaves you free at almost human hours to enjoy your life, family, and whatever else you may choose. The job does have built-in late nights and early mornings, but generally you can take off most holidays and weekends. Knowing this, you should of course take advantage and live your life with normalcy. The amount of time you spend on the job usually indicates your level or position in the company. For instance, as your status grows, you will be more involved in opening new operations and guiding the company as a whole.

# Earnings and Outlook

| 2003 Earnings in B&I Foodservice | | |
| --- | --- | --- |
| | **CHEFS AND HEAD COOKS** | |
| | **Mean (Average) Hourly** | **Mean (Average) Annual** |
| Business and Industry | $25.13 | $52,280 |

**SOURCE:** *2003 OES National Industry-Specific Occupational Employment and Wage Estimates,* Bureau of Labor Statistics, 2003

The outlook for jobs in B&I foodservice depends, of course, on the economic health of business and industry itself. Total employment for all industry sectors is projected to grow about 15% from 2002 to 2012. Data indicate that employment in the professional and business service sector will increase 30.4% in that period, the second-highest rate of growth of all sectors. Education and health services industries, transportation and warehousing, and information industries are others anticipating rapid growth. Employment in manufacturing is expected to decline slightly.

# Professional Organizations

General culinary associations, such as the American Culinary Federation, are discussed in Appendix A. This section mentions associations for culinary and foodservice professionals working in business and industry.

## SOCIETY FOR FOODSERVICE MANAGEMENT (SFM)

### Who Are They?

The Society for Foodservice Management (SFM) serves the needs and interests of executives in on-site foodservice industry (predominantly B&I). SFM provides member interaction, continuing education, and professional development via information and research.

### Where Are They?

304 West Liberty Street, Suite 201
Louisville, KY 40202
502-583-3783
**www.sfm-online.org**

### Publication

Monthly email newsletter: *FastFacts*

# Interview

**BILL CHODAN, Vice President of Culinary Operations, Flik International**

**Q / What is your current position, and to whom do you report?**

*A: My job is Vice President of Culinary Operations with Flik International. I report to the president of Flik.*

**Q / How many people do you supervise?**

*A: In my department, there are six corporate Chefs, a nutritionist, and someone who handles the paperwork of our department. So there are eight direct reports, but in theory all the culinary employees of the company and Executive Chefs report to the corporate Chefs and myself.*

**Q / How many accounts does Flik have?**

*A: We have about 220 accounts in 188 locations. Some accounts we count as more than one. For instance, in a large law firm, we may count the cafeteria as one account and catering and executive dining as two additional accounts.*

**Q / What's a typical day of work?**

*A: That's a rough one. If there is ever a typical day, it is just trying to return calls and respond to people and respond to problems. Of course, a lot of times I am like a firefighter who gets 20 to 30 voice mails a day. Some of them need immediate attention, such as a Chef resigned or a client wants a special party—things that just need particular attention. Another hat I wear with Flik is that I am in charge of all the account renovations, designs, construction, so a lot of my time is taken up by meeting with clients and architects about building new facilities, new food programs, etc. There really isn't a typical day.*

**Q / How many hours do you work in a day?**

*A: The minimum is probably 12 hours a day. I leave my house by 6:30 AM and get into the office (if I'm going to the office that day) by 7:30. I usually don't leave there or wherever I am until 7:30 to 8:00 PM, and get home around 9:00 PM. Very rarely do I get home before 8:30.*

**Q / How many days a week?**

*A: I would say five is the average. We don't do a lot of weekend work, except when there are special events, we open a new account, or we have renovations in an account. For instance, we are taking over a new account in New Jersey next Friday, so we have to go in there Friday afternoon. Once the old company leaves, we move in and get everything prepped by Monday to open. So the amount of weekend work depends on how many openings and new pieces of business we take on. If I'm renovating an account, it's always done on weekends. You go on Friday afternoon, do the demolition work, and then it all gets put back together during the weekend so you can open up on Monday. It depends how much growth we're doing as a company and how many projects are going on.*

**Q / How long have you been with your current employer?**

*A:* *I've been with Flik for 15 years. I started as a corporate Chef with Flik. In its infancy, Flik only had one account in Manhattan. I knew people in food-service, and Flik was an up-and-coming company. So I called them, and Mr. Flik said the company's growing and is going to need a corporate Chef. He said if I did a good job for him, that he would eventually make me corporate Chef. So I took the risk, and went to work in this law firm in Manhattan. I did a couple of years there, and proved what I could do and turned the account around. Mr. Flik was true to his word, and made me his corporate Chef about three years later. I was the only corporate Chef, and then the company started to really grow. We started landing law firm after law firm, and we continued to grow. I would open the new accounts, train people, and my role just sort of evolved. Then as the company really grew, we needed more corporate Chefs to assist.*

**Q / What were you doing before you joined Flik? Did you plan out your career?**

*A:* *There was no master plan for me. I wasn't a great student in school; I was probably quite the opposite. I just didn't apply myself. I knew I'd do well at whatever I did, but I didn't know what I wanted to do. I just happened to be cooking in a restaurant when I graduated high school. So over the course of three to four years, I worked in some different restaurants and then I got an opportunity to go work for a hotel. I thought that would be interesting because at that time you had a lot of different trades in the hotel. You had a Pastry Chef, a Garde-Manger, a Butcher. I thought that could be good for I had no formal schooling, so I went into the hotel and started as a breakfast cook. I got to work in banquets, with the butcher, with the pastry department, and really learned from some good Chefs and took a little something from each one. Over a course of five to seven years, I started working my way up the hotel to Banquet Chef, Executive Sous Chef, and eventually Executive Chef. I did that for ten years with Stouffer Hotels.*

*At that point, after ten years in the hotels, I knew the only way I could move up was to relocate to other hotels around the country, and I didn't want to get into that. That's when I decided to get out of hotels and look for something else. That's when I found out about Flik.*

**Q / Is there something you would do differently if you could go back now and plan?**

*A:* *I probably would have gotten more formal training because I really had none. It was more just learning from other people. So I may have missed certain elements of a formal training. Actually, I used to go home and practice skills I was weak in. I know that sounds silly but pastries, for instance, I never really spent a lot of time in the pastry shop, and we had a good Pastry Chef in the hotel, so I would ask him for recipes. Then I would go home and bake something to learn that technique, such as make a great cake or marzipan. But I guess a little more formal training would have helped early on.*

**Q / How did you pick up things like speaking and working well with others?**

*A:* It was a long, drawn-out process for me. I'm a streetwise kid from the Bronx. I think it was just constant development over the past 20 years. Probably the last ten years with Flik really helped my development. Rudy and Julie Flik taught me that how you present yourself, both physically and verbally, is very important. It was rough going at first. There were actually points in my career, even with Flik, when I felt I may not make it. I didn't express myself as clearly as I wanted, nor was my appearance at the corporate level. But I had good role models who liked me and saw I had skills, which just needed to be polished. So, really it was just a development process over 20 years.

**Q / What's your management philosophy?**

*A:* I really think it is treating people fairly and making sure they have fun coming to work. And that's probably another side of me—I try to have a little too much fun at times. If people have fun coming to work, I think they will contribute more. I think it all comes down to respect. You deal with everyone on the same playing field. You can't treat different levels of the kitchen or the hierarchy of the kitchen different; you treat them the same. You say good morning to everybody, you thank them for coming in every day, thank them when they do a good job. If they don't do a good job, tell them and be honest with them. I think that's the roughest part of a job is when you have to tell someone they're not doing a good job and here's why. I can honestly say, in all my years, out of all the people I've mentored, even those I had to terminate, they usually thanked me for treating them fairly. I told them when they made mistakes, told them what to correct, and warned them what would happen if they kept coming in late or whatever the problem was.

It's also important for people to like you. When people like you, you'll get more out of them. Let's say you have two individuals and they both have equal skills, but one is a little ornery and sullen at times, comes in with a bad attitude, or is not happy to be here. The other guy is happy to be here, always has a positive attitude, and never says no. If you had to choose between those two people, it's always going to be that person with the better attitude. It has to be sincere; it can't be fake. You can see I want to be here, I like being here, and I'm having fun being here. People have to like you. I would not have moved up as much in this company if people didn't like me and if I wasn't open to change and critiques.

**Q / What do you like most about your job?**

*A:* I like being able to control my job. One thing about Flik is that we have always had an entrepreneurial spirit. If we see a great trend out in a little store down in Greenwich Village, I may want to try it out. I can do that and be very creative. I guess I like the freedom and the variety too. My personality likes to be continually challenged. We are always landing new accounts, new pieces of business, and diverse pieces of business, so I am continually challenged. I prefer not to stagnate in one position, and [don't like] not being able to move out of it.

**Q / What do you like least about the job?**

*A: One of the harder things of the job is keeping up with voice mails, cell phone messages, emails—just getting back to all those people and responding to them on a timely basis. Obviously, the best thing is to prioritize them. Even if a vendor calls me and he wants to sell me some bizarre product that I have no interest in, I like to get back to him and say thanks, but no thanks. Because somewhere down the road, you may cross paths again. It's back to respect and treating everybody equal and giving them your time. But it's getting hard to give everybody that amount of time they deserve. At the end of the day, you see you haven't gotten back to a person and feel guilty. I think that's the hardest thing. You have to prioritize your day, and try to get back to everybody.*

**Q / Speaking to a young person who wants to end up where you are, what advice would you give them?**

*A: Build a good foundation; really learn all aspects of the business, culinary and otherwise; get good training; understand a little bit of everything in the kitchen. You really have to be a jack of all trades. Make sure people like you and get their respect. Be trustworthy and dependable. That's probably the two biggest things I would say that's been helpful for me. And try not to lie. Tell people the truth. You'll get caught in the lies. You have to be honest with people, but they'll appreciate it in the long run. Be dependable, too.*

**Q / What are some of your thoughts about trends in your industry? Where do you think it's going?**

*A: Keeping up on ethnic trends is really important. The great thing here in Manhattan, you can walk around the neighborhoods and you'll see the trends out on the street. I also try to read a lot—I'll read a magazine in 25 or 30 seconds. As I flip through a magazine, if something catches my eye, I cut it out and put it in a folder. Then when I'm challenged or I need new ideas, I reach into that folder. I actually have folders labeled Spring, Summer, Fall, and Winter, and if I see something I like, I'll throw them in those folders. And then going into the Fall, I'll go into the folder and say that's great, I saw that in the food show or in the little store in the mall. We do what we call retail tours, where we travel around New York and see some hot shops and new things out there. I just visited a store that sells 40 types of rice pudding with great roasted nut toppings and fruit toppings, and it's all natural. I ask myself if I can take an element of that and apply it to my business. Everybody says there's no new ideas, you just steal the old ones and put your spin on them. I think there is some truth to that. I think it's just rediscovering old cuisines and tweaking them to make them work for you. I think the importance is the authentic use of the ingredients.*

# Organization and Job Description

## JOB DESCRIPTION

### Chef, Private Executive Dining

*Position Type:* Full-Time/Permanent

*Pay Basis:* Salary

*Reports to:* Executive Chef

### RESPONSIBILITIES:

*Supervise and coordinate all related culinary activity for executive dining rooms.*

*Develop daily custom menus for the CEO's office and board events.*

*Develop and train waitstaff to understand menu concepts.*

*Prepare scratch baked goods for catering and elsewhere.*

*Establish presentation techniques and quality standards.*

*Plan and price menus.*

*Estimate food consumption and requisition food.*

*Select and develop standardized recipes.*

*Control inventory.*

*Ensure proper safety and sanitation in kitchen.*

*Ensure proper equipment operation and maintenance.*

*Maintain a commanding presence in the dining room and interact well with client.*

*Be able to work some nights and weekends.*

### QUALIFICATIONS:

*Bachelor's degree preferred, or related culinary degree.*

*Certified Executive Chef preferred.*

*Seven+ years of culinary management experience.*

*Ability to focus on client services.*

*Experience with food and labor cost control, profit and loss accountability, demonstration cooking, menu development, pricing, and development of culinary team.*

*Contract-managed service experience desirable.*

## JOB DESCRIPTION

**Chef Manager** | *Position Type: Full-Time/Permanent*
*Pay Basis: Salary*
*Reports to: District Manager*

**RESPONSIBILITIES:**

Oversee all front- and back-of-house operations.

Supervise food production and all kitchen activity, including cleaning and sanitation.

Plan and implement all catering events, including cooking.

Interact with customers and clients daily.

Manage and control food costs.

Prepare budgets.

Interview and hire cooking and kitchen staff.

Train cooking staff monthly on appropriate topics.

Complete and carry out performance appraisal system for all employees.

**QUALIFICATIONS:**

Bachelor's degree preferred or related culinary degree.

Two+ years of management experience.

Previous experience with food production, purchasing processes, and on-site foodservice experience required.

## EXERCISES

1. The 2003 mean annual wage for Chefs and head cooks in all industries is $32,620. Compare that to the 2003 mean annual wages for Chefs and head cooks in B&I.

2. Go to the website of Sodexho/Marriot, Aramark, or Compass. What types of culinary jobs do they have in B&I? Print out a description of at least one job.

3. How would you rate the job outlook for jobs in B&I foodservice: excellent, good, fair, or poor? What does the job outlook depend on?

4. Would you consider working in B&I foodservice? Why or why not?

5. How is being a Chef in B&I foodservices like being a restaurant Chef? How is it different?

**Figure 7-1**

Sample Business and Industry
Organizational Chart

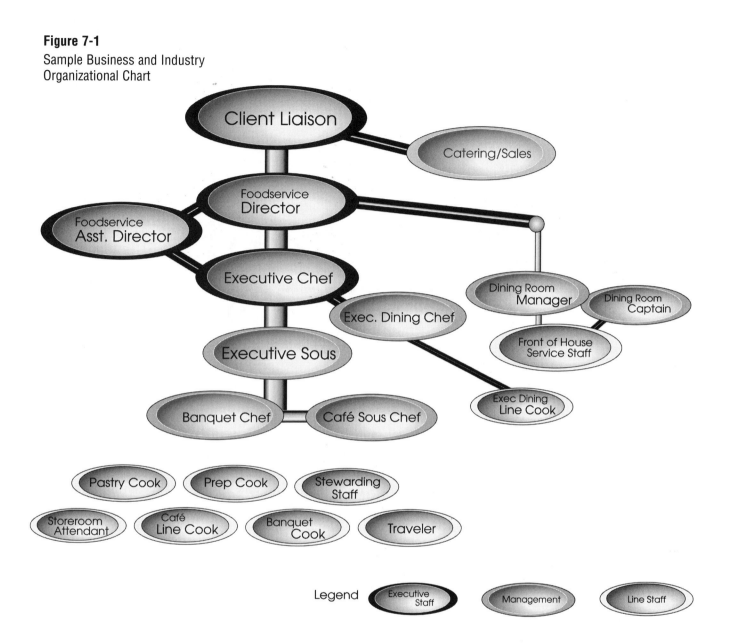

6. Using the Career Path Guide (page 122) and Education Path Guide (page 124), map out a possible career and education path for an Executive Chef in B&I, including:

   - number of months in each job
   - each place of employment
   - how and when you would complete your formal education
   - how and when you would complete skills and competencies

7. Go to the library and find five periodicals that relate to on-site foodservices. List the names of each one, and compare and contrast them.

# Culinary Careers in Universities and Schools

## Introduction

**STUDENTS ON COLLEGE AND UNIVERSITY CAMPUSES** have such sophisticated tastes these days that many creative Chefs have found a home there, cooking comfort foods such as pizza and ethnic foods from around the world. With this impetus, college and university foodservices have changed dramatically from 20 or 30 years ago. Gone are the days of mystery meats in the dining halls and dorms loaded with students who were required to buy full board, meaning three meals a day, seven days a week. Nowadays, less than half of students are likely to live on campus. The rest live off-campus or commute from home. Working as a Chef in the ivory towers is all about keeping residents happy with food selections and quality and working to attract off-campus students, commuters, faculty, and staff employees.

## The Feel

### UNIVERSITIES

University foodservice has three segments that emphasize fresh food, made-to-order food, variety, and authenticity.

◆ Residential dining
◆ Retail outlets
◆ Catering

Residential dining encompasses the dining halls primarily designed for resident students. No more cold breakfasts or unimaginative menus in the dining hall

| The Feel | A Day in the Life | The Reality | Earnings and Outlook | Professional Organizations | Interview | Organization and Job Description |

**145**

these days! At one college, the dining hall has over ten stations offering traditional menu selections, deli sandwiches and panini, salad bar, soups, pizza, pasta bar, stir-fry and sauté dishes, ethnic favorites, whole fresh fruit, bakery and other desserts, ice cream and frozen yogurt. Some resident students are vegetarian (using dairy products and eggs) or vegan (no dairy or eggs), so both options are available at every meal.

Retail foodservices include restaurants (sit-down and quick service), convenience stores, coffee shops, bakeries, and delivery service. For example, Pennsylvania College of Technology operates eight retail services: a sit-down restaurant, two convenience stores, two snack bars, a café, a coffeehouse, and a pizza delivery service. Retail foodservices blend well-known national brands with signature brands, meaning brands you create or other Chefs have created.

Last, university foodservices do lots of catering that can be compared to a neighboring hotel or conference center. The busiest time of the year is often around graduation, when the foodservice caters many events, some of which are outdoors. Year-round, the service caters for faculty seminars, presidential fundraisers, socials, trustee meetings, and many other events. Catering often requires the foodservice to abandon the normally casual feel and move on to linens, three-course plated dinners, and exquisite cooking marked by superb technique and flavoring. After all, the customers are the powerhouses of academia, the business world, and elsewhere.

The major source of university foodservice income is resident students. Students have a variety of meal plan options when they arrive at college, and the information is encoded on their college identification card. Here are typical options:

**Board plans**—Board plans are used almost exclusively by students living on campus. Many board plans offer a certain number of meals per week (such as 10 or 14) and flexible spending dollars (often $100). The meals must usually be taken in the residence cafeterias, and the flexible spending dollars can be spent anywhere. Students contract for the number of meals they expect to eat, taking into account factors such as "I never eat breakfast" and "I'm not on campus during weekends." Unused meals are not normally refunded at the end of the semester.

**Declining balance plans**—With the declining balance plan, residents or commuter students can set up accounts in specified amounts (often $500) to cover purchases. When food is purchased, the purchase amount is deducted from the account. Balances left at the end of the year may or may not be refunded.

**Non-semester plans**—Commuter students may buy a meal plan that includes a given number of meals and flexible spending dollars; the plan lasts until the total number of meals and the flexible dollars are exhausted.

The variety of plans is endless, but one thing is clear: College and university foodservice has a much more predictable volume of sales, and therefore production volume, than many other foodservices. This is a plus for the Chef.

## SCHOOLS

Whereas college and university students will probably love your latest Thai menu creations, don't be surprised when second graders will not eat your kid-friendly vegetables. Children can be picky eaters, which is something the U.S. Department of Agriculture (USDA) acknowledged when they launched Team Nutrition. Team Nutrition is a program that provides training and technical assistance to schools to ensure that personnel have the education, training, and skills necessary to provide healthy meals that appeal to children.

For students in kindergarten through 12th grade, the National School Lunch Program (NSLP) is a federally assisted meal program operating in almost 100,000 schools and residential childcare institutions. It provides nutritionally balanced low-cost or free lunches to more than 26 million children each school day. The USDA administers the program at the federal level. In school districts, Chefs often plan menus and develop recipes, and they may supervise food production. Each school district usually decides whether to offer the NSLP and additional programs, and whether to use a contract services company or operate the foodservice themselves.

School lunches must meet the recommendations of the Dietary Guidelines for Americans, which recommend that no more than 30% of an individual's calories come from fat and less than 10% from saturated fat. USDA regulations also establish a standard for school lunches to provide one-third of the Recommended Dietary Allowances of protein, vitamin A, vitamin C, iron, calcium, and calories. To provide local foodservice professionals with flexibility, four menu planning approaches yield healthful and appealing meals.

The choice of what specific foods are served and how they are prepared and presented is made by local schools and/or the managed services company they hire. Many school districts involve students in discussing what foods they would like to be served at mealtime. Chefs may develop new recipes and test them with groups of students for acceptance.

As children grow, their tastes start to expand beyond the familiar comfort foods of chicken nuggets, pizza, hot dogs, and hamburgers. Middle schools and high schools often offer à la carte purchases in addition to the school lunch meal. Some schools also get involved in catering and other opportunities to bring in revenue.

## A Day in the Life

Planning menus for students at this state university in upstate New York has certainly forced me to be right up to date on cooking trends and techniques. Most students are more adventurous and know more about ethnic and healthy foods than I remember being when I was in college. Their tastes have become much more sophisticated. Of course, they love the standards: pizza, hoagies, burgers, fries, and so on. But they also love the sushi bar on weekends, foods from Japan or Morocco, and really good vegetarian foods.

On a typical day, I go into work early only if there's a special catering event, such as a breakfast involving the college President. Otherwise, I come in around 8:00 and work through lunch and dinner because that is when our outlets are the busiest. We have two cafeterias, one restaurant, a specialty coffee and bakery shop, and two convenience stores (c-stores) with lots of grab-and-go foods. At night, we have a pizza delivery business, and we also make food baskets for students that we market to their parents.

Every morning, I oversee lunch production and get most of my ordering done. All food production for the cafeterias, catering, and the c-stores is done in one unit—the large cafeteria in which I work. Hot and cold foods are transported to the other cafeteria and c-stores for service. The restaurant uses a different menu and has its own cooking staff, as does the coffee and bakery shop. As colleges have added different venues for eating, the Chef's job of supervision and quality control has grown a bit more difficult. No matter how hard I try, I still haven't figured out how to be in two places at the same time.

During either lunch or dinner, I talk with customers in any of our units. A lot of being a good Chef in a college is keeping customers happy. And what a diverse group of customers we have. Good people skills and just general communication skills are very important.

Afternoons and evenings are often busy catering times, and these are excellent opportunities to do a little public relations with college administration, as they are usually the ones who request the catering. Almost every catered event is unique, and I find them all exciting to design and cook for. I usually work directly with administration on the menu for catered events.

By dinnertime, when I'm getting a little tired, the students are just gearing up to eat dinner, study, party, and more. They love the endless variety of bars, such as pasta bars and sushi bars, we offer. They also love having dinner cooked to order in front of them. It's a big challenge to offer new services and programs, such as exhibition cooking, while at the same time restrain expenses, particularly labor expenses. Oh well, every segment has its challenges—but like I said, this job environment keeps me thinking young!

## The Reality

### UNIVERSITIES

A university foodservice, when school is in session, is a 24/7 operation. Food is served morning, afternoon, and night. Catering and special events often take place at night and on weekends. Luckily, though, business goes down tremendously during breaks (holidays, spring, etc.) and during the summer, so you have some time to recharge your batteries, evaluate your programs, and test new recipes and ideas.

Whether you work for a managed services company or self-operate, it is critical to work with staff and the dining services director to increase sales and maintain or reduce costs, especially labor costs. To increase sales, sometimes you get involved in building new outlets or renovating. You must be up to date with the latest in equipment and renovations to successfully attract off-campus students, commuters, and faculty and staff to your foodservice outlets.

In general, food production in universities is varied and challenging from a culinary point of view. To be successful, you need to consult regularly with students to find out what they like about the food, what they don't like, and what they would like to see in the future. Good communication and people skills are essential for dealing with students, working with your staff, and arranging and producing catered events for the President and others.

## SCHOOLS

School foodservice is not as demanding from a time perspective as working in colleges and universities. Only one or two meals are served, and work is often done by the middle of the afternoon. From time to time, events will crop up that require evening hours or part of a weekend. During summer, you are usually off, although that varies. Some Chefs working for managed service companies may be assigned to other accounts for the summer.

As a Chef in a school district, your goals are challenging: increasing student participation, boosting student satisfaction, making healthy meals, having only 30 minutes to get 100 children through the lunch line and eat lunch too, and achieving cost reductions. The profit margin in school foodservice is not nearly as high as in many other types of foodservice, so cost control is vital. With constrained funding for the school lunch program, it might seem easy to just raise prices, but when you do so, participation decreases.

Along with the challenges, you also get the satisfaction of making a difference for the kids you serve, whether they are in first grade or high school seniors. If you enjoy working with children, a career in school foodservice can be fun and satisfying.

## Earnings and Outlook

| 2003 Earnings in Universities and Schools | | |
|---|---|---|
| | **CHEFS AND HEAD COOKS** | |
| | **Mean (Average) Hourly** | **Mean (Average) Annual** |
| Universities and Colleges | $18.28 | $38,010 |
| Schools (Elementary and Secondary) | $12.65 | $26,230 |

**SOURCE:** *2003 OES National Industry-Specific Occupational Employment and Wage Estimates*, Bureau of Labor Statistics, 2003.

One reason for the lower pay in schools is that you normally work about ten months out of the year. Many college and university Chefs continue to work during the summer, so they work year-round. Also, the industry has only in the past ten years started to hire more Chef-level employees.

Through 2012, overall K–12 enrollments are expected to rise more slowly than in the past. As the children of the baby boom generation get older, smaller numbers of young children will enter school behind them, resulting in average employment growth for people working in school foodservice. Projected enrollments will vary by region. Fast-growing states in the South and West—particularly California, Texas, Georgia, Idaho, Hawaii, Alaska, and New Mexico—will experience the largest enrollment increases. Enrollments in the Northeast and Midwest are expected to hold relatively steady or decline.

College and university enrollments are expected to grow over the next decade due, in part, to an expected increase in the population of 18- to 24-year-olds. Adults returning to college and an increase in foreign-born students also will add to the number of students, particularly in the fastest-growing states of California, Texas, Florida, New York, and Arizona. In addition, workers' growing need to regularly update their skills will continue to fill university and college classrooms.

# Professional Organizations

General culinary organizations, such as the American Culinary Federation, are discussed in Appendix A. This section mentions organizations for culinary and foodservice professionals working in schools and universities.

## SCHOOL NUTRITION ASSOCIATION (FORMERLY THE AMERICAN SCHOOL FOOD SERVICE ASSOCIATION)

### *Who Are They?*

The School Nutrition Association (SNA) has over 55,000 members involved in some way with the National School Lunch Program. SNA works on making sure all children have access to healthful, tasty school meals and nutrition education. SNA does this by providing education and training, setting standards, and educating members on legislative, industry, nutritional, and other issues.

### *Where Are They?*

700 South Washington Street, Suite 300
Alexandria, VA 22314
703-739-3900
**www.schoolnutrition.org**

## Publications

Monthly magazine: *School Foodservice and Nutrition*
Semiannual journal: *Journal of Child Nutrition and Management*

## Credentials

School Foodservice and Nutrition Specialist (SFNS)

# NATIONAL ASSOCIATION OF COLLEGE AND UNIVERSITY FOODSERVICES (NACUFS)

## Who Are They?

The National Association of College and University Foodservices (NACUFS) is the trade association for foodservice professionals at nearly 650 institutions of higher education in the United States, Canada, and abroad.

## Where Are They?

1405 South Harrison Road, Suite 305
Manly Miles Building
Michigan State University
East Lansing, MI 48824-5242
517-332-2494
Fax: 517-332-8144
**www.nacufs.org**

## Publication

Quarterly magazine: *Campus Dining Today*

# Interview

**DONALD R.MILLER, CEC, CCE, AAC, Executive Chef, University of Notre Dame**

**Q / What is your current position at the University of Notre Dame?**

**A:** *I'm currently the Executive Chef of the Morris Inn, a hotel and conference center, which is a $4 million food revenue a year operation. But in the next couple of weeks, I'm going to be promoted to the Executive Chef of Notre Dame foodservices. The food revenues are around $40 million.*

**Q / How many meals for students are there on an average day?**

**A:** *Just for the students, it's 28,000 meals a day. And that's actually relatively small. What people don't understand is the University of Notre Dame has only 12,000 students, not including graduate students.*

| The Feel | A Day in the Life | The Reality | Earnings and Outlook | Professional Organizations | Interview | Organization and Job Description |

**151**

**Q / And how many venues are there for them?**

*A: We have two student dining halls: North and South. You go into South Dining Hall and it is set up in an international format, everything from Mongolian wok to pasta. We have American, Italian, South American, and other cuisines. With their meal plans, students can get their meals at North or South dining halls, or go into any of our satellite operations.*

**Q / How many satellite operations are there?**

*A: Quite a few. In the Student Center is the Huddle Food Court, which includes branded concepts such as Starbucks and Burger King. We have three restaurants on campus and seven express eateries, each with a different theme and menu. All are open to the students, faculty, staff, and public.*

**Q / How is the foodservice in the hotel organized?**

*A: At the hotel, there is an Executive Chef and an Executive Sous Chef. There is a Pastry Chef and then, of course, there is the structure of AM and PM Sous Chef and then all the line staffing cooks, like a normal hotel operation would be structured. The hotel makes 95% of the menu items from scratch. The kitchen has been renovated and now includes a refrigerated butcher room which we also use for Garde Manger work, pastry specialties, and cold hors d'oeuvre presentations. Now, in the college foodservices, there is a structure of an Executive Chef and then what we call Unit Chefs, which would be much like Sous Chefs. But in some of the satellite outlets they don't necessarily need the culinary training that one would need at the hotel level or our satellite restaurants, just cooks and pantry Garde Manger levels. Universitywide, we employ in the area of 165 to 175 cooks.*

**Q / Who does all the baking for the foodservices?**

*A: We have a centralized operation that includes a bakery, butcher shop, and a vegetable processing kitchen. It also includes the central warehouse for all food products. It's a tremendous, fascinating facility that make all their own breads, rolls, danishes, as well as thousands and thousands of cookies. And there is an area where they do cake decorating, finer à la carte desserts, chocolate work, and the more creative pastry workmanship.*

**Q / So what is a typical day like for the Executive Chef of Notre Dame's dining services?**

*A: Nothing in this job is typical, but let me tour you on what I perform as a daily routine. The Executive Chef is responsible for the culinary integrity, unit operations, culinary operations, and their costs. Once a month there is a mandatory meeting of the culinary staff leads, all the Unit Chefs, Sous Chefs, Pastry Chef, and of course the Executive Chef. We talk about everything that is going on within our units so we can create a synergy in our areas. The problems are brought up, and everyone can help each other out. So that would be something that would be a regular weekly routine along with the department management meetings, which would include all the satellite operations.*

*On a daily basis, I report in the morning to the office, which is above South Dining Hall, and read the daily reports, check and answer email, look at the daily costs. From then on it's a situation where 30 to 35% of the time I will be out at the different units, 20 to 25% of the time I am involved in training, and then the rest of the time I will be working with menu development. And you know, of course, that there's always more meetings to attend.*

**Q / Let's step back a little bit and talk about how and why you got into this profession.**

*A:* People have asked me that question before, and I think this career picked me. But if you go back further to when I was a kid, I was influenced by my grandmother, who came here from Austria. She was like the cook of an estate in Austria and came here before World War II. In her big home in Kenosha, Michigan, she would be doing all the cooking, and I would do things like crush the black walnuts to go into her traditional Austrian bread.

While I was in high school, I worked as a dishwasher in a restaurant. I also got to prepare salads, shred cheese, and perform other simple food preparation jobs. Then, one day one of the cooks didn't show up, and the Chef asked me if I would help work on the line. That was fun and exciting. I worked there for three years when I moved with the Chef in his position at a new Holiday Inn. I worked at the opening, which was round-the-clock setup, as the Banquet Chef. I was really in over my head. The food wasn't unsafe, I just wasn't doing basic cooking techniques—I was young and clueless.

After high school, I applied to go to the Culinary Institute of America and got in, but I was put on a waiting list until a spot came open, which was close to two years. In the meantime I went to the University of Wisconsin for two years. In retrospect, I wish I had graduated from Wisconsin and then gone on to Chef school after that. Also, when I dropped out of the University of Wisconsin to go to culinary school, my parents did not talk to me for about a year.

The two years that I was at the University of Wisconsin, I took a lot of business classes, and I fall back on a lot of those classes, especially the systems that they taught me. And of course, it's not good to have a system unless you have the discipline to implement it and follow up on it.

At culinary school, the first semester was so hard for me, I almost flunked out. Unfortunately I went there with the attitude that I'd been cooking seven years, so I knew it all. I struggled a lot because I went in there thinking it was going to be easy. But I was wrong. Cooking is a science and a discipline that needs to have a foundation of basics. By the last three semesters of school, everything finally clicked for me, and I got it.

**Q / So after your culinary training, where did you go to work?**

*A:* I worked in a huge Holiday Inn hotel at Holiday Inn's corporate headquarters in Memphis. I worked for them for a year, then I found out Opryland Hotel was opening in Nashville, Tennessee. It was an exciting place—over 2,000 rooms and lots of restaurants. So I had an interview there. I walked through *that place and it was incredible. Absolutely incredible. So I took the job. I was the Second Saucier in the gourmet dining room, called Old Hickory, which is all copper service. We all worked 17-hour days to get the kitchen in shape; it was incredible. I was Second Saucier, and the First Saucier was Peter Fousch, who was on the 1968 Olympic team. Now all of a sudden I'm being introduced to a whole level of focus that was a fabulous culinary experience. It was like an awakening for me. I was so excited, all I could think about was food and cooking. I was working with Peter Fousch, who taught me how to set up the station as he displayed his European discipline within his own station. We were doing 450 people a night, copper service, that was taken out and plated tableside.*

*I trained for months as the First Saucier and took it over, and then later was promoted to Sous Chef, which was an honor because they normally filled that position with Europeans because of their training. I worked there for another year and then a resort, called Sandestin, opened up in Florida, and they needed an Executive Chef to open up the hotel. I got the job, and it was my first Executive Chef's job ever. I stayed with that company for several years before I found I was burned out from the long hours and tiring work.*

**Q / How do you define "burned out"? Why did it happen?**

*A: I was very passionate, but I had nothing left to reach into. I had never rested—I worked every day of the week. So here is how I got out of it. I applied for a job teaching—one opened in Joliet Junior College. I thought I'd never get it because I didn't finish my bachelor's degree, but I got the job and taught for five years, which got tenure. Teaching helped my burnout and helped me resolidify my base so I could go out again into an industry position.*

**Q / And then from there you went to the University of Notre Dame?**

*A: Yes, there was something I liked about the academic atmosphere of being at that junior college. But here I was, I think I was 33 years old at the time, and I felt like I wanted to be back. I missed the life. It's in your veins. But I wanted that academic background. I looked, waited, and watched, and when this job came up at Notre Dame, I came here, interviewed, and I've been here ever since.*

**Q / Now tell me a little bit about your management philosophy.**

*A: I'm open to constructive ideas. I like taking those ideas and working on them independently and also working on them together as a group. I like challenging myself and others, and I like other people challenging me. With a great rapport, you can raise the caliber of the kitchen to amazing levels.*

**Q / As you talk about nurturing your staff and really developing them, is there an apprentice program here that is very intense?**

*A: We have a wonderful apprentice program here. There are four apprentices at the university who rotate through the hotel, all the foodservice outlets on campus, which gives them a well-rounded experience. They go to school at our local community college to gain the academic classes required by the ACF.*

*We have strict guidelines and demand the best from our students, as they receive the best from us. In the apprentice program here at the University of Notre Dame, you start in stewarding. You go through every aspect of sanitation, cleaning, and equipment maintenance needed to run a kitchen. The next step is to become a breakfast cook and go through that entire segment, rotating through all the necessary stations. This continues through all stations until they have completed their 6,000 hours of kitchen training. This program is sanctioned by the American Culinary Federation apprenticeship program, which awards them a Certified Cook certificate upon completion. The program is also recognized of course by the U.S. Department of Labor.*

### Q / What do you love most about your job?

*A: I'm passionate about my work. I've heard this before, that the culinary field is the only profession in which you have to use and develop all your senses. You have to educate all your senses. This is a hands-on profession where you also use academic abilities. For instance, I have to crunch numbers to analyze financial progress. If you were to tell me that I need to keep food cost at 31% next year, I know how to position my kitchen structure to achieve that goal. And yet at the same time I've developed my taste buds, my sense of smell, as well as all the dexterity that comes with physical skills.*

### Q / What advice would you have for a young culinarian who is choosing this profession?

*A: When you come out of school going into this industry and have your mind set to become an Executive Chef, you have to pace yourself first of all. You have to be patient, put in your time, and build your foundation by developing your technical skills and your personal repertoire. I'm not suggesting by any stretch of the imagination that you do what I did, because working every day was what we had to do back then to prove ourselves to the European market. Keeping in mind dedication, commitment, and perseverance are still your number-one priorities.*

### Q / What do you see as trends in the industry?

*A: I feel that the industry is going toward fresh, wholesome, well-balanced (nutritionally) foods, not adulterated by poor cooking skills. The customer demands a perceived value that is not based on huge portions. They are based on well-prepared foods, fairly priced, with a new focus on presentation. Our clients are looking for a relaxed atmosphere and great service complimented by creative, well-prepared foods. This is what I have based my career on, and it has worked for me.*

# Organization and Job Description

## JOB DESCRIPTION

**Executive Chef**

Reports to: Director of Dining Services

Position Type: Full-Time/Permanent

Pay Basis: Salary

SUMMARY

Oversee a large multi-unit campus dining facility, including both retail and catering operations.

JOB DUTIES

1. Plan menus for residence cafeterias, retail foodservices, and all catering events.

2. Supervise production of food for all cafeterias and retail foodservices.

3. Prepare and supervise the production of all catering events.

4. Order all products and supplies and conduct inventory.

5. Interview, hire, train, and evaluate food production staff.

6. Ensure proper merchandising of food and facilities.

7. Maintain service and sanitation/safety standards.

8. Prepare budgets, update forecasting, and maintain costs within budget.

9. Obtain regular feedback from customers and adjust menus accordingly.

10. Work with many campus administration and employees to ensure high-quality catered events.

QUALIFICATIONS:

*Bachelor's degree preferred.*

*Certified Executive Chef or Certified Master Chef.*

# EXERCISES

1. How would you rate the job outlook for culinary jobs in universities and schools: excellent, good, fair, or poor?

2. Would you consider working in universities or schools? Why or why not? If you would, what type of university or school would you like to work in and where?

3. Look at the job listings for university and school jobs at websites of managed services companies such as Sodexho-Marriott and ARAMARK. Print one out that sounds exciting to you.

**Figure 8-1**
Sample University
Organizational Chart

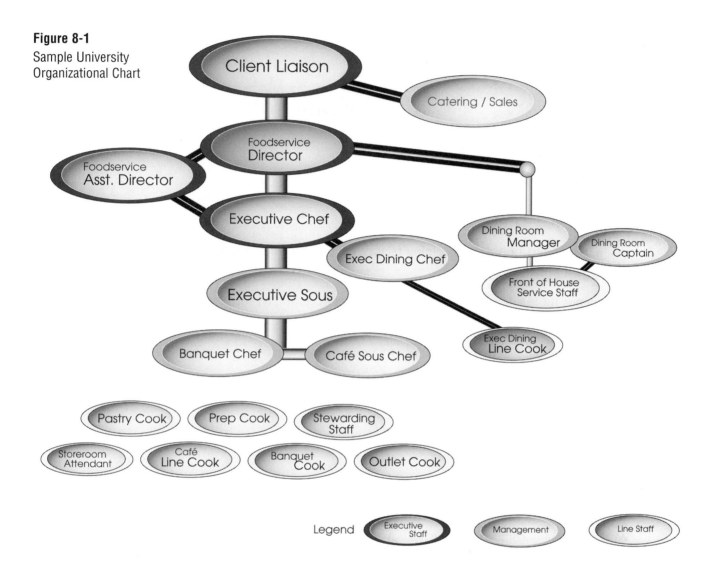

4. Using the Internet, find a school lunch menu and a university menu. Compare and contrast the menus from a culinary point of view.

5. If you are enrolled in a college, find out about what types of catering the dining services provides.

6. Interview someone who works in a school or university foodservice. Find out what he or she likes about the job and the challenges.

7. Go to the NACUFS website and list the benefits of becoming a member.

8. Using the Career Path Guide (page 122) and Education Path Guide (page 125), map out a possible career and education path for an Executive Chef in a university foodservice, including:

   • number of months in each job

   • each place of employment

   • how and when you would complete your formal education

   • how and when you would complete skills and competencies

# Culinary Careers in Healthcare

## Introduction

**HEALTHCARE FOODSERVICE HAS ITS OWN** excitement, challenges, and avenues for creativity. Imagine preparing food every day in a hospital for:

- 250 hospital patients who use a room service menu to order meals
- Close to 1,000 hospital employees looking for meals and snacks in the Café
- Dozens of catering affairs that routinely include dinner service for the hospital President, Board of Directors, and donors.

Hospital food is changing, and foodservice directors are constantly looking outside healthcare to hotels, business and industry, and colleges for new ideas. Hospital foodservice sales are growing too, and employment benefits are great, including Chefs and Managers being off many weekends and holidays.

The healthcare field also encompasses long-term care, an area that is becoming more important as the population over 65 years old continues to increase. About one in every eight Americans is over 65, and this group will more than double over the next 25 years. The elderly frequently have disabilities or chronic medical conditions such as high blood pressure, so there will be lots of job opportunities in this area. Long-term care choices include:

**Community services**—In many communities, services and programs are available to help seniors. These programs often include meals available at senior centers or delivered to the home (like Meals-on-Wheels). Over 3 million Americans receive these types of services every year.

**Assisted-living facilities**—These are facilities in which residents often live in their own room or apartment within a building or group of buildings and take some or all of their meals together. Although they can be quite costly, assisted-living facilities provide older or disabled adults with the help they need as well as social and recreational activities.

**Nursing facilities**—Commonly called nursing homes, these facilities provide care to people who need a wide range of personal care and health services, as well as 24-hour skilled nursing care.

**Continuing Care Retirement Communities (CCRC)**—These communities, also called life care communities, extend different levels of care based on the resident's needs. The same community may include individual homes or apartments for residents who still live on their own, an assisted-living facility for people who need some help with daily care, and a nursing home for people who require higher levels of care. Residents move from one level of care to another based on their needs but remain in the community. CCRCs generally charge a large payment before you move in (called an entry fee) and then charge monthly fees. Many guarantee lifetime shelter and care.

This chapter discusses careers in hospitals and CCRCs, as these segments offer the most job opportunities.

# The Feel

## HOSPITALS

Working as a Chef in a hospital foodservice is a satisfying job. You are helping the patients get better by making great food that meets their nutritional requirements. Most of the patients get a regular diet, but some need less sodium, less refined sugar, and so on. No, you don't interact with patients every day like the nurses do, but many Chefs visit patients frequently. Just like in a restaurant, you need to get the input and feedback of your customers. Of course, patients are a very different kind of customer because they don't want to be in the hospital; they would much rather be home. In any case, I was a nervous wreck the first day I went to visit some patients with the dietitian. Of course, I survived, and I am much more comfortable now on the floors with my Chef's hat on, doing public relations with the patients and the nurses.

In a hospital, the foodservice is often called the food and nutrition services department. The reason for this name is simple: The department not only prepares and serves food but also provides nutritional care and education to patients. Food and nutrition services is one of the support services of a hospital. Other support services are housekeeping, laundry, and maintenance. The Chef normally reports to the Director of Food and Nutrition Services, who in turn reports to a hospital administrator (usually the administrator in charge of support services).

| The Feel | A Day in the Life | The Reality | Earnings and Outlook | Professional Organizations | Interview | Organization and Job Description |
|---|---|---|---|---|---|---|

**159**

The organization of the food and nutrition services department varies depending on size and function, but it often includes a Director and two or more Assistant Directors (Figure 9-1). One of the Assistant Directors is usually the Nutrition Manager, or Chief Dietitian, who oversees the nutritional care activities, and the other is the Chef or Executive Chef. The Chef in a hospital foodservice has not just one but several groups of customers: patients, hospital employees, and hospital visitors.

Before going any further, you need to know a few things about the environment in which hospitals operate. Government agencies and insurance companies reimburse hospitals for many expenditures, so the healthcare environment is rather regulated. With pressure from government and insurance companies to reduce escalating costs, hospitals have worked for many years to cut costs and increase revenue in new ways. In addition, the federal government and other entities pressure hospitals to provide quality services and ensure patient satisfaction with those services. The hospital Chef knows and understands these trends and takes actions to reduce costs and increase revenue and satisfaction.

### Patient Service

Hospitals vary in how patients order meals, how the food is prepared, and how the food is assembled and delivered. Service is not as simple as in most restaurants. Patients order meals using a paper menu, spoken menu, or room service menu. Traditionally, patients receive paper menus on their breakfast trays. The menu contains the food selections for the next day and, after the patient fills it out, is picked up and returned to the kitchen. The menu is usually a one-week cycle menu. Because the length of a patient's stay in most hospitals is less than one week, the one-week cycle suffices for variety and is easier and less expensive than a two-week cycle. This method of sending up menus for the next day on the breakfast tray sounds simple but has its share of problems: The menu may not be picked up; the patient may not realize they are selecting for the next day; the patient may not have a pencil or may have other difficulties completing the menu; the patient may change his or her mind, and so on.

With the spoken menu system, a host or hostess visits the patient just before meals to review the menu options for that day. The patient can ask questions and make selections. The host or hostess then returns to the kitchen, assembles the meals, and delivers them.

Room service menus are similar to those in hotels, except that patients are encouraged to call for meals during certain time periods and the service is normally not available throughout the night. Meals are usually delivered within 30 to 60 minutes of ordering. If the patient is not capable of calling down to place a meal order, a nurse or family member may do so.

In most hospitals, food is prepared just prior to service. The amount of scratch cooking and convenience cooking varies from hospital to hospital. In an effort to save money, some facilities use cook-chill systems (a food-production system in which large quantities of foods are cooked, chilled, and then reheated) or even get cooked foods from central commissaries.

Assembly of the individual meals is often centralized in the main kitchen or may take place in smaller kitchens closer to the patient floors. In hospitals where patients fill out their menus a day ahead of time, trays are commonly assembled on a trayline. Trays, dishes, and all foods are positioned along a conveyor belt, and employees are stationed to place specific items on each tray as the trays flow past. The first employee on the line usually puts the tray mat, silverware, and completed menu on the tray. As the tray starts down the line, hot and cold foods are added, and the last person checks the tray contents against the menu to see if it is correct. All food items are covered, and the trays are stacked into carts or trucks and taken to the hospital floors, where they are delivered directly into each patient's room.

If the hospital uses the spoken menu, each host or hostess assembles the trays of the patients from whom they have taken menu orders. Stations are set up in the kitchen to make it easy for the host or hostess to assemble the meals quickly. For example, he or she may start at a condiment station; continue to cold units to pick up juices, salads, desserts, and milk; and pick up hot food from the cooks and coffee last.

As you can see, the newer systems involving the spoken menu and room service are closer to the services available in hotels and restaurants. Patients have been quite satisfied with these options.

### Café

With fewer inpatients and more outpatients, hospital foodservices have jumped into the Café business, as well as other nonpatient services, to increase departmental revenues. In the old-fashioned hospital cafeteria, the food was boring, the atmosphere was stale, and the décor was unappealing. Nowadays, it is common for Chefs and directors to take a retail approach, meaning that they use restaurant strategies to target and compete against a specific group of concepts such as fast food, independent, or chain restaurant. By serving more upscale food items different from what customers expect in a hospital cafeteria, Chefs have created a retail setting and built perceived value.

ARAMARK, which has more than 1,200 healthcare clients nationwide, calls their hospital cafeterias by new names: cafés, on-site restaurants, or retail foodservices. The cafés often resemble food courts, fast-casual restaurants, or high-end buffets with display cooking stations and recognized brands. At the 5151 Café at Zale Lipshy Hospital in Texas, the independently run foodservice changes its self-serve bars, such as Asian Noodle Bar, Greek Sandwich Bar, and the Pasta Bar, *every day.* Most cafés use a cycle menu, to ensure lots of variety.

### Catering

Catering is an integral part of most hospital foodservices, especially in medium and large facilities. The activities hosted within hospitals—meetings, seminars, receptions, and parties—often call for catered services. Chefs develop standard menus for most catering but create one-use menus when needed.

## Additional Retail Services

Hospital foodservices have branched into many other retail areas such as the following:

**Convenience stores**—For example, Zale Lipshy Hospital has a store off the hospital lobby called the 5151 Market. They bake cookies, pop popcorn, and serve other grab-and-go foods such as soups, salads, sandwiches, and to-go plates of the daily specials from the 5151 Café. Sundries like toothpaste and other personal products are also available.

**Kiosks or cart service**—As you enter York Hospital in Pennsylvania, you will immediately see (and smell) a Starbucks cart offering a variety of drinks, baked goods, and sandwiches. One employee staffs the cart, which can be moved to other locations as needed.

**Home meal replacement**—With so many hospital employees running home after work to make dinner, some Chefs have capitalized on this by offering dinners to go. For example, Zale Lipshy Hospital offers what they call Friday Night Feasts. Their specialty is gourmet cooking for two. Employees simply order, pick up their meals from the Nutrition Services office, and follow the heating directions at home for a delicious meal. Most items can be reheated with a microwave.

**Bakeries**—Another convenience to offer employees in a hospital is baked goods such as birthday cakes, cookies for Mother's Day, decorated cakes for Halloween, and so on.

**Branded fast-food restaurants**—Some hospitals, typically larger ones, have contracts with fast-food restaurants to operate units on their campuses. For example, on the first floor of Children's Hospital of Philadelphia is both a café and a McDonald's.

**Vended foods**—Vending is yet another retail possibility to feed employees and visitors. Vending is often used to feed the night shift in hospitals.

## Dietitians

In the hospital setting, Chefs work with dietitians to develop patient menus and recipes for modified diets. The over 50,000 registered dietitians (RDs) in the United States constitute the largest and most visible group of professionals in the nutrition field. Individuals with the RD credential have specialized education in human anatomy and physiology, chemistry, medical nutrition therapy, foods and food science, the behavioral sciences, and foodservice management. Registered dietitians must complete at least a bachelor's degree from an accredited college or university, a program of college-level dietetics courses accredited by the Commission on Accreditation for Dietetics Education (CADE), a supervised practice experience, and a qualifying examination. Continuing education is required to maintain RD status. Most RDs are members of the American Dietetic Association and many are licensed or certified by the state in which they work. In addition to working in hospitals, they also work in other healthcare settings, private practice, sales, marketing, research, government, restaurants, fitness, and food companies.

In the hospital setting, you may also work with registered dietetic technicians (DTR). To become a DTR, one must complete an associate degree and curriculum requirements of a program accredited by CADE, then take the registration examination for dietetic technicians to become credentialed. DTRs often work assisting registered dietitians or in supervisory positions in the food and nutrition department.

### Accreditation

Yet another characteristic that distinguishes hospitals from restaurants is that most hospitals comply with the standards of a regulatory agency called the Joint Commission on Accreditation of Healthcare Organizations (JCAHO). JCAHO evaluates and accredits thousands of healthcare facilities and programs in the United States. JCAHO publishes standards relating to food and nutrition services, and checks to see if your department is compliant through site inspections that last several days on average. JCAHO does not substitute for the local health inspection. You get those visits too.

## CONTINUING CARE RETIREMENT COMMUNITIES (CCRC)

### Levels of Care

CCRCs are unique in that they provide various levels of care within one community for elderly residents whose needs change over time. This is referred to as a continuum of care. There are three stages of care:

**Independent living**—Virtually all CCRCs offer independent living. The main purpose of this unit is to provide a sense of independent living for older adults who are capable of doing the basic chores of everyday life but who may need occasional help. Units come in a variety of forms including studio apartments and one-bedroom, two-bedroom, and even single homes such as cottages. Services such as meals, housekeeping, and laundry service are usually available.

**Assisted living**—Available in over 80% of CCRCs, assisted living is an intermediate step between independent living and nursing care. It provides assistance for residents with chronic care needs that do not require 24-hour skilled nursing care. Assisted-living services include helping a resident with bathing, dressing, taking medications, and other daily activities.

**Nursing facility (nursing home)**—Almost all CCRCs offer skilled nursing care in a nursing home setting. Most nursing homes offer round-the-clock nursing care for those who require it. Care is provided under the supervision of licensed nurses and other medical professionals such as physical therapists under the supervision of a licensed physician.

| The Feel | A Day in the Life | The Reality | Earnings and Outlook | Professional Organizations | Interview | Organization and Job Description |
|---|---|---|---|---|---|---|

**163**

Some CCRCs also have special units such as those for residents with Alzheimer's disease.

CCRCs combine a variety of services to improve the wellness of residents: good meals and nutrition, exercise, social activities, and educational activities. Independent living residents are usually offered dinner every evening in a dining room setting with waitstaff and may receive other meals as well. Different communities offer different styles of service. For instance, a CCRC in a major city may serve meals in a classy restaurant-style setting to cater to residents who are used to city life. A community in the Amish areas of Pennsylvania, on the other hand, might serve family-style meals where the food is passed around the table as if you were at home. Distinctions like these make an enormous difference in the overall atmosphere of a community.

The kitchen prepares and serves three meals a day to residents in the assisted-living unit and nursing home. Most of these residents also eat in a dining room setting.

Menus for independent and assisted-living residents are often restaurant-style, with lots of daily specials to add needed variety. Buffets and food bars such as salad bars are popular, as they offer lots of choices and self-service.

### Accreditation and Inspections

Some CCRCs voluntarily ask for accreditation from the Continuing Care Accreditation Commission (CCAC), an organization that is the nation's only accrediting body for CCRCs. Accreditation from CCAC helps market the retirement community as a well-run operation.

The assisted living and nursing facility units may be accredited by agencies such as JCAHO, which publishes standards relating to food and nutrition services and checks compliance during site inspections.

If the nursing facility accepts Medicare payment, the state health department will come in, at least once a year, to inspect for compliance with Medicare regulations.

## A Day in the Life

It's almost 6:30 AM as I open the back door of the 600-bed hospital on the New Jersey shore and hustle up the hall past Maintenance and Purchasing to get to the Food and Nutrition Department. Although I can enter my office directly from the hall, I always walk through the department first to say good morning to everyone I see and to check in with the supervisors and cooks to see if there are any pressing problems. Luckily, there are no big problems this morning. Only one employee called in sick, and a replacement was found.

After I put my things away, wash my hands, and get my Chef's hat on, I check on breakfast preparation for the patients and the café. The café is already open; the breakfast bar is in full swing, and the short-order cook is keeping up with orders for eggs and other breakfast foods. Next, I watch the setup of the trayline for breakfast and taste the foods. Everything is fine, and the trayline starts humming with swift production. Two trays are ready, and only 376 to go! I review with the cooks and baker the prep sheets for today's needs and remind them to pull out frozen items for later this week. The baker gives me a fresh danish that he made for a special 9:00 meeting of the hospital administration. The danish melts in my mouth.

Next, I do a quick walk-through of the food storage areas, including the walk-ins, to see that the receiver is putting orders away properly and that everything is dated according to policy. I also check the temperature logs of the refrigerators and freezers to make sure they are up to date and at appropriate temperatures. Lastly, I check the refrigerators of the cold prep cooks because, as usual, there are odds and ends of food in their refrigerators that are either not covered properly or not dated. After speaking with them, I head off to do my purchase orders, including dry goods that will come in later this week from the foodservice distributor.

By the time I finish my orders, it's time to check on and taste the foods going out for lunch for the patients and in the café. One of the homemade soups for the café tastes a little too salty, so I talk to the cooks and remind them how a physician screamed at me last month because he actually measured the sodium in his soup and it was sky-high. The cooks laugh with me about this incident and assure me it was a mistake and won't happen again. During lunch service, I remain in the kitchen or café to supervise and talk to employees and café customers. Customers and employees tell me what they like and don't like about the menu. Café customers love all the bar ideas we try out—and then the employees tell me how hard it is to keep the bars clean and stocked at all times. I praise the café employees for doing a great job of keeping everything clean and making the foods look appetizing. They love making the garnishes we've been working on in the new garnish program.

Once lunch service slows down, I eat lunch with the Director of Food and Nutrition Services and her administrative boss. I learn about how the hospital is doing and get positive feedback about the foods we make for the new vending machines that feed the night shift. It's nice to finish lunch with administration and not have indigestion.

Next, I run to Human Resources to meet and interview two candidates for a food preparation position that has been open for weeks. I pray that one of the two will meet the requirements. After the interviews, I get back to the recruiter and say I want to offer the job to one of the candidates. Luckily, she has already checked his references and agrees that he is a good prospect. She says she will call him later and get back to me.

When I am back in my department, the nutrition manager asks me to go visit a patient who is complaining about the food. After being briefed, I head up to the floors and visit not only the complaining patient but others as well. The complaining patient is asleep, but luckily a family member is there and tells me what the concerns are. The patient is not getting the exact foods he orders, so I get all

| The Feel | A Day in the Life | The Reality | Earnings and Outlook | Professional Organizations | Interview | Organization and Job Description |

**165**

the details and tell the family member I will take care of the problem. For some reason I was petrified to visit patients when I first came to work here. I really didn't enjoy going into patient rooms because I was scared of what I might find. Now I know that patients are people too, and most try to be nice. I've also learned to talk to the nurses. They give me lots of information that can be helpful when dealing with patients.

By the time I get back to my office and get today's problems squared away, it's 3:15 PM. I still have to do my milk and bread orders for tomorrow and check the dinner meal. That's fine. If I were still working at my restaurant job, I would just be starting work!

## The Reality

### HOSPITALS

As in restaurants, your hospital customers can be demanding, but patients have the potential to be *very demanding*. After all, they are sick, so don't be surprised about how fussy and unhappy they seem. Add to the mix that many patients are on modified diets for the first time, and you can imagine their unhappiness. It's not a pretty sight when salt-lover Mr. M. in Room 318 learns that the salt packet on his tray has been replaced by a seasoning mix!

Café customers can be demanding too; after all, there are not a lot of lunch choices outside the hospital when employees normally get just 30 minutes for lunch. And, yes, that includes the time to get to the café and return to work. So lunch must be quick! Expect serious complaints if hospital employees have to wait at the cashiers' stations.

Staffing is almost always a concern in hospitals. Finding personnel such as café servers, cooks, and dietary aides and then training them is a priority and a challenge. You can make great food, but you need good support employees to get it to your customers.

Working conditions may include some long days, and you do have to work some weekends and holidays, but say good-bye to working every Friday and Saturday night. Salaries in hospitals also tend to be as good or higher than in restaurants, and benefits, such as vacation time and insurance, are often better, or at least competitive. Another benefit is that helping people can be very fulfilling.

### CONTINUING CARE RETIREMENT COMMUNITIES

The graying population demands the best of culinary care in their communities. The residents see food as the highlight of the day and expect good value for the money they are spending. After all, it is coming out of their pocket, not their insurance.

The menu for independent living must offer variety, style, and flexibility because, after all, the residents eat in your restaurant every day! Due to health concerns, residents ask about the ingredients in many of your recipes because they must watch their sodium intake or simply avoid foods that cause gastric distress. Along with healthy selections, residents also want decadent desserts and the occasional filet mignon.

Being a Chef in a CCRC is a challenge because in addition to satisfying the desires of the fine diner, you have to feed the sick too. The residents in the assisted-living units and nursing home are often on modified diets to limit sugar, sodium, and so on. Older adults also don't taste food as well as they did when they were younger, and they often have trouble chewing.

Having multiple dining areas is a challenge, as is finding the time to chat with your residents. Continually surveying the residents and knowing what they want (luckily, most speak up about their needs), and then following through with the changes, is necessary for survival.

Staffing is almost always a concern. Finding personnel such as cooks and servers and then training them to work with older adults is essential, but not easy.

On the bright side, the salary and benefits in a CCRC are generally good. You won't have to work every weekend or holiday. You can be creative with the independent-living menus and go all out on holidays for special meals. It's nice to hear the praise of the residents and to know you helped take care of someone's mom and dad today.

## Earnings and Outlook

| 2003 Earnings for Chefs and Head Cooks | | |
| --- | --- | --- |
| | **WAGE ESTIMATES** | |
| | **Mean (Average) Hourly** | **Mean (Average) Annual** |
| General Medical and Surgical Hospitals | $18.43 | $38,320 |
| Nursing Care Facilities | $16.70 | $34,740 |

SOURCE: *2003 OES National Industry-Specific Occupational Employment and Wage Estimates*, Bureau of Labor Statistics, 2003.

Salaries generally increase as the number of patients/residents increases.

Wage and salary employment in the health services industry is projected to increase 28% through 2012, compared with 16% for all industries combined. Employment growth is expected to account for about 3.5 million new wage and salary jobs—16% of all wage and salary jobs added to the economy over the 2002–2012 period. Projected rates of employment growth for the various segments of the industry range from 12.8% in hospitals, the largest and slowest-growing industry segment, to 34.3% in nursing and residential care facilities.

Employment in health services will continue to grow for several reasons. The proportion of people in older age groups, with their greater-than-average health-care needs, will grow faster than the total population between 2002 and 2012, increasing the demand for health services, especially home healthcare and nursing and residential care. Advances in medical technology will continue to improve the survival rate of severely ill and injured patients, who will then need extensive therapy and care. New technologies will enable conditions not previously treatable to be identified and treated. Also contributing to industry growth will be the continuing shift from inpatient to less expensive outpatient care, made possible by technological improvements and consumers' increasing awareness of, and emphasis on, all aspects of health. All these factors will ensure robust growth in this massive, diverse industry.

Employment growth in the hospital segment will be the slowest within the health services industry, a result of efforts to control hospital costs and of the increasing utilization of outpatient clinics and other alternative care sites. Hospitals will streamline health services delivery operations, provide more outpatient care, and rely less on inpatient care. Job opportunities, however, will remain plentiful because hospitals employ a large number of people.

CCRCs and assisted-living centers are growing about six times faster than nursing homes, in part simply because nursing homes have been around much longer. There are currently over 18,000 nursing homes. Community care facilities for the elderly are expected to be among the ten fastest-growing industries and will add many new jobs between now and 2012.

# Professional Organizations

General culinary organizations, such as the American Culinary Federation, are discussed in Appendix A. This section mentions organizations for culinary and foodservice professionals in healthcare.

## AMERICAN SOCIETY FOR HEALTHCARE FOOD SERVICE ADMINISTRATORS (ASHFSA)

### *Who Are They?*

The American Society for Healthcare Food Service Administrators is an affiliate of the American Hospital Association. Members include food and nutrition service management professionals in hospitals, continuing care retirement communities, nursing homes, and other healthcare facilities. Members work as Directors of Food and Nutrition Services, Directors of Dining Services,

Café/Catering/Vending Managers, Clinical Nutrition Managers, and Dietitians. ASHFSA welcomes food and nutrition service professionals from both independent and contract operations.

### Where Are They?

304 West Liberty Street, Suite 201
Louisville, KY 40202
800-620-6422
**www.ashfsa.org**

### Publication

Quarterly magazine: *Healthcare Food Service Trends*

### Recognitions

ASHFSA has a professional recognition program called APEX (Actions for Professional Excellence). The three levels of achievement are:

**Level 1.** Accomplished Healthcare Foodservice Administrator (AHCFA)
**Level 2.** Distinguished Healthcare Foodservice Administrator (DHCFA)
**Level 3.** Fellow Healthcare Foodservice Administrator (FHCFA)

The ASHFSA Professional Recognition Program is designed to recognize those factors that are indispensable to true professionalism—basic and continuing education, experience, and participation in professional and society activities. After successfully completing the requirements for Level 1, you can then move on to Level 2. Each level requires additional education, work experience, and participation in society activities. Members may put the acronym for the level they have accomplished after their name (for example, John Hall, DHCFA).

## DIETARY MANAGERS ASSOCIATION (DMA)

### Who Are They?

Dietary Managers Association is a national association of over 15,000 professionals dedicated to the mission of "providing optimum nutritional care through food service management." Dietary managers work in nursing homes and other long-term care facilities, hospitals, schools, correctional facilities, and other settings. Responsibilities may include directing and controlling menu planning, food purchasing, food production and service, financial management, employee hiring and training, supervision, nutritional assessment, and clinical care. Dietary managers who have earned the Certified Dietary Manager (CDM) or Certified Food Protection Professional (CFPP) credential are also specially trained in food safety and sanitation. Dietary managers may work as foodservice directors, assistant foodservice directors, supervisors, and in other positions.

| The Feel | A Day in the Life | The Reality | Earnings and Outlook | **Professional Organizations** | Interview | Organization and Job Description |

**169**

### Where Are They?

406 Surrey Woods Drive
St. Charles, IL 60174
800-323-1908
**www.dmaonline.org**

### Publication

Monthly magazine: *Dietary Manager*

### Certification

To become a CDM or CFPP, you have to meet certain education and experience requirements as well as take an exam. If you hold a two-year or four-year college degree in foodservice management, nutrition, culinary arts, or hotel-restaurant management, you can submit your transcripts to DMA to determine your eligibility to take the exam. If you pass, you must complete continuing education requirements to maintain the credentials.

## NATIONAL SOCIETY FOR HEALTHCARE FOOD SERVICE MANAGEMENT (HFM)

### Who Are They?

The National Society for Healthcare Foodservice Management is a professional association representing healthcare foodservice operators and their suppliers. HFM accepts only members who operate independent operations and are not contracted. HFM offers advocacy for independent healthcare foodservices as well as management tools to decrease costs, increase patient and staff satisfaction, and define successful operational performance. Members are mostly from hospitals.

### Where Are They?

204 E Street NE
Washington, DC 20002
202-546-7236
**www.hfm.org**

### Publication

Monthly magazine: *Innovator*

## AMERICAN ASSOCIATION FOR HOMES AND SERVICES FOR THE AGING (AAHSA)

### Who Are They?

The American Association for Homes and Services for the Aging represents 5,600 not-for-profit nursing homes, continuing care retirement communities, assisted-living and senior house facilities, and home- and community-based service providers. The organization seeks to advance affordable, ethical aging services and enhance the health of seniors.

### Where Are They?

2519 Connecticut Avenue NW
Washington, DC 20008
202-783-2242

### Publication

Monthly magazine: *Best Practices*

## AMERICAN HEALTHCARE ASSOCIATION (AHCA)

### Who Are They?

The American Healthcare Association (AHCA) is a federation of affiliated state health organizations, together representing nearly 12,000 assisted-living, nursing facility, developmentally disabled, and subacute care providers. AHCA provides information, education, and administrative tools to its members and represents the long-term care sector to the nation. Their focus is to provide quality care for the elderly and disabled. Within AHCA is the National Center for Assisted Living (NCAL), which specializes in helping members in assisted-living units.

### Where Are They?

1201 L Street NW
Washington, DC 20005
202-842-4444
**www.ahca.org**

### Publication

Monthly magazine: *Provider*

## Interview

**BRENT RUGGLES, CEC, Corporate Executive Chef, St. Paul and Zale Lipshy University Hospitals**

**Q / What is your current position, and to whom do you report?**

**A:** *I am the corporate Executive Chef, and I report to Mary Kimbrough, RD, LD, who is the corporate Director of Nutrition and Hospitality Services. I work in St. Paul and Zale Lipshy University Hospital, which is a private hospital.*

**Q / What services does your department provide?**

**A:** *We provide foodservice for patients, employees, and visitors to Zale Lipshy University Hospital and St. Paul University Hospital. We also do the catering* *for both those hospitals and the University of Texas Southwestern Medical School.*

*At Zale Lipshy University Hospital, we offer a number of retail services. First, we have the 5151 Café, which is located on the first floor adjacent to the main lobby. A wide variety of choices is available for breakfast and lunch. The café offers home-style favorites such as country-fried steak with mashed potatoes and gravy, roasted vegetable chili and cheese enchiladas, roasted chicken with homemade macaroni and cheese, and meatloaf with Creole sauce. We also feature savory self-serve bars that change every day, such as Asian Noodle Bar, Greek Sandwich Bar, and the incredibly popular Friday Pasta Bar. Gourmet sandwiches and salads as well as fresh soups and breads are also featured daily. The 5151 Market is our c-store and is open every day until 7 PM. The c-store makes its own popcorn and cookies, and also offers to-go plates of the same daily specials that we serve in the 5151 Café. Soups, salads, sandwiches, pizza, ice cream, cold beverages, coffes, and cappuccino are available as well. We also offer to-go services for employees, including holiday meals, all-occasion cakes, and what we call the Friday Night Feasts—a gourmet meal for two that is picked up on Friday afternoons and reheated by the customer.*

### Q / What is a typical day of work, and how many hours do you work in a week?

*A: Report time is usually 7 AM. When I first arrive, I check with the Chef to see how the day is progressing so far. After talking with the Chef, I then go to the office to check messages and answer any email and do some paperwork. I work in the kitchen helping cooks with lunch and catering for about six hours. After lunch, I do more paperwork, writing menus, recipes, etc. I check with dinner cooks on dinner production. If I am not needed for dinner service, I will usually leave around 4:30 or 5:00.*

### Q / What was your career track to your position?

*A: I started working in kitchens after school and while going to college. I worked my way up from prep cook to lead dinner cook to Sous Chef. I decided to quit college and pursue a career as a Chef. I joined the Texas Chef Association of the American Culinary Federation and started working on my certification. I was hired as working Chef at a small resort hotel property on the Texas coast. Then I moved to Dallas to work as the Executive Chef at the Dallas Market Center. I left the Market Center to be the Executive Chef for the Dallas Cowboys and Texas Stadium. I received my Executive Chef certification from the ACF in 1998. Then I went to work for a small independent restaurant in Dallas, City Café, achieving a four-star rating. Then I moved to my current position with the hospital.*

### Q / What do you like most about your job?

*A: I enjoy the challenge of running two kitchens and managing 24 cooks, and producing wholesome, quality food. We try hard to make our food delicious and nutritious—see our website to see how we do that (www.zluh.org/deptSpe/nutri/10Things.htm).*

## Organization and Job Description

**WOOD COUNTY HOSPITAL**
**CHEF/MANAGER 8050-027**

# Job Description

### STATEMENT OF PURPOSE

Provides personal care and professional excellence in dietary services under the direct supervision of the Foodservice Administrator. The Chef/Manager supervises and coordinates activities of the dietary staff who prepare, cook and serve hospital patients, staff, and visitors.

### MAJOR TASKS, DUTIES AND RESPONSIBILITIES

1. Supervise and develop to high standards the practices of the kitchen production staff.
2. Teach and supervise the production staff in grilling, roasting, poaching, charcuterie preparation, breads, pastries, and garde manger.
3. Must appreciate the important elements of foods—their honest flavors, pleasing textures, comforting colors, and nutritional values.
4. Prepare proper mise en place to facilitate food production and service.
5. Supervise all areas of the kitchen.
6. Responsible for sanitation guidelines and strict HACCP controls.
7. Prepare late trays as required or assign others to the task.
8. Assist with physical inventories in all areas.
9. Perform any or all duties in the kitchen, trayline, cafeteria, and dish rooms as required by staffing needs.
10. Close or open shifts including all assigned paperwork and cash control.
11. Attend inservices, supervisory meetings, and Education Day. Actively pursue professional development by attending meetings, seminars, and classes, or by reading.
12. Assist in public relations by making promotional presentations to public and civic groups or organizations as deemed appropriate by the hospital.

### EDUCATION AND TRAINING

Prefer two-year culinary arts school graduate and/or certification by the American Culinary Federation plus three to five years' experience in foodservice production with verifiable track record and healthcare experience.

### PHYSICAL DEMANDS AND WORKING CONDITIONS

**1.** In an eight-hour workday, employee stands/walks (hours at one time):

[ ] 0–2     [ ] 2–4     [X] 4–6     [ ] 6–8

**2.** In an eight-hour workday, employee sits (hours at one time/hours during the day):

   [ ] 0–2     [X] 2–4     [ ] 4–6     [ ] 6–8

**3.** Employee lifts:

   [ ] up to 10#   [ ] 10–20#   [X] 20–50#   [ ] over 50#

**4.** Lifting as indicated in item 3 above is performed during the workday:

   [ ] occasionally   [X] frequently   [ ] continuously

**5.** Employee uses hands for repetitive:

   a. Simple grasping:   [X] yes [ ] no
   b. Pushing/pulling   [X] yes [ ] no
   c. Fine manipulation   [X] yes [ ] no

**6.** Employee uses feet for repetitive movement in operating foot controls:

   [ ] yes [X] no

**7.** Employee must be able to:

   a. Bend       [ ] frequently     [X] occasionally     [ ] not at all
   b. Squat      [ ] frequently     [X] occasionally     [ ] not at all
   c. Kneel      [ ] frequently     [X] occasionally     [ ] not at all
   d. Climb      [ ] frequently     [X] occasionally     [ ] not at all

**8.** Good mental and physical health required.

**9.** Visual and aural acuity required.

Source: Tim Bauman, Food and Nutrition Services Director, Wood County Hospital, Bowling Green, Ohio. Reprinted with permission.

# EXERCISES

1. Find a hospital and a continuing care retirement community in your town, city, or state. What are their names, and where are they located? How big is each facility? Do the facilities have websites? What do the websites say about food or nutrition services?

2. Go to the website of St. Paul and Zale Lipshy University Hospitals, www.zluh.org. Briefly describe their retail foodservice concepts.

3. The 2003 mean annual wage for Chefs and head cooks in all industries is $32,620/year. Compare that to the 2003 mean annual wages for Chefs and Head Cooks in hospitals and nursing facilities.

**Figure 9-1**
Sample Hospital
Organizational Chart

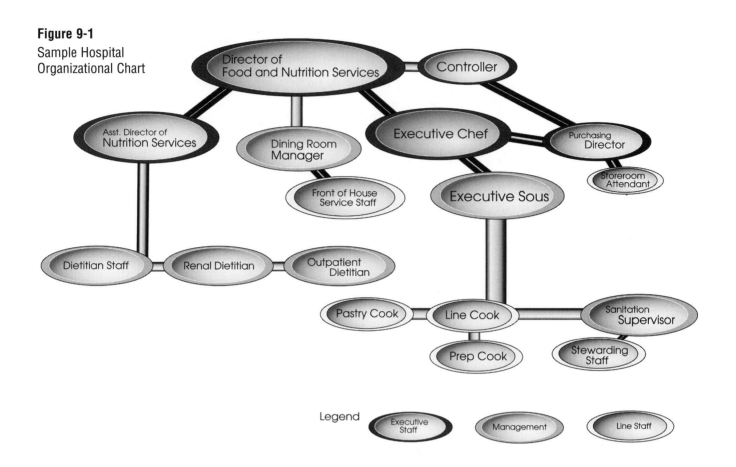

4. Go to the website of Sodexho-Marriot, Aramark, or Morrison Healthcare Food Service. What types of culinary jobs do they have in hospitals or CCRCs? Print out a description of at least one job.

5. How would you rate the job outlook for jobs in foodservices in healthcare: excellent, good, fair, or poor?

6. Would you consider working in healthcare foodservice? Why or why not?

7. What different responsibilities would a hospital Executive Chef have from an Executive Chef in a CCRC?

8. Visit the websites of HFM and ASHFSA. What is different about the memberships of these organizations? Which organization seems to offer more resources for its members?

9. Using the Career Path Guide (page 122) and Education Path Guide (page 126), map out a possible career and education path for an Executive Chef in a hospital, including:

   • number of months in each job

   • each place of employment

   • how and when you would complete your formal education

   • how and when you would complete skills and competencies

# Culinary Careers in the Armed Forces

## Introduction

**EVERY DAY, OVER ONE MILLION MEALS** are prepared in military kitchens. Some kitchens prepare thousands of meals at one time, while others prepare food for small groups of people. Culinary careers are possible in the all of the Armed Forces.

- **U.S. Army** — Today's Army is a highly trained team. Soldiers build bridges, calibrate and operate computers, and apply state-of-the-art tools and methods to solve critical problems. They operate tanks, fly helicopters, and launch missiles. Working together, they enable the Army to accomplish its mission to deter war, but they are prepared to fight and win when war is not deterred. This is important to know because, as a military Chef, you must have this training as well.

- **U.S. Navy** — The Navy operates throughout the world to preserve peace. Navy cruisers, destroyers, frigates, submarines, aircraft carriers, and support ships are ready to maintain the freedom of the seas. Navy sea and air power are available to assist in the defense of allies or engage enemy forces in the event of war.

- **U.S. Air Force** — The men and women of the Air Force fly, maintain, and support the world's most technically advanced aerospace vehicles, including long-range bombers, supersonic fighters, and many others. These forces are used whenever and wherever necessary to protect the interest of the United States and its allies.

- **U.S. Marine Corps** — The Marines are a part of the Department of the Navy and operate in close cooperation with U.S. naval forces at sea. The Marine

Corps' mission is unique among the services. Marines serve on U.S. Navy ships, protect naval bases, guard U.S. embassies, and provide a quick, ever-ready strike force to protect U.S. interests anywhere in the world. All Marines can move on short notice to match up with equipment stored on floating bases on the world's oceans.

◆ **U.S. Coast Guard**—The Coast Guard regularly performs many functions vital to maritime safety. The Coast Guard's most visible job is saving lives and property in and around American waters. The Coast Guard also enforces customs and fishing laws, protects marine wildlife, fights pollution in lakes and along the coastline, and conducts the International Ice Patrol. The Coast Guard also monitors traffic in major harbors, keeps shipping lanes open on ice-bound lakes, and maintains lighthouses and other navigation aids.

In 2003, more than 1.4 million individuals were on active duty in the military—about 490,000 in the Army, 377,000 in the Navy, 368,000 in the Air Force, and 179,000 in the Marine Corps.

Whether you are on a base, ship, or submarine, your culinary efforts will be appreciated by the personnel you cook for. After all, your skills are key to keep the morale of the troops high. In addition to cooking for troops, you may be selected to prepare gourmet meals for formal dignitaries and distinguished guests, both within and outside of the United States.

In the regular service, personnel serve on a full-time basis. After enlisting in a service, they are sent to basic training. Upon graduation, they are sent to specialty job training schools, after which they are assigned to a station for duty. After 20 years of regular service, they qualify for a military retirement.

The reserves are part-time military soldiers. Personnel serve an initial period on active duty after attending basic training and job training. After the training period, which usually lasts several months, reservists are free to return to civilian life, but for the remainder of their service obligation they attend training sessions and perform work in their job specialty one or two days a month with their local unit. Once a year, they participate in an active-duty training session for 14 days. When reservists have completed 20 years of service and have reached age 60, they are entitled to retirement based on reserve pay.

The military distinguishes between enlisted and officer careers. Enlisted personnel begin at the lowest rank in the military and serve as the main workforce, making up about 85% of the Armed Forces. With education and service, advancement can be expected. Officers begin at a supervisory rank and are the leaders of the military, supervising and managing activities. They must have a four-year college degree from an accredited institution before being commissioned. The military has several programs that lead to becoming a commissioned officer, such as officer candidate schools, Reserve Officer Training Corps (ROTC), and direct appointments.

Some service members enter the military as enlisted and convert to the Warrant Officer Corps. The Warrant Officer Corps is made up of specialized technicians with solid leadership experience who serve an entire career as a

| The Feel | A Day in the Life | The Reality | Earnings and Outlook | Professional Organizations | Interview | Organization and Job Description |

**177**

Warrant Officer in their chosen field. Warrant officers are also commissioned and receive the same privileges as senior ranking enlisted and commissioned officers of all ranks. The Army and the Marine Corps both have Warrant Officers in the foodservice career field.

In order to enlist, one must be between 17 and 35 years old, be a U.S. citizen or an alien holding permanent resident status, not have a felony record, and possess a birth certificate (some requirements vary depending on the service). Applicants must pass both a written examination and meet certain minimum physical standards, such as height, weight, vision, and overall health. All branches require high school graduation or its equivalent for certain enlistment options. Most active-duty programs have first-term enlistments of four years, although there are some two-, three-, and six-year programs. The contract also states options such as sign-on bonuses and the types of training to be received.

## The Feel

Each service branch has a different feel. In the Navy and Coast Guard, you are frequently based on ships or at shore bases. You operate kitchen and dining facilities, known as galleys, and do the usual jobs of preparing menus, ordering foods, and keeping records and financial budgets. The Army, Marines, and Air Force are more land-based, so foodservice is usually centered around permanent dining facilities in the United States and installations around the world. In addition, the land forces operate military field kitchens in support of operational contingencies.

Foodservice specialists in all branches of the military are enlisted personnel who prepare all types of food according to standard recipes and modified recipes for special diets. They also order and inspect food supplies and keep kitchens and dining halls clean. Foodservice managers are officers who perform some or all of the following duties:

◆ Manage the cooking and serving of food at mess halls.
◆ Direct the operation of officers' dining halls.
◆ Determine staff and equipment needed for dining halls, kitchen, and commissaries.
◆ Set standards for food storage and preparation.
◆ Estimate food budgets.
◆ Maintain nutritional and sanitary standards at foodservice facilities.

These are just a few examples of military culinary jobs.

Working in the culinary field in the services can be very similar to and also very different from working as a civilian. The greatest difference between a service job and a civilian job is culture. The military culture is noted for being more formal, and it is based on rules and regulations, discipline, and rank. The chain of command is very important, as it provides the framework on which each service is built. In addition, each service has its own customs, courtesies, and terminology. For example, when Army cooks talk about preparing meals in the field, they are cooking in a mobile kitchen where the troops are doing maneuvers.

## A Day in the Life

In the military, culinary personnel can work in a number of positions, so their daily routine varies. For example, some Army foodservice soldiers are Dining Facility Managers. This is the equivalent of a Chef running an operation in the Army. They write the menus and implement nutritional menus, short-order menus, and the main line menu. They order the food, oversee production, and manage the disposition of the resources. They must account for all of the food and the money as well as provide a good training program for the cooks. They write recipes and enforce standards in foodservice. They host market research boards, called dining facility council meetings, to find out what the diners would like to see changed in their dining facility. They also maintain an energy conservation program and an equipment replacement program.

Cooks and Chefs in the services prepare meals for wherever the troops are: in the field, in permanent dining facilities, on ships, in submarines, or at shore bases.

When troops are in wartime environments, they receive a mix of individual feeding rations (called Meals, Ready-to-Eat, or MRE) and group meals. The MRE comes in 24 varieties and includes an entrée, sides, condiments, and dessert. Group meals may include food prepared with hot water or more sophisticated cooking methods.

It is challenging to get the right meal to the soldiers on the battlefield at the right time. Soldiers are fed around the clock in shifts. Many of the challenges are in the supply chain. Once the cooks get the food and equipment, they can prepare the meals, but the supply chain can be broken because of competition between ammunition and food for the transportation assets. The Army has recently come up with a new meal concept where the troops get food options, including many of the short-order items they miss so much, such as cheeseburgers.

## The Reality

In return for your commitment, the services offer a lot of training (sometimes including college courses) and career advancement opportunities you won't find everywhere. Generally, the first few promotions come easily; subsequent promotions are more competitive. Many service people get college credit for the technical training they receive on duty, which, combined with off-duty courses, can lead to an associate degree through programs in community colleges. In addition to on-duty training, military personnel may choose from a variety of educational programs. Most military installations have tuition assistance programs for people wishing to take courses during off-duty hours. The courses may be correspondence or online courses or courses in degree programs offered by local colleges or universities. At this time, the military provides 100% tuition at any accredited college. Solders can attend college online or in the classroom, fully funded, while they serve on active duty.

Veterans who participate in the New Montgomery GI Bill Program receive educational benefits. Under this program, Armed Forces personnel may elect to deduct up to $100 a month from their pay during the first 12 months of active duty, putting the money toward their future education. Those who enlist and serve for two years will receive $429 a month in basic benefits for 36 months. In addition, each service makes its own contributions to the enlistee's future education. The sum of the amounts from all these sources becomes the service member's educational fund. Upon the member's separation from active duty, the fund can be used to finance educational costs at any institution approved by the Veterans Administration (VA). Among approved institutions are many vocational, correspondence, certification, business, technical, and flight training schools; community and junior colleges; and colleges and universities.

The services also offer structure, a daily routine, and opportunities to see new places. If you like to travel and have an adventurous life, one of the services might be for you. Navy cooks travel all around the world on ship preparing regular galley meals, "steel beach picnics" on the ship's flight deck, meals for admirals, and more. Army cooks might haul their kitchen to a field site in Hawaii and prepare dinner for the troops on top of a mountain, or celebrate the Army's birthday by preparing a special menu.

Life on a military installation is like living in a close community. Facilities — such as gyms, tennis courts, golf courses, bowling allies, stores, libraries, and movie theaters —are available at many military installations.

Being in the services means you must make a commitment of two years or more. During that time, you will move to new locations. You could even be transferred to a war zone to feed troops. Wherever you are, you have to work with the military culture and obey its rules and regulations. Your work hours are less flexible than they would be in a civilian job.

## Earnings and Outlook

Earnings vary by whether you are enlisted, a Warrant Officer, or a commissioned officer. Officers make more than enlisted personnel. Pay also varies by total years of service as well as rank.

In addition to receiving basic pay, military personnel are provided with free room and board (or a tax-free housing and subsistence allowance), free medical and dental care, a military clothing allowance, military supermarket and department store shopping privileges, and 30 days of paid vacation a year (referred to as leave). In many duty stations, military personnel receive a housing allowance that can be used for off-base housing. This allowance can be substantial, but it varies greatly by rank and duty station. Other allowances are paid for foreign duty, hazardous duty, and submarine duty.

The outlook for culinary professionals in the military is good. Each year, the armed forces need new foodservice specialists and foodservice managers due to changes in personnel and the demands of the field. Opportunities should be good for qualified individuals through 2012.

The exact career path varies widely depending on whether you are enlisted or a commissioned officer. For an enlisted person in the military, a sample career path might be as follows. The years shown represent typical time in service before advancement to that level. Actual career advancement depends on individual experience and performance.

### Cook    0–7 years

Cooks work under the supervision of experienced Chefs. They prepare ingredients and cook basic dishes. Cooks maintain the kitchen and dining areas in a clean, orderly fashion.

### Chef    7–13 years

Chefs plan and prepare food menus and recipes. They direct kitchen staff in food preparation and develop work schedules. Chefs determine food and supply needs and prepare order forms and records. They also conduct training classes and assign trainers to new cooks.

### Foodservice supervisor    15–21 years

Foodservice supervisors set foodservice standards, policies, and work priorities. They prepare standard operating procedures and administrative reports on foodservice activities. Supervisors plan budgets, monitor foodservice expenses, and determine personnel, equipment, and food supply needs.

# Professional Organizations

The American Culinary Federation (ACF) is an important affiliation for Chefs who work in one of the service branches. Information on the ACF is in Appendix A.

## MILITARY HOSPITALITY ALLIANCE (MHA)

### *Who Are They?*

The Military Hospitality Alliance (MHA) is the affiliate of the International Foodservice Executives Association that focuses on the needs of the military. Projects include a military culinary competition, Enlisted Aide of the Year awards, and more.

### Where Are They?

836 San Bruno Avenue
Henderson, NV 89015
888-234-3732
**www.mhaifsea.com**

### Certifications

Registered Military Culinarian (RMC)
Registered Military Hospitality Manager (RMHM)

## Interview

**TRAVIS W. SMITH, CEC, AAC, Chief Warrant Officer Three, United States Army**

Our Armed Forces interview is with Travis W. Smith, CEC, AAC, Chief Warrant Officer Three, United States Army. A career for an Army Foodservice Warrant Officer is professionally challenging and personally rewarding. Warrant officers manage or direct foodservice programs at all echelons from brigade, group, regiment, division, corps, and installation to major commands and Department of the Army staff. These leaders and technical experts provide valuable guidance to commanders on the Army Foodservice Program. Candidates who successfully complete Warrant Officer Candidate School are appointed in the grade of Warrant Officer One. When promoted to Chief Warrant Officer Two, Warrant Officers (who were originally enlisted personnel) are commissioned by the president and have the same legal status as their traditional commissioned officer counterparts. Unlike commissioned officers, Warrant Officers remain single-specialty officers whose career track is oriented toward progressing within their career field rather than focused on increased levels of command.

### Q / What is your current position in the Army?

*A: I'm chief of the Craft Skills Training Branch at the U.S. Army Quartermaster Center and School in Fort Lee, Virginia. I am the manager of the U.S. Army Culinary Arts Team. I led the Army team to the Culinary Olympic Championship in 2000 and to a second-place finish at the 2002 Culinary World Cup. I revised the annual culinary competition to include live cooking competition in public view, the Baron H. Galand culinary knowledge bowl, the nutritional hot food challenge, the team buffet table concept, and an emphasis on hot food competition formats to foster the growth of cooking skills throughout the Army.*

### Q / What is the main focus of your position?

*A: My job overall is managing resources and personnel, budgeting for the foodservice program for the culinary arts training program for the Army, focusing on the development of cooks from the very basic level to the most advanced level that we offer in the military in any branch of the service.*

*Our basic course teaches the fundamentals of cooking and baking. As personnel progress to middle management, we focus on the trainer program because our middle-level managers are our trainers for the on-the-job training program, something I'm working on improving now. Through our Advanced Culinary Skills Training Course (ACSTC), we train selected personnel in advanced culinary techniques and skills required for menu planning, cost comparison, food purchasing, advanced gourmet food preparation/production, menu evaluation, and meal service. My job is to lead and manage the culinary team, the culinary program, the culinary arts competition, and develop programs and initiatives like cook certification. That's something I'm working on right now for the Army. Ultimately, my job is about improving the quality of food that's available to soldiers and improving training that's available to the cooks.*

**Q / To whom do you report?**

**A:** *I report to the Chief of the Culinary Institute Skills Training Division. I also report to the Director of Training for the Army Cook School and the Director of the Army Center of Excellence Subsistence. And then it goes on up the line to the Quartermaster General.*

**Q / How many people do you supervise?**

**A:** *I supervise around 30 people on average. They're instructors, the actual hands-on cooking instructors from basic to middle level to the advanced culinary level.*

**Q / What is a typical day of work for you?**

**A:** *Typically, I'll come in just to make sure that I don't have anything that really needs my attention immediately. I will make sure that the people are where they need to be, doing what they need to be doing. My day starts early in the morning with an hour workout Monday, Wednesday, and Friday. Then I'll take off, come back after breakfast and get cleaned up, and then go through and check email and messages. Then I just proceed with whatever projects I have going on. It might be planning for the Annual Army Culinary Arts competition. It might be planning for a training event with the U.S. Army culinary team, getting them ready to go somewhere. We go to the Chicago National Restaurant Show every spring. We also go to the American Culinary Federation national convention. We do training sessions at both major events and then occasionally, if we're getting ready for an international culinary competition event, then we'll do training sessions at Fort Lee. I make the arrangements for all the lodging, the travel, the budgeting. I manage the budget for the team, and I manage the training that we conduct to perfect our plan.*

**Q / How many hours do you spend every day or every week on work?**

**A:** *It probably averages about, I would say, 50 hours a week.*

**Q / What was your career track to this position?**

**A:** *I was in high school and I wanted to be a Chef, so I started working in restaurants as a dishwasher and then I bused tables and then eventually I got into the kitchen to work. It was a kind of steak and seafood restaurant.*

### Q / Where was that?

**A:** *That was in the suburbs of Denver, Colorado. It was interesting because, during my high school senior year, I got into the vocational-technical school for culinary arts. At that school, I started realizing that this is really a neat profession and there's a lot to learn, but I thought that I needed some structure in my learning process. I didn't really know that there were cooking apprenticeships in existence then, but one of my Chefs at the school told me about the American Culinary Federation (ACF) apprenticeship program. He recommended the Chef at the Sheraton, Michael Campe, CEC, AAC. I felt like I would go around and meet Chefs and decide who I would like to work for. But I liked the Chef at the Sheraton, and he was willing to take me under his wing —so that's what I did. I went through the rotation of the ACF apprenticeship with him, starting at the Sheraton and then moving on to the Radisson in downtown Denver. And I finished it up in November of 1989. The day after graduating from my apprenticeship program, I left for the Army.*

*Nearing the end of my apprenticeship program, I was looking for a place to go to—I was looking for which direction I wanted to go in. I wanted to see the world. I wanted to continue to cook. And I had done a couple of culinary competitions and got more into it. Then I heard about the Army Culinary Arts Team, so I checked it out and it seemed like a secure way to go about it, and it had good benefits. I also needed a way to support my daughter, Rachel, and provide medical benefits. I also considered how hard my Chef was working. He was working 80 hours a week. And I thought, if I go in the Army, I can have a good career in the cooking field, provide structure in my life, and in 20 years I can get out and still be a young man.*

### Q / Where did you go first in the Army?

**A:** *I didn't have to go to a cook's training school because of my ACF apprenticeship training. So I did my basic training in Missouri, which was basic combat training, lifesaving skills, and even shooting a rifle. I went from there to Korea for the first three years, where I learned that the Army foodservice programs are a little bit different than I had thought. So I wasn't really all that crazy about it. But I stuck with it. I was obligated to stick with it. I signed a contract. And so I tried out for the Culinary Olympic Team in 1991 and made the team. Every year, I just stayed with Culinary Arts competition and continued to improve my skills by doing them. I did this in addition to my regular job, which would be just cooking food and baking for the Army soldiers and officers at my unit. The competitions were an avenue for me to get some release and be artistic. Hot food competitions became more popular. The Army started its Army Chef of the Year competition in 1993, and I looked forward to that every year. At that time it was a planned menu. You'd have to cost out your menu, send in a menu, come in, and prepare a planned menu. And so I went in 1993, 1994, 1995, and I finally won it in 1995. I was thankful that I didn't win it the first year because I learned so much getting ready for it in subsequent years.*

**Q / Where else have you been stationed?**

*A: I've been stationed in South Korea, New York (Fort Drum in Watertown), Hawaii, and my next assignment is South Korea (again), where I'll have about 20 dining facilities to oversee and provide feedback. I'll be like a regional type of corporate Chef, traveling, not actually having to go into the kitchen to cook. I will host incentive award contests among the cooks so that they can receive recognition for excellence at various levels, and I will head up the Culinary Team from Korea at next year's culinary competition at Fort Lee, Virginia.*

**Q / Why did you change over from the enlisted branch to become a Warrant Officer?**

*A: I changed over from the enlisted branch to become a Warrant Officer because it was a better way of life. You get more benefits as an officer with regard to respect and the way that you're treated by people. You also have the opportunity to impact a greater number of people and improve the Army food program. I got into my current position as Chief of the Culinary Branch by staying active in culinary competitions and positioning myself for the job.*

**Q / Do you plan to retire from the Army?**

*A: Yes. As you know, you can retire with a pension and benefits after 20 years. In only six years, I will be able to retire.*

**Q / What's your management philosophy?**

*A: Treat people with respect. Demand high standards. And set goals, set objectives, work toward those goals, build the team, get people on the team. My management style is to get people on the team and make them a part of the planning process, and whatever the project is to help them make it their project, too, and take everybody to the finish line together and just try to do great work that is self-fulfilling and builds self-esteem.*

**Q / What's the most valuable part of your education?**

*A: I think the most valuable lesson that I've learned is, the more that I learn, the more I know that there's more to learn. Never stop learning, and never think that you know it all.*

**Q / What do you like most about your job?**

*A: What I like most about it is...I can create my own destiny with my job. When I came to Fort Lee, Virginia, as the manager of the culinary program, I made changes that I knew needed to be made with the culinary arts competition, the annual competition. I got to have fun. I got to be creative and involved in the ACF more, and some of the things we were doing.*

**Q / What advice would you give someone telling you that they want to be an Army cook?**

*A: I would advise them to establish a goal. If they wanted to be an Army cook, I would really want to know why do you want to be an Army cook? I just believe that structure is important and goals are important.*

| The Feel | A Day in the Life | The Reality | Earnings and Outlook | Professional Organizations | Interview | Organization and Job Description |
|---|---|---|---|---|---|---|

185

**Q / How about if someone said he or she wanted to be an Army cook because he or she wants to work in varied foodservices, stay fit, have structure, and retire in their forties. What do you say to them?**

*A: I say it's a great career with great potential. And, if you have the right attitude and you're willing to buckle down when times are not as good as you want them to be, your potential is really unlimited. Like any other job, the Army has its high and low moments. I mean part of my job as an Army cook when I first came in, and even now, you're out there training in the field. You actually leave in an Army truck pulling your kitchen, which is enclosed in a trailer, out to the field somewhere, and you set that up, and you set up your kitchen site. Or you could be in Hawaii sitting on a mountaintop. I'm serious. I did this in the field, watching hang gliders and stuff come down, and I'm out there cooking pizza. I'm making meals out of whatever food I've got. And I'm cooking up some good food that you don't normally associate with the Army. It can be really good, or you could be somewhere in Alaska doing the same thing in below-zero weather. But it's an exciting life. I mean it's an adventurous life. Even in Alaska, you can catch salmon and halibut on your days off. Embrace each day and make the most of it.*

**Q / What types of foods are you cooking these days?**

*A: We focus on what our customers prefer to eat. Our customers tend to be young, often male, between 18 to 22 years of age. Favorite foods include cheeseburgers and fries. We stay attuned to what our customers like to eat, and as their eating habits change, so will we. But there will always be those old-time favorites like creamed beef on biscuits. We serve that every morning at every Army dining facility. We also serve biscuits and gravy, grits, and other foods because we have such a diverse population. You see, people like comfort foods to make them feel at home when they are deployed and away from family. We keep up their morale the best we can.*

**Q / Do you actually cook outside of the military?**

*A: Yes, I used to be the Executive Chef of the Chester Plantation, a cozy bed and breakfast built in 1845, located in Prince George County, Virginia. They were open to the public Thursday through Saturday evenings for fine dining, and Sunday for an upscale plated five-course brunch. I have also operated my own ice sculpting and culinary consultation business in Richmond, Virginia, and surrounding areas. I teach cooking to home cooks and food enthusiasts at local foodie stores.*

**Q / What are the future hot topics in your area?**

*A: Our training programs are improving for our cooks. We're talking about certification now. There's going to be an explosion for the military—for the Army on certification, also for the Navy. It's going to be a huge. I can see, in the future, there's going to be higher morale for the cooks when they start realizing what the benefits are, what they can learn as they travel up the certification ladder. Another hot topic within the military is that some foodservice and cooking functions have been outsourced to managed service companies.*

## Organization and Job Description

### JOB DESCRIPTION
#### UNITED STATES ARMY

**Chief, Foodservice Operations**

RANK: *Warrant Officer 3*

UNIT: *Headquarters and Headquarters Company, 19th Theater Support Command, Eighth United States Army, Taegu, Republic of Korea*

RESPONSIBILITIES:

- Supervise and coordinate foodservice operations for the 19th Theater Support Command ranging from the demilitarized zone of North Korea to the port of Pusan in the southernmost regions of South Korea.

- Responsible for the operation of 34 dining facility operations supporting foodservice for over 3,500 soldiers in peacetime, 35,000 civilians in emergency evacuation, and potentially 75,000 soldiers in contingency operations.

- Evaluate and monitor the overall readiness posture of foodservice teams in areas of food preparation, food safety and sanitation, personnel strength, equipment and facility maintenance and replacement program, food ration accountability, and internal foodservice training programs.

- Advise commanders on all foodservice matters that could impact war fighting and humanitarian missions. Team Captain of the Eighth United States Army Culinary Arts Team.

- Advisor, United States Army Culinary Arts Team.

- Member, World Association of Cooks Societies Culinary Guidelines Committee.

- Instructor of Lieutenants, Captains, Warrant Officers, and Senior Noncommissioned Officers for foodservice materials.

- Develop, write, review, and approve foodservice lesson plans and curriculum for quarterly training initiatives.

- Develop culinary training and deliver presentations at regional quarterly food management meetings to impact food and food management quality throughout the peninsula.

## EXERCISES

1. Would you consider working in armed forces either full time or part time? Why or why not? If so, which branch would be of interest?

2. Talk to someone who has been in the military. Find out which service branch they were in and for how long. Ask them what they liked and didn't like about their service.

3. What benefits does a military job offer that a civilian job doesn't offer? What obligations does a military job have that a civilian job doesn't have?

# Part 3

# INTRODUCTION TO HOME ON AND OFF THE RANGE

Part 3 talks about nontraditional careers. In other words, these Chefs are not usually cooking in a kitchen for guests in the dining room. Instead, they might be doing any of the following.

- *Teaching the culinary arts in high schools, postsecondary school, or cooking schools for nonculinary professionals.*
- *Developing new products for restaurants, retail, and manufacturing.*
- *Planning and cooking meals for families.*
- *Purchasing all food and beverages for a foodservice.*
- *Styling or photographing food for newspapers, magazines, cookbooks, advertisements, movies, television shows, or theater productions.*
- *Writing about food and cooking for a newspaper, magazine, or other publication, or authoring cookbooks.*
- *Promoting a foodservice operation, food company, or other company as a public relations manager.*

Chapter 11 discusses culinary educators, and chapter 12 delves into the other careers noted here. Nontraditional careers for Chefs are available more now than ever before. Depending on your interests and skills, one or more of these careers is sure to interest you.

# Culinary Careers in Education

## Introduction

**TRAINING EMPLOYEES IS A PART** of every Chef's job. Chefs who find they really enjoy teaching can go on to become educators, either part time or full time. Chef educators work in many exciting work settings.

- **Vocational-technical high schools**—Most vocational-technical high schools offer training in the culinary arts. These high schools may be part of a regular high school or in a separate location. Nearly two-thirds of all high school graduates from vocational high schools enter some form of postsecondary program.
- **Professional culinary schools**—Many of these postsecondary institutions offer associate degrees or certificate programs in the culinary arts.
- **Colleges and universities**—These include both two-year colleges that award associate degrees and four-year institutions that award bachelors degrees.
- **Cooking schools**—Schools offer cooking classes mostly to adults who are not Chefs but simply want to improve their culinary knowledge and skills for use at home.

Most teaching careers (except within cooking schools) involve teaching students who are working or plan to work in the culinary field.

In all of these settings, educators often perform these job duties:

- Assess the learners' educational needs.
- Design and prepare classes.
- Teach classes, including food labs, using a variety of teaching methods and media.

- Demonstrate food preparation, cooking, and presentation techniques to groups of learners.
- Establish and maintain an effective learning environment.
- Respond to individual, cultural, and age differences in students.
- Work with learners individually and in groups.
- Motivate learners and inspire their trust and confidence.
- Advise learners in choosing courses, colleges, or jobs.
- Evaluate learners by preparing and grading tests and other assessment tools.
- Work cooperatively and communicate effectively with other learners, support staff, parents, and members of the community.
- Keep up to date with trends and developments in the culinary field.

This chapter introduces the life of a Chef educator.

# The Feel

### THE FIRST MONTH

As Chefs in a professional kitchen, we are predominantly teachers, mentors, and daily role models for our staff. After some time, many of our staff members move on to expand their careers and experience. Nothing is more rewarding than the phone calls you receive from past employees bringing you up to date on their lives and accomplishments. They feel like extensions of your family, as many of them worked for you in their early years, when they were still kids. The conversation usually ends with, "Hey, Chef, thanks for everything you taught me," or "Chef, you were right. I needed those basics to be successful, and now at work I blow everyone else away." Words like that are like music to our ears, for we are all teachers at heart. Our success is based on the success of those around us, trying to emulate what, when, and how we do our job. When the opportunity came my way to teach, I thought, "What a fine way to transfer my years of experience to the next generation."

The first month of being a Chef instructor in a professional culinary school was one of the toughest I have ever worked. Somehow or other, teaching didn't look so hard when I was a student. Teaching cooking and techniques in the food lab isn't nearly as hard as classroom teaching, yet I am constantly praying that everyone leaves the lab with all ten fingers intact and that my boss doesn't show up and see how disorganized and messy the kitchen has become. It's mind-boggling that some of my students have never cooked anything except microwave popcorn. But despite their primitive cooking skills, many of them seem to have a passion for foods and cooking, and that will help them (and me) along this path called education.

Teaching in the classroom is rough. I spend hours preparing my classes and putting together good visuals, yet I'm not entertaining enough to stop some students from instant messaging each other, playing computer games, or watching DVDs on their laptops. I'm supposed to be using new technology in the classroom to improve teaching, not developing ways to counter students' uses of technology that disrupt my class!

I know deep down that if I can run a professional kitchen, I can get a class of 19 students under control. I like the students; they are all good kids who are just looking for some direction. They keep me young in my thinking and constantly challenge my knowledge with their never-ending questions. If I don't have the answer, we either research it together or I have the answer for them the next day. I prefer to show them how to find answers than to tell them everything.

## MIDDLE OF THE SCHOOL YEAR

I survived the first semester, and thank goodness for the long break between semesters. The break allowed me to recharge my batteries and reflect on what worked and what didn't work, as well as to network with other Chef educators about how to deal with classroom disruptions and students who really don't want to be there. The school has rules, and there is a right and wrong way to properly handle a student not participating in a class. I want to make sure from the get-go that I do things correctly. All in all, the majority of students, now that I have gotten to know more of them, are good students who work and go to school. They may not get A's, but they are committed to cooking and at least have a sense of humor.

I'm not an academic student myself in the traditional sense. In fact, I chose this profession so I could be more hands-on. The truth is there is a portion of our profession that is academic and can expand with the areas we choose. For the most part, as your commitment and dedication grow for your profession, your academic skills develop from the passion you possess for cooking. If we try to incorporate hands-on activities in the midst of the classroom, fewer students drift off. One of the reasons we went into this work is because it is fast-paced, always changing, interactive—which is what people like us love. We definitely aren't loners. We enjoy stimulating conversations, interfacing with other people, and, most important, we need to be busy. Luckily, I've discovered that the problems I have in the classroom occur in every corner of this institution, so I've gotten lots of ideas from the other instructors here. They are a great resource, especially if you make them lunch or dinner.

## END OF THE SCHOOL YEAR

Somehow all the blood, sweat, and tears I put into my first year of teaching disappeared as I watched my students graduate. I felt a sense of accomplishment I have never felt before. I know I have made a positive impact on the lives of many of these young culinarians, and I have a great feeling of pride whether they thank me or not.

There is a significant parallel between teaching and running a professional kitchen. You must use the same patience and understanding when instructing a student as you would with a new employee. The realities of a lab are not much different from a typical day of kitchen instruction. The pressures may be different, but not the people. The students as well as the employee must have a sense of your expectations of them. They need to understand that the Chef-educator is there to guide and inform them as they learn and progress, but responsibility for the result is ultimately their own. Lecturing is another story altogether. You must recognize the different learning styles of the students. Some are visual learners; others are hands-on learners or verbal learners.

A teacher is like a conductor of an orchestra; he or she must be gentle and instructional to the violins, then turn with command and urge on the drums while confidently creating a flow and balance for the trumpets. This is the most challenging aspect of the job, while at the same time it is the most rewarding. I now see that teachers can have a long-lasting influence on their students. We can affect how students learn, what they learn, the quantity and quality of what they learn, and the ways they will interact in the future.

Having so much power can be a little scary, but I believe my understanding, passion, and commitment in being a Chef has had a positive influence on these students. I'll be coming back to teach next year. Yes, it will still be a challenge, but I also consider it a privilege.

## A Day in the Life

Arriving for work by 7:30 AM every day at the community college, the first thing I do is check my email, and then by 8:00 I check with the lab assistant to make sure I have all the foods needed for my 1:00 lab class. Everything is in, so I finish preparation for both the lab and my 10:00 lecture on baking yeast breads. By 9:00 I have a few students in the office. Some have questions; some just want to say hi and chat for a few minutes. I love the interaction with the students. At 9:30 I have an appointment with a student in another major who wants to switch to culinary arts. I review his transcript, tell him what additional courses he will need, and give him an idea of what cooking and being a Chef is like.

Now it's on to my 50-minute class on bread baking. Teaching lecture classes these days is a challenge because you can't lecture in the traditional sense without losing most of the class. To be successful, I have to come up with great visuals for my presentations, find ways to get students to interact with the material and each other, and shift gears when I see I've lost their attention. Luckily, I spend many more hours in the kitchen for food labs doing hands-on activities. Most students enjoy the hands-on work best. In any case, they get to apply what we talked about in the classroom setting.

After answering student questions at the end of class, I dash to an 11:00 department meeting. We meet once a month to discuss college issues, but we also update each other on what each program is doing. As in some other community colleges, the culinary arts program is in the business department, so in addition to interacting with the one other full-time Chef instructor, I work with people who teach legal assistant, administrative assistant, and other business classes. My boss, the Dean of the Business Studies Division, runs the meeting and shows us how to fill out the forms requesting new equipment for our programs. As always, money is tight, and you have no guarantee you will get what you need.

The meeting goes until 12:20, at which time I grab something to eat in the student center, change into my Chef's outfit, and get into the kitchen lab. The lab assistant has laid out all the supplies we need for the day, which includes both ingredients to make bread dough from scratch and prepared bread dough. To start the lab, I demonstrate how to make the yeast bread dough and then supervise the students in making their own dough. While the dough rises, I use bread dough that has already been fermented to demonstrate scaling, rounding, benching, makeup, and panning. The students keep busy the rest of the class by using the prepared bread dough to make loaves, and then use the dough they prepared themselves to make a variety of hard rolls. Toward the end, they work in teams to make sure the lab is clean and everything is put away. At 5:05, I release the students to go home. They are proud of their new bread-making skills.

Once I finish responding to the emails and phone messages, it's time to check what's on my calendar for tomorrow. Luckily I have no college meetings, so I should be able to finish grading a test I gave yesterday. But the high point of the day will be when I watch the Quantity Foods class make and serve the foods at a college event. I can't wait!

## The Reality

In any setting, seeing learners gain knowledge and develop new skills is highly rewarding. However, teaching has its share of frustrating moments, whether working in cooking facilities that need to be updated or dealing with unmotivated or disruptive students. Students, and teachers too, come from varied backgrounds. With growing minority populations in most parts of the country, it is important for teachers to work effectively with a diverse student population.

This section discusses the reality of teaching in cooking schools, vocational high schools, and postsecondary schools.

### COOKING SCHOOLS

Working as a cooking instructor in a cooking school can be a lot of fun. Learners pay to be there, so they are motivated to learn and are invested in getting something out of your class. Most cooking school jobs are filled by free-lancers who are paid by the class. Cooking school instructors may specialize in certain types of classes, such as baking or cooking for children. The highlight of cooking classes is the recipes, so recipes must be carefully chosen, concisely written, and absolutely correct in measurements, directions, temperatures, and timing. The instructor or the school may be responsible for ingredient shopping. Before class starts, instructors must have the lab, including food and equipment, ready for the learners. During class, instructors have a lot to do: explaining and demonstrating foods and cooking, supervising food preparation of different dishes by various groups, answering learners' questions, having tastings, setting up meals.

## VOCATIONAL HIGH SCHOOLS

To teach in a vocational high school, you must be licensed by the state in which the school is located. In many states, vocational teachers (also called career technology teachers) must meet many of the same requirements for teaching as other secondary school teachers. However, because knowledge and experience in a particular field are important criteria for the job, some states license vocational education teachers without a bachelor's degree, provided they can demonstrate expertise in their field. A minimum number of hours in education courses may also be required. For example, in Pennsylvania you can get an intern certificate by enrolling in an approved college program and meeting program requirements. To receive your permanent certification, you must complete three years of satisfactory teaching in a vocational high school as well as 60 college credits.

Most states have tenure laws that prevent teachers from being fired without just cause and due process. Secondary teachers may obtain tenure after they have satisfactorily completed a probationary period of teaching, normally three years. Tenure does not guarantee a job, but it does provide some security.

Including school duties performed outside the classroom, most teachers work more than 40 hours a week. Most vocational teachers work traditional ten-month school years, with a two-month summer vacation. Teachers in districts with a year-round schedule typically work eight weeks, are on vacation for one week, and have a five-week midwinter break.

Area employers frequently provide input into the curriculum and offer internships to students. Culinary arts teachers must be able to play an active role in building and overseeing these partnerships. For example, teachers may work with the National Restaurant Association Educational Foundation to use its ProStart® program. This state-based program for high school juniors and seniors links two years of classroom curriculum with a mentored worksite experience. Over 40 states currently have schools that use the ProStart® program.

## POSTSECONDARY SCHOOLS

Postsecondary schools include professional culinary schools, colleges, and universities. In the past, students entered postsecondary schools right after high school and attended full time. Nowadays, postsecondary faculty work with an increasingly varied student population made up of growing shares of part-time, older, career change, and culturally and racially diverse students.

Postsecondary instructors prepare classes, grade papers, attend departmental and faculty meetings, and keep abreast of trends and developments in the culinary field. The teaching load is usually heavier in two-year and culinary schools than in four-year institutions. Faculty members are often involved in setting up and supervising internship programs for students, and they provide students with information about prospective employers. Most faculty members also serve on one or more academic or administrative committees that deal with the policies of their institution, academic issues, curriculum, budgets, equipment purchases, and hiring.

Postsecondary teachers usually have somewhat flexible schedules. They must be present for classes and for meetings. Most establish regular office hours, usually three to six hours per week. Otherwise, they work on course preparation, grading, study, research, and other activities.

Some postsecondary teachers teach night and/or weekend classes. This is especially true for teachers at schools with large enrollments of older students who have full-time jobs.

Depending on the school and the position, full-time contracts may require you to work 9 months each year or 12 months each year. About three out of ten college and university faculty worked part time in 2002. Part-timers, known as adjunct faculty, often teach in addition to doing their regular job. Many come into the college or university one evening a week during the semester to teach one course. Part-timers generally get no benefits; their pay varies from $1,000 to over $3,000 or more for a three-credit course.

Training requirements for culinary schools vary, but in general, teachers need at least an associate degree plus work experience. More schools are recommending a bachelor's degree or higher, and this may become a standard requirement. Certification by the American Culinary Federation may also be required.

In two-year colleges, master's degrees are preferred and often required, although some jobs, especially part-time jobs, may accept a bachelor's degree. Many two-year colleges increasingly prefer job applicants with some teaching experience.

Four-year colleges and universities usually consider doctoral degree holders for full-time, tenure-track positions, but they may hire master's degree holders or doctoral candidates. Most college and university faculty are in four academic ranks (from highest to lowest): professor, associate professor, assistant professor, and instructor. Most full-time faculty members are hired as instructors or assistant professors. A smaller number of additional faculty members, called lecturers, are usually employed on contract for a single academic term and are not on a tenure track.

For faculty, a major step in the traditional academic career path is attaining tenure. New tenure-track faculty are usually hired as instructors or assistant professors, and must serve a period—usually seven years—under term contracts. At the end of the period, their record of teaching, research and writing, service, and overall contribution to the institution is reviewed. Tenure is granted if the review is favorable. Those denied tenure usually must leave the institution. Tenured professors cannot be fired without just cause and due process. Tenure protects the faculty's academic freedom—the ability to teach and express opinions without fear of being fired for advocating unpopular ideas. It also provides financial security for faculty. Some institutions have adopted posttenure review policies to encourage ongoing evaluation of tenured faculty.

The number of tenure-track positions is expected to decline as institutions seek flexibility in dealing with financial matters and changing student interests. Institutions will rely more heavily on limited-term contracts and adjunct faculty. In a trend that is expected to continue, some institutions now offer limited-term contracts to prospective faculty—typically two-, three-, or five-year full-time contracts. These contracts may be terminated or extended when they expire. Institutions are not obligated to grant tenure to the contract holders. In addition, some institutions have limited the percentage of faculty who can be tenured.

For most postsecondary teachers, advancement involves a move into administrative and managerial positions, such as departmental chairperson or dean. At most four-year institutions, a doctoral degree is necessary to advance. Figure 11-1 (page 204) shows the organizational structure of a large culinary school.

## Earnings and Outlook

Cooking school instructors are normally paid by the class. The exact amount varies based on what the school can charge for the class, your background and reputation, the popularity of the class, the cost of the cooking ingredients, and the amount of time required to prepare and clean up. An instructor may receive from $100 to over $1,500 (for a big-name Chef) for a two-hour class. There are more opportunities for cooking school instructors in cities and rapidly growing communities where many home cooks want to take cooking courses.

The mean (average) annual salary in 2002 for culinary arts teachers in secondary schools was $45,850. Keep in mind that most of these jobs are ten-month positions. For secondary schools, job opportunities for culinary arts instructors are strong. Through 2012, overall student enrollments, a key factor in the demand for teachers, are expected to rise more slowly than in the past, when large numbers of children meant the building of new schools. Many job openings will be attributable to the expected retirement of a large number of teachers.

Median (middle) annual earnings of all postsecondary teachers in 2002 were $49,040. The middle 50% earned between $34,310 and $69,580. The highest 10% earned more than $92,430. Earnings for postsecondary culinary school teachers vary widely by academic credential, experience and reputation, and region of the country. Part-time instructors usually receive few benefits.

Earnings for college faculty vary according to rank and type of institution, geographic area, and field. According to a 2002–03 survey by the American Association of University Professors, salaries for full-time faculty averaged $64,455. By rank, the average was $86,437 for professors, $61,732 for associate professors, $51,545 for assistant professors, $37,737 for instructors, and $43,914 for lecturers. Faculty in 4-year institutions earn higher salaries, on average, than do those in 2-year schools. In 2002–03, average faculty salaries in public institutions—$63,974—were lower than those in private independent institutions—$74,359—but higher than those in religiously affiliated private colleges and universities—$57,564.

Many faculty members have earnings in addition to their base salary from consulting, teaching additional courses, and even owning or working in restaurants. In addition, many faculty members enjoy benefits that may include access to campus facilities, tuition waivers for dependents, and paid sabbatical leave.

Because one of the main reasons students attend postsecondary institutions is to obtain a job, the best job prospects for postsecondary teachers are in fields such as culinary arts and foodservice where job growth is expected to be strong over the next decade. Overall employment of postsecondary teachers is expected to grow much faster than the average for all occupations through 2012. A significant proportion of these new jobs will be part time. Good job opportunities are expected as the retirement of current postsecondary teachers and continued increases in student enrollments create numerous openings for teachers at all types of postsecondary institutions.

Projected growth in college and university enrollment over the next decade stems also from the expected increase in the population of 18- to 24-year-olds. Adults returning to college and an increase in foreign-born students also will add to the number of students, particularly in the fastest-growing states of California, Texas, Florida, New York, and Arizona.

## Professional Organizations

### AMERICAN CULINARY FEDERATION (ACF)

*Who Are They?*

The American Culinary Federation (ACF) is the largest and most prestigious organization dedicated to professional Chefs in the United States today. ACF offers its members many opportunities to keep up to date with the latest in knowledge and skills through its journals, seminars, workshops, national conventions, and regional conferences. ACF accredits culinary programs at the secondary and postsecondary levels. Local chapters of ACF offer members opportunities to network with nearby culinary professionals.

### Where Are They?

180 Center Place Way
St. Augustine, FL 32095
800-624-9458
**www.acfchefs.org**

### Publications

Monthly magazine: *The National Culinary Review*
Monthly newsletter: *Center of the Plate*

### Certifications (Education-Related)

Certified Culinary Educator (CCE)
Certified Secondary Culinary Educator (CSCE)

## ASSOCIATION FOR CAREER AND TECHNICAL EDUCATION (ACTE)

### Who Are They?

The Association for Career and Technical Education is the largest national education association dedicated to the advancement of education to prepare youth and adults for successful jobs and careers. Its membership consists of educators, administrators, and others involved in career and technical education programs at the secondary and postsecondary levels.

### Where Are They?

1410 King Street
Alexandria, VA 22314
703-683-3111
800-826-9972
**www.acteonline.org**

### Publications

Monthly magazine: *Techniques*
Semimonthly email: *Career Tech Update*

## FOODSERVICE EDUCATORS NETWORK INTERNATIONAL (FENI)

### *Who Are They?*

Foodservice Educators Network International (FENI) is a group of foodservice educators who work in high schools and postsecondary programs. FENI works with educators to help advance their professional growth. A key element of FENI is to assist culinary educators in sharing teaching techniques and other information with colleagues and industry partners to enhance high standards of culinary education.

### *Where Are They?*

20 West Kinzie, 12th Floor
Chicago, IL 60610
312-849-2220
**www.feni.org**

### *Publications*

Quarterly magazine: *Chef Educator Today*

### *Certification*

Certified Culinary Instructor (CCI)

## INTERNATIONAL ASSOCIATION OF CULINARY PROFESSIONALS (IACP)

### *Who Are They?*

The International Association of Culinary Professionals (IACP) is a group of approximately 4,000 food professionals from over 35 countries. IACP provides continuing education, networking, and information exchange for its members who work in culinary education, communication, or in the preparation of food and drink. Its mission is to "help its members achieve career success ethically, responsibly, and professionally." Many of its members are cooking school instructors, food writers, cookbook authors, Chefs, and food stylists.

### *Where Are They?*

304 West Liberty Street, Suite 201
Louisville, KY 40202
502-581-9786
**www.iacp.com**

### *Publications*

Quarterly: *Food Forum*

### *Certification*

Certified Culinary Professional (CCP)

## INTERNATIONAL COUNCIL ON HOTEL, RESTAURANT, AND INSTITUTIONAL EDUCATION (I-CHRIE)

### *Who Are They?*

I-CHRIE is the advocate for schools, colleges, and universities that offer programs in hotel and restaurant management, foodservice management, and culinary arts.

### *Where Are They?*

2613 North Parham Road
Richmond, VA 23294
804-346-4800
**www.chrie.org**

### *Publication*

Quarterly journal: *Journal of Hospitality and Tourism Education*

## Interview

**THOMAS RECINELLA, CEC, AAC, Assistant Professor, State University of New York At Delhi**

**Q / How many students are in the Hospitality Management Program?**

**A:** *There are over 300 students in hospitality management. Within hospitality management, there are majors in travel and tourism, culinary arts, hotels and resorts, and restaurant/foodservice management. About 140 are in culinary arts, and they are split between those going two years for an Associate in Applied Science (AAS) and those going four years for a Bachelor in Business Administration (BBA). All majors offer AAS and BBA degrees. We offer the only BBA in culinary arts in the state of New York.*

**Q / How is the department you teach in organized?**

**A:** *We have a division Dean who splits her time between the hospitality department and the business department. All faculty report directly to her. We have two secretaries, and our Dean reports to the Vice President of Academic Affairs.*

**Q / Besides preparing and teaching classes, what other responsibilities do you have?**

**A:** *I do a lot more besides teach classes. I:*

◆ *Coordinate all culinary student activities, culinary competitions, fundraisers, and any dinners that we put on for the department.*

◆ *Advise the culinary club, which is a philanthropist organization called the Delhi Escoffier Club.*

◆ *Coach the culinary teams and chair all culinary competitions/contests held on campus.*

| The Feel | A Day in the Life | The Reality | Earnings and Outlook | Professional Organizations | Interview | Organization and Job Description |
|---|---|---|---|---|---|---|

**201**

- *Order all of the food for culinary arts, including developing specifications and negotiating prices.*
- *Supervise the making and harvesting of the ice blocks from the Clinebell machine (for making large ice blocks) to be used for ice carving.*
- *Advise 45 students in their academic careers.*
- *Handle job placement for summer internships.*
- *Design and spec new lab spaces and equipment as needed.*
- *Work on search committees when needed.*
- *Serve on other committees when needed.*
- *Serve as the division's representative on the college senate.*
- *Write new courses and revise old curriculum as necessary.*
- *To date, I have written five new courses which are used in our program. I stay pretty active in virtually every department endeavor.*

**Q / What is it like to be a teacher in the college classroom these days?**

*A: It is challenging, especially for Chef educators. With career changers and traditional students, we need to be able to relate to both. All students learn differently, so it is very challenging, to say the least, when speaking of a lecture. Lab is a little different. It is easier for me anyway to guide a struggling student through lab than it is in the classroom setting. The lab is really no different from a normal kitchen situation. The patience and the willingness of the Chef to work with the employee is the same in a lab situation. The student must know that there are expectations and what they are. Then they must know and feel that the Chef will help to guide them to those expectations, but the responsibility is ultimately theirs. In lecture, it is quite different. Some students are visual learners, while others learn by doing. There seem to be more students with special needs these days than there were when I went to college, or maybe we are just doing a better job of identifying and helping them than we used to. Learning disorders and attention disorders are very prevalent. Diversity of course is a big issue. A teacher in today's classroom is a lot like a maestro. The key is to treat everyone with respect and consideration and hope that your example is transferred to the entire class.*

**Q / Describe a typical workday.**

*A: In between 7:00 and 7:30 AM, I prep for class. I also compile orders for the restaurant class. Each day is different depending on the semester load. I teach 26 contact hours currently, meaning that I am in class or lab 26 hours each week. Labs are four hours long. My club meets once a week. I leave school by 7:30 PM, except on Fridays when the restaurant is open, I am not done until 11:00 PM. I guess my days are not typical. But that is what I like. I am often in on Saturday, especially when our team is practicing.*

**Q / What do you like most about your job?**

*A: I like the students. They keep me thinking young, and they challenge me not to grow stale and uninformed. I like being at a small school where I can do it my way. And my way has worked. (I like that best.) I like knowing my students' names, all of them. I like knowing that Mike comes from Binghamton and has a dog named Buddy and a sister in the army. I like the feeling when I leave at night or watch one of my students graduate that they are better prepared than I was when I left school, and they got there with my help. I like making a difference. I love to work, and the students give me a lot of work to do.*

**Q / What are the challenges in your job?**

*A: I am not an academic person in the traditional sense. I am a Chef and a cook. I like to make things happen. In academia, sometimes things move way too slow. I am a 6 AM to 10 PM guy working in a predominantly 9-to-5 world. That can be frustrating at times. A recent challenge has arisen with our continued success. We are constantly asked to take more students. I refuse to do it. Our strength is our instructor-to-student ratio. I will not compromise the integrity of that ratio. My boss supports me in this as well. The pressure comes from above her. But we deal with it. The only way we can take any more is to build a new facility. I sometimes get frustrated with those professors that I encounter that take the students for granted. They forget why they are here and why they have a wonderful job.*

**Q / What led you to teaching? What was your career path?**

*A: I started at 14 as a dishwasher and line cook, working with my sister in a greasy spoon in Livonia, Michigan. I worked there and a similar place for six years and then went to college for culinary arts. I have been a Technical Chef in a large university dining service and a Chef de Cuisine for a private club as well as a 2,200-seat banquet facility. I worked in large-volume à la carte restaurants doing 900 to 1000 covers.*

*I am really a line cook at heart. My son's birth led me into teaching part time in Michigan to make some extra money, and I took a liking to it. I pursued it full time here at Delhi by taking over the struggling culinary program five and a half years ago. The program was in jeopardy, and I was their last resort.*

*I also owned and operated several of my own businesses. Most recently, my wife and I had our own restaurant for three years here in New York. I have much experience in off-premises catering for 100 to over 5,000 guests. I also worked in several smaller family Italian restaurants. I now run a very private and exclusive club in the Catskills in addition to my teaching. I do all of the cooking and baking, and my wife works with me. It is very important to stay in tune with cooking and running a kitchen. I think it is a disaster when Chef educators leave the kitchen entirely. They cannot possibly be effective teachers without engaging in some kind of cooking outside of school. Labs are fine, but they just do not cut the mustard for keeping an instructor up on their own cooking ability.*

**Q / What professional associations for teachers do you belong to?**

*A: I belong to the American Culinary Federation and the National Ice Carving Association. I follow Council on Hotel, Restaurant, and Institutional Education, and I read a great deal.*

# Organization and Job Description

## Job Description
### Culinary Arts Program at Fox Community College

**ASSISTANT PROFESSOR**

**PROGRAM:** Culinary Arts

**POSITION TYPE:** Full Time/Permanent

**PAY BASIS:** Salary

**REPORTS TO:** Dean, Business Studies Department

### RESPONSIBILITIES:

1. Teach coursework within the culinary arts program.

2. Show evidence of planning classes, care in preparation of teaching materials and handouts, and reflective and conscientious feedback on students' work.

3. Teach using a variety of teaching methods that allow students to be actively involved as much as possible. Offer learning experiences that help students attain high levels of knowledge, a repertoire of core thinking skills, and culinary skills they will need to draw on in the workplace.

4. Foster understanding of diversity through educational experiences.

5. Maintain four office hours each week.

6. Recruit students.

7. Advise 40 students each year.

8. Evaluate transcripts of college students who want to enter one of the culinary programs.

9. Establish and maintain collegial relationships with local hospitality employers and internship sites. Secure and maintain contractual agreements.

10. Participate in continuing education activities.

11. Promote high standards of ethical and professional conduct.

12. Participate in yearly evaluation process. Set yearly goals.

### QUALIFICATIONS

Education: Master's degree required.

Experience: ACF certification required. At least five years' experience as a Chef. Teaching experience preferred.

**Figure 11-1**
Sample Culinary Education
Organizational Chart

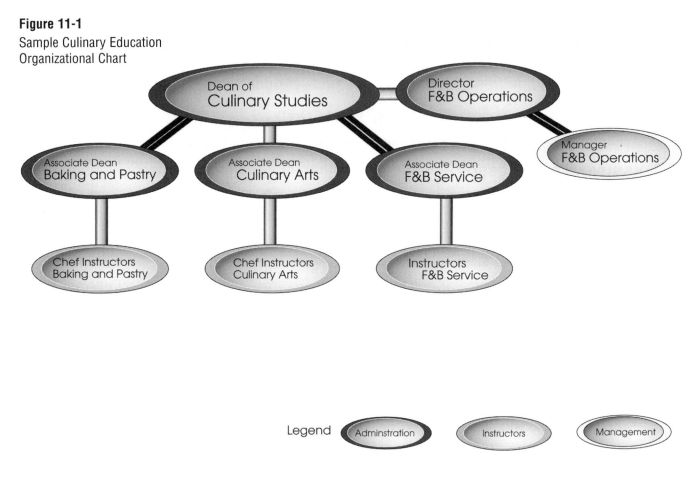

Legend

Adminstration    Instructors    Management

## EXERCISES

1. Would you consider working in education either full time or part time? Why or why not? If so, which type of school or college would you like to work in and where?

2. Visit the website of a cooking school such as Tante Marie's Cooking School (www.tantemarie.com) and look at the classes offered. Do they offer day, evening, or weekend classes? What types of classes seem popular—basic skills classes, international cooking?

3. If you worked full time in a postsecondary school, what benefits would professional organizations such as ACF, I-CHRIE, or FENI offer you?

4. Explain why you agree or disagree with Tom Recinella's following statement: "I think it is a disaster when Chef educators leave the kitchen entirely. They cannot possibly be effective teachers without engaging in some kind of cooking outside of school. Labs are fine, but they just do not cut the mustard for keeping an instructor up on their own cooking ability."

5. Find out from one of your instructors why he or she became an educator and what he or she likes and dislikes about the field.

# Additional Culinary Careers

## Introduction

**THERE ARE MANY INTERESTING CAREERS** for culinarians in addition to those discussed in the last 11 chapters. This chapter discusses what it is like to work in the following fields.

◆ Research and development Chef
◆ Personal and private Chefs
◆ Food stylist and food photographer
◆ Food communications
◆ Public relations
◆ Food purchasing managers
◆ Celebrity Chef

Without a doubt, we didn't cover all the jobs that a Chef could venture, but we have tried to show you the venues where the large majority of Chefs work.

## The Feel

Research and development (R&D) Chefs are getting a lot of attention these days, and understandably so. These professionals, who are part Chef and part food scientist, use their cooking and marketing know-how to develop and evaluate new menu items for restaurants, food manufacturers, and other companies. Most large food processing companies have test kitchens staffed with home economists, food scientists, and often professional Chefs. More Chefs are becoming involved in research and development in part because consumers have more sophisticated palates and want food that tastes great.

The process of creating new menu items involves a lot of time and work. To develop prototype menu items, research Chefs look broadly at marketing research, customer data and input, food trends in restaurants of all types as well as the media (television, cookbooks, magazines), information on ingredients, and more. Once a prototype is developed, research Chefs:

◆ Cost out the recipe.
◆ Determine what price can be charged.
◆ Analyze the nutritional content.
◆ Work with vendors to develop specifications.
◆ Develop equipment, methodology, and presentation of item.
◆ Make the recipe replicable in any location.
◆ Present the product to tasting groups to gather feedback and insight.
◆ Test the item with real customers.

Multi-unit restaurants then test promising new items in selected markets before determining whether to roll them out throughout the company. Many prototype menu items fail to get to this step.

So who are these research Chefs, and where do they work? According to a survey by *Nation's Restaurant News* (September 2003), most research Chefs have worked in the culinary field for more than 20 years. The research Chef needs strong culinary knowledge and skills and extensive restaurant experience. Operational experience is important so the Chef understands menus, customers, and how new menu items are integrated into a restaurant. Because so much of the job involves communication, the Chef also needs excellent oral, presentation, and writing skills. Research Chefs work in many settings: all types of restaurants, food manufacturing companies, hotels, cruise ships, healthcare, schools and universities, business and industry, ingredient supply houses, and even flavor and spice manufacturers. Page 207 shows a job description for a research Chef position.

Research Chefs are often members of the National Restaurant Association, American Culinary Federation, and the Research Chefs Association (see Appendix for more information). The Research Chefs Association (RCA) was founded in 1996 and now has about 2,000 members, including research Chefs, food scientists, and other food professionals in sales, marketing, manufacturing, distribution, and the media. As you can read on their website (www.culinology.org), RCA members are "the pioneers of the discipline of Culinology™—the blending of culinary arts and the science of food." RCA is helping colleges develop culinology bachelor's degree programs that offer coursework in culinary arts, food science, business management, nutrition, processing technology, and government regulations.

# Job Description

## Research and Development Culinary Manager

*POSITION TYPE: Full Time/Permanent*

*PAY BASIS: Salary*

### SUMMARY

This individual will support the business strategies of the foodservice division by providing expertise and experience in food science, foodservice operations, culinary, and product knowledge as an integrated business team member. This individual will build relationships with customers via their culinary teams in order to identify customer needs and possible solutions. These solutions will be found in our current product lines through menu/recipe applications and through new product concepts. This individual will be a communication link with research and development by providing customer insights and jointly working on new product concepts. Also, this individual will provide industry and culinary expertise to photography projects and communications related to food and menu trends, and orchestrate product samples for special events or specific customers. This individual will also be responsible for recipe development for advertising, public relations programs, brochures, and promotions, and will provide culinary presentations to customers and internal groups. Travel will be approximately 25–35%.

### POSITION REQUIREMENTS AND SPECIFICATIONS:

*Bachelor's degree in food science, nutrition, or business area strongly preferred; culinary degree required.*

*Minimum of ten years' experience within the foodservice industry, in particular with a variety of national full-service family-style restaurant (FSR) organizations; corporate culinary experience desired.*

*Proven expertise in food science, foodservice operations, and culinary arts as they relate to FSR operations.*

*Strong presentation skills to both internal and external customers.*

*Experience with creative recipe and menu development.*

*Strong planning and consulting skills specifically to uncover and build new food applications or product lines.*

*Ability to manage multiple projects as well as bring projects to successful completion.*

*Effective time management, communications, and decision-making skills.*

### IN ADDITION, THE FOLLOWING CRITICAL SUCCESS FACTORS ARE NECESSARY:

*Proven passion for food and the desire to build on expertise and industry knowledge.*

*Strong business orientation, with an understanding of the costs and labor involved in product development.*

*Confidence in culinary abilities and willingness to share insights and opinions with internal and external customers.*

*Strong team orientation, including the ability to lead a project team and partner with a variety of personalities and styles.*

*Ability to handle multiple projects and adapt to new situations.*

*Customer service orientation and ability to respond to customer needs.*

*Strong written and oral communication skills, including presentation skills.*

RCA also offers two certification programs.

◆ Certified Research Chef (CRC)—Certified Research Chefs are experienced Chefs who have worked in food product research and development.
◆ Certified Culinary Scientist (CCS)—Certified Culinary Scientists are experienced food scientists who are also knowledgeable about the culinary arts and use this knowledge in food product research and development. Food scientists are trained in science and engineering and work in product development as well as quality control, food processing and packaging, and nutrient analysis.

CRC and CCS professionals offer the unique advantage to their employer of having the skills of both Chef and Food Scientist.

Being a research Chef has certain benefits. It allows more time for creativity because the job does not entail running a business, dealing with customer and employee problems, and worrying about food and labor costs. The hours are longer than 9 to 5, but they allow for more flexibility than many other businesses. The salary is financially competitive. The survey from *Nation's Restaurant News* found that about half of respondents make between $50,000 and $125,000 annually.

After working for many years as a Cook and Chef in well-known hotels, restaurants, country clubs, yachts, and more, Roland Schaeffer started with the Heinz Company as an Experimental Chef and worked his way up to Senior Experimental Chef. Along the way, he learned that most products developed in R&D kitchens never get to the restaurant or supermarket. Besides developing new products, he was also called on to refine product costs. Imagine how much money you can save a company that produces millions of bottles of ketchup if you can save a penny for each bottle produced!

## INTERVIEW

**LORI DANIEL, Chief Inspiration Officer/Chef, Two Chefs on a Roll**

**Q / What is your current position?**

*A: My current position is Chief Inspiration Officer/Chef, and my role is new business development, marketing, and ambassador of culture for my company.*

**Q / And your company is?**

*A: The name of the company is Two Chefs on a Roll, Inc., and we are a custom food manufacturer. I cofounded the company in 1985 with Eliot Swartz, my husband. We create products for retail, food service, and manufacturing. The product range is from dips, sauces, pestos, puddings, to a full range of desserts. Customers are unique retailers or the top-tier supermarkets such as Trader Joe's, and restaurant chains such as California Pizza Kitchen. For manufacturing clients, we co-pack recipes that they supply to us, or we will create a specific product for their needs. Currently we co-pack several labor-intensive food items for different companies. Often the opportunities that come our way are products that they're not geared up to do, such as a new product launch for which the equipment is a significant investment or something that is extremely labor-intensive, particularly on the dessert side. On the dessert side, we are very hands-on. On the savory side, we're highly automated. We are at a turning point right now in our company, trying to decide what our new identity is going to be.*

**Q / How big is Two Chefs on a Roll?**

*A: We have 230 hands-on employees and a full office support staff. We relocated our facility from 33,000 to 100,000 square feet, building the plant, negotiating the lease, getting the building done, the contracting. It was a major undertaking, and then moving into the new plant with a new overhead, a new breakeven, new responsibilities, so obviously we want to grow it much larger. We've been working on putting the organization together and bringing on a Vice President of Operations and a Chief Financial Officer. We even have a Chief Information Officer. For a food manufacturer that is our size, we have quite a team, and that's what we need to be in place to continue to grow.*

*The challenge is we only have so many resources. We have some bank loans and we have an SBA [Small Business Administration] loan, but we haven't been able to get any more money than what Eliot and I put against it. So, if we put in 250, they put in 250. It does inhibit your growth, and it kind of chokes your resources a little bit. And now that we're spread over those three markets—retail, food service, manufacturing—it's hard to market the company in those three arenas. Trade publications are different. Buyers are different. Products are different. People are all different.*

**Q / What was your career path up until Two Chefs on a Roll?**

*A: How did I choose the food industry? Well, for me it was sort of by default in some ways. I wasn't a very good student, and I found that I loved to work and did well. So, when I became a senior in high school, there were these work-related occupational programs that you get out of school at 11 and go to work. It made all the sense in the world to me. One of the work programs was called home economics–related occupations. Well, I figured I could probably get into that one, so I signed up for it. I had to find a job in a kitchen even though I didn't know how to cook. I found a job as a prep cook at the nicest steak and lobster house in town, but only after convincing the interviewer that I could lift 50-pound bags of onions, put up with the heat, and so on. They told me that "we don't hire girls," but somehow or other I got a phone call from them to start work. I fell in love with the work. I loved being on my feet. I loved the excitement. I had been involved in drama in high school. So this was like being in theater.*

*I was 16 years old, and I just knew I had found a place for myself that combined a lot of the things that I liked. But understanding a career in it,*

*I had no clue. But part of this program was you had to spend time in the college resource center. I knew with my grades I was never getting into college. So here I was just flipping through brochures. I didn't even know what the word* culinary *was, and I fell into it. I was looking under cooking, I guess, because I didn't know what else to do. There was this brochure for the Culinary Institute of America, and I had to look up the word* culinary *in the dictionary. I couldn't believe it. There was a whole school on this. I stole it from the college resource center. I took it home. I told my parents that this is what I want.*

**Q / After you graduated, did you think right away that you were going to be a Chef?**

*A: I thought I was going to be a great saucier; that's what I wanted to do. I fell in love with that idea. During my externship, I worked at the sauce and soup station at the Hilton on Michigan Avenue in Chicago. All we did every day was support all the other venues by doing the soups and the sauces for every restaurant in this giant—I mean the hotel did thousands of covers a day. So there were a lot of sauces to be made and soups. And so it made sense that you could actually pursue a career in that type of specialty. That's what I wanted to do. I knew I wouldn't be a Chef right away. This was in the days that you would never call yourself a Chef until you were really a seasoned professional.*

*One of the reasons that I fell in love with going to culinary school was because I could live anywhere in the world and support myself. It was also a very big part of the romance of this industry. I could go to a Club Med and work, and I could travel the globe and get a job here and stay for a year then move on. And that was a very big part of the romantic aspect of being in the food industry—I could live anywhere and support myself.*

**Q / So what work experience did you get after graduating?**

*A: I was looking to build skills. I worked in a Scandinavian restaurant in Atlanta. I worked in a Russian restaurant in Atlanta. And then I worked in a continental cuisine restaurant. Smaller restaurants offered me the best opportunity to build skills. When I went to work for the smaller restaurants, they really appreciated me. I got to do the specials of the day. I cooked the line at night. I was a very loyal employee. That's where I really learned to cook. They gave me the opportunity to do all the cooking—the side dishes, the sauces, the entrées, the appetizers, everything. I even got a little exposure to desserts.*

*So, after working in Atlanta, I moved to Los Angeles. Los Angeles is a very big city, a very difficult place to find your way. And I had some very crappy jobs. I hit rock bottom when I was running a cookie store at a mall. I got a phone call from the woman who was a Pastry Chef at a restaurant where I had worked. She remembered that I wanted to bake and she told me that her Sous Chef just left for the summer, but he would be back in the fall. She asked me, "Do you want to learn how to bake right now?" And I said, you bet. So I quit the cookie store in a heartbeat and went to work for her. It was a frustrat-*

ing summer. I didn't pick it up very easily but, by the end of the summer, she said I was a great Baker and a great Sous Chef. That was a big deal for me.

After that, I went to a brand-new restaurant in Redondo Beach which was about to open. I sold myself as a Sous Chef making simple desserts. And they were American-style desserts because that's what I love to create. I made specialties like chocolate-strawberry shortcake, linzer cookies, chocolate truffle, and seven-layer chocolate decadence. It was in the early 1980s, when there was a big renaissance in desserts. And I couldn't make enough of them. Within six months, I had a pastry assistant, and we were doing wine dinners every month with the pastries and special anniversary parties with pastries and getting written up in every newspaper, magazine, and guide in Los Angeles. It was a real exciting time for me. And the owners of the restaurant kept saying they wanted to open a bakery, but they clearly didn't have the money to do it.

So Eliot and I saved $20,000 and decided to do something on our own. I didn't really want to do retail bakery and sell cookies and wedding cakes. I wanted to sell my desserts to restaurants and businesses, and that was the premise for the business. I thought I was the first one to ever think of it, not realizing there was such a thing as a wholesale dessert company. Then during my first few weeks out there I started to learn that there was competition, so I was terrified. I thought, I can't do this. We would never survive.

**Q / Did you rent a little store to start your business?**

*A: Actually, it was a restaurant. But first, I baked six cakes in the apartment that we had at the time, and I took a slice out to my favorite restaurateurs. I asked them if I were to make these things, would they buy them? In those days, everybody had a Pastry Chef. And every single restaurant I went to said, "Yes, I would buy them, and I want two of those and three of those." And I said, well, I don't really have anywhere to make them because I was adamant that I could never cook professionally out of my house. So one of the people I visited said they had a restaurant in a nearby town that's not doing very well and is only open for lunch. They let me go in there at 1 AM, just be out by 10 AM. So I agreed, paid the rent, and that's how I got started. I would bake through the night, deliver the cakes in the morning, buy ingredients, and work on sales to build the business.*

**Q / Did you leave your job at that point to start your own business?**

*A: Yes, and actually before we determined what direction to take the business, we traveled to Europe. Neither one of us had been to Europe, and I thought, well, I'm doing American-style desserts, and all the customers of this restaurant were saying have you ever had a strudel in Germany, or have you ever had a petit four in France, have you ever tasted any of these great desserts. And I hadn't. So I felt I didn't know what direction to take the company in. So we said we need to go to Europe. We're not going to have a vacation for a long time, and we never had a honeymoon. So we took a four-month trip to Europe and ate our way across the 11 countries of Western Europe. We went to every chocolate shop in Brussels, every strudel shop in Germany, and we went to the Hotel Sacher in Vienna. We just tasted all the regional pastries in every country. It was a magnificent trip.*

*I enjoy the American-style desserts with all the French and German techniques. I did make strudel, but it was my own style. We started the business with a whole renewed sense of pride in our own skills and style.*

**Q / Tell me about how you moved from supplying local restaurants with desserts to going national?**

*A: A very big opportunity that changed our business was with a foodservice account where we walked in at just the right moment when a competitor had presented them with a chocolate dessert but couldn't come through with the pricing and the packaging. I was sitting in the lobby being stood up by the same Chef for the fourth time. And I'm sitting there with my box of cake—Pastry Chefs never travel empty-handed. So I had a cake in my hand, and it was a type of chocolate cake made like a soufflé with nowhere to fall. His assistant comes to tell me that he wasn't going to see me when she noticed the box and asked what was in it. I showed her my chocolate cake, and she asked me to sit right down. She went away and came back with a little tiny chocolate truffle cake and asked if I could make something like this. So I looked at it, and I could tell what it was by looking at it. I tasted it too. Because I had been baking cakes how many thousands of hours, I told her I could make the cake for her. I have a really good taste memory. I can taste something and figure out how to make it. She wanted the cake by Friday, which I did. Actually I think mine was better than hers. We agreed on details such as price and packaging design. We went national with that product in six weeks. We shipped pallet loads of product for the very first time. With Trader Joe's, we were doing fresh daily deliveries unwrapped—shrink-wrapped little cakes with labels on top. For foodservice, this had to be tray-packed, case-packed, marked, palletized, wrapped, and listed up, then brought to the distributor dock.*

*That changed the business. It was a struggle because I never knew what it meant to make a profit or keep books or hire employees or manage production schedules or any of that. I didn't know any of it. Eliot and I didn't know that part of the business. But we both had our own skills which sort of sifted out. We made a lot of mistakes. I did finally take classes at a local college on how to start a business. Taking this one small course at a community college taught me some very important basics that kept me in business—by taking liability insurance, by getting workers' comp, etc., by doing the right thing from day one.*

**Q / How many employees did you have at that time?**

*A: At that time, we probably had 35 employees. And that was a big deal. I was amazed at how fast we had grown.*

**Q / What skills are you looking for in a Chef that comes to work for you?**

*A: For Chefs to work in R&D, they need to have good knife skills, understand how to cook, have experience working a line and working with customers. I'm looking for people who actually have the foodservice experience, because how else can they help me create products for a retailer or a restaurant chain if they really don't understand the customers, the people who eat, how they eat, what they're eating, as well as how food is made? We bring in R&D Chefs who are great with the customers. They need to be able to do presentations demonstrating their culinary skills as well as to use PowerPoint and have photographs of menu items to give to customers. R&D Chefs are actually salespeople—they need to sell the products that they develop. They can't just stand there and cook the whole day and hope that maybe something hits.*

*The challenge of coming from a culinary background into our environment is also that you need to be very organized and detail-oriented. Those skills in a Chef can be hard to find. What makes it exciting for a Chef to work here is the opportunity to create new products for brand-name customers.*

**Q / What paths could a culinary student take to get into the R&D field?**

*A: I think there are a lot of different paths that one could go into to get into R&D. If a culinary graduate just wanted to go out and get a food science degree and do R&D, that path is open because someone with a culinary degree and a food science/technology degree would have no problem finding a job. That being said, in my opinion, I would recommend spending time learning the foodservice industry, because understanding how people eat, where they eat, what they're going to be eating, is a great way to connect to that customer. And you're going to need those skills in R&D and manufacturing. Additionally, running an organization where you have profit and loss responsibility, and you have an opportunity to manage people, all of that directly relates to a manufacturing facility, because you will need to control costs and supervise people.*

*I have two senior-level scientists and technologists who can't solve problems any better than our culinary creative team, because the Chefs' approach to things is really grass-roots and practical. R&D people are always focused on how to make that product hold that color, how to make it stay that thick, how to make it have that flavor. The Chefs and food technologists work great together. We're looking for the pure culinarian to work hand in hand with the pure technologist and personally we get a better result.*

**Q / What have you loved the most about growing your own business?**

*A: What I have loved the most in the past is building the organization, finding the right people for the right job. That has always given me my greatest pleasure, as well as being able to be out networking with the industry and understanding the trends.*

**Q / And maintaining the quality that you started by yourself?**

*A: Right. That hasn't given me pleasure though; that's been brutal. Now that the organization is built, I don't need to be involved. I have excellent people who can find their own people. The Chef can find more Chefs, and the CFO can manage his own finance department. So I'm not needed to do that recruiting anymore. Now I'm needed to do the new business development, thinking carefully about our strategies for the future. That is more my focus now.*

# Personal and Private Chefs

Personal Chefs run their own businesses to provide customized home-cooked meals for busy individuals and families. They help individuals and families who:

◆ Don't have enough time to plan and cook meals.
◆ Want to eat healthier meals.
◆ Are giving a party but don't want to cook.

The number of personal Chefs has been growing. Some individuals are going into this segment because they want to be their own boss, or they want to make a comfortable living while working flexible daytime hours.

What does a personal Chef do? Most personal Chefs own and operate their own business and have a number of clients. Once a week, they usually prepare Monday through Friday meals (often just entrees and side dishes) for one client each week. During a typical week, he or she is involved with any or all of these activities:

◆ Conducting an assessment with potential clients to learn about what foods they like, whether or not they have any allergies or special dietary needs, and so on.
◆ Planning weekly and other menus.
◆ Food shopping.
◆ Setting up and cooking for clients, including preparing containers and labels with reheating instructions.
◆ Preparing bills for clients and depositing checks.

Although it is possible to start your business with clients you already know, you will at some time have to market your services to gain more clients.

Whereas a personal Chef has a number of different clients and is self-employed, a private Chef has only one client, and is usually an employee of the client. Private Chefs prepare three meals per day and often live with the client, as well as travel with the client.

## INTERVIEW

### CANDY WALLACE, Owner/Operator, The Serving Spoon

To learn more about being a personal Chef, read the following interview with Candy Wallace, Executive Director of the American Personal Chef Association (APCA).

**Q / In addition to being the Executive Director of APCA, do you run your own personal Chef business?**

*A:* Yes, I do. I am the owner and operator of *The Serving Spoon,* a personal Chef service in San Diego, California.

**Q / What types of job duties do personal Chefs do each day, each week?**

*A:* We counsel clients and design menus specifically to meet their dining and nutritional needs. If a personal Chef is amply qualified and trained, he or she may discuss medically specific diets. Generally speaking, meals are prepared in the clients' homes. This is helpful from a sanitation and transportation point of view because the food is prepared and immediately stored at the proper temperature.

**Q / How many hours do you work each week in your business?**

*A:* Like many other personal Chefs, I usually work with one client each day. I generally work five hours each day, about 25 hours weekly.

**Q / What can a personal Chef expect to earn by cooking for one client each day, Monday through Friday?**

*A:* About $60,000 to $75,000 annually, but there's a potential of $100,000.

**Q / What was your career track to this position?**

*A:* I worked in restaurants and have a lot of experience in catering businesses. I also went to college and got a degree in political science. I moved to San Diego, and owned a catering business called *Taste,* which did primarily dinner parties. In 1992, I realized that the catering grind was too much and I wanted to control more of what I was doing. I am a big entertainer, so I got all my friends together and told them, "I will support you and your families every day with fresh meals," and I did just that. I got a great deal of exposure and started to get calls from people who wanted my services as well as calls from people who wanted to start this type of business.

**Q / What would you have done differently in your career path if you could do it again?**

*A:* Since my only formal culinary training was continuing education courses, perhaps I would have gone to culinary school rather than university.

**Q / Has any investment made in your career paid off?**

*A:* What has helped me a lot is the fact that I have traveled a lot internationally and domestically.

**Q / What do you like most about your job?**

*A:* I make an enormous contribution to the quality of life of my clients by providing great nutritious meals and removing stress from their lives. Personal Chefs bring families back to the dinner table.

**Q /** **What do you like least about your job?**

*A:* *Paperwork and accounting.*

**Q /** **What advice do you have for new graduates in choosing a career like yours?**

*A:* *Go to school, learn the skills, and get some varied foodservice experience, because once you are in this business, you are on your own. This business should be entered once you are ready, individually and professionally, to operate in a self-directed environment.*

**Q /** **Is it true that personal Chefs can become certified through the American Culinary Federation (ACF)?**

*A:* *Yes, there are several certifications available: Personal Certified Chef (PCC) and Personal Certified Executive Chef (PCEC). Also, Certified Culiniarian is being added for personal Chefs. Requirements for these certifications are available on the ACF website, www.acfChefs.org.*

## Food Stylists and Photographers

Look at the front cover of any food magazine and you'll see the work of food stylists and food photographers. Food stylists and photographers work together to produce those beautiful, glossy food photos. But you need to be cautious about eating the food they photograph. Food stylists are like makeup artists: They use all types of tricks to make food look fresh and delicious, including some that make it inedible. For example, instead of using powered sugar on top of a quick bread, a food stylist may use talcum powder because it won't melt. Another trick is to substitute motor oil for pancake syrup because it looks thicker.

Food stylists arrange food to be photographed professionally. This is a creative job that involves working with food photographers, photo editors, art directors, and others involved in print publications such as newspapers, magazines, and cookbooks. Food stylists also prepare food for advertisements, movies, television shows, and theater productions.

Here are tasks food stylists typically do:

◆ Meet with clients to discuss what is to be photographed and how, including props.

◆ Shop for food and equipment needed for the shoot or presentation.

◆ Cook or bake the food items.

◆ Arrange the food over the course of the shoot or presentation.

◆ Maintain a toolkit to use when styling foods, including dishes, glasses, and other items that can be used as props.

A stylist's toolkit also includes the tools needed for most of the tricks that keep food looking delicious and fresh—paintbrushes, skewers, tiny spatulas, and eyedroppers. Also, a food stylist may spray ArmorAll® on a tortilla so it doesn't dry out.

A culinary background, meaning both schooling and cooking experience, is crucial to succeed as a food stylist. In addition, food stylists need training in art and photography. An artistic background helps a stylist use colors, props, camera angles, and lighting to make food look great. Good communication skills help a stylist interact with clients, editors, producers, and other people he or she will work with to set up the stage. Because many food stylists work on a freelance basis, they also need some business background so they manage their business and finances well.

Besides having a background in foods and cooking, art, and business, a food stylist must be able to work well with clients and handle deadline pressure. Food stylists have to be flexible when the client wants to change something or has a strong opinion about a look. A food stylist may prepare a dish ten times before a good shot is taken, or freshen up a dish many more times, so patience is important. On photo shoot days, hours are usually quite long—ten hours or more.

To break into this field, most people start by getting a job as an assistant stylist. This is a good way to learn food styling techniques as well as photography. Most culinary schools offer classes in food styling; these are helpful, but they cannot replace hands-on experience with food stylists and photographers.

To start a career as a food stylist, you must do a lot of networking, because much of this business is by word of mouth, and you need to develop a strong portfolio of sample shots. Full-time food stylist positions are rare, so most stylists work as freelancers. Potential employers of both full-time and freelance stylists include advertising agencies, food companies, restaurant chains, food magazines, and television networks. Los Angeles and New York are two of the biggest places where food stylists are needed because that's where many advertising agencies, films, and television shows and commercials are produced.

Although the job outlook for food stylists is steady, there is a lot of competition for the jobs, in part because the work is fun and the hours are good. Full-time food stylists make from $30,000 to $50,000 or more each year. Freelancers make anywhere from $300 to $1,000 or more per day for a photo shoot, depending on the type of job, location, and their experience. Self-employment allows for greater autonomy, freedom of expression, and flexible scheduling. However, income can be uncertain and the ongoing and time-consuming search for new clients can be stressful.

Like food stylists, food photographers work most often as freelancers. Less than half their time is normally involved in photographing foods. Other tasks include researching new clients such as art directors and magazine editors, talking to clients about upcoming assignments, preparing for a shoot, estimating and negotiating fees, getting supplies, developing and delivering prints, bookkeeping, paying bills, return phone calls and emails, and so on.

The food photographer uses a variety of tools to create a mood or feel in a picture. He or she uses natural and commercial lighting to help paint a picture, almost like an artist does. For example, a photographer might use a wooden block to create a shadow for the feel of a candlelit meal.

To break into this exciting field, you might start work as an assistant to a photographer or even assist a food stylist. Like food stylists, most food photographers work as freelancers, and the field is very competitive. Freelance photographers can make upwards of $1,000 per day for a shoot, but keep in mind that they need a lot of money for equipment to start up their business. They also need to be well versed in digital photography and software programs like PhotoShop. Photography is a highly technical field, and attention to details is crucial.

Food stylists and food photographers often belong to the International Association of Culinary Professionals (IACP). IACP is a group of approximately 4,000 food professionals from over 35 countries. IACP provides continuing education, networking, and information exchange for its members. See Appendix A for more information about IACP. Food photographers may also be members of the Advertising Photographers of America or the American Society of Media Photographers.

## Food Communications

Culinary professionals can work in communications as food writers, food editors, and cookbook authors. Writers and authors develop original writing for books, magazines, trade journals, online publications, company newsletters, radio and television broadcasts, motion pictures, and advertisements. Journalists write articles mainly for newspapers and magazines. Editors select material for publication or broadcast. They review and revise a writer's work for publication or dissemination.

Major newspapers and newsmagazines usually employ several types of editors. The executive editor oversees assistant editors who are responsible for particular subjects, such as food and cooking. The food editor determines the subject or content of the food pages, which appear once a week in many newspapers or in each issue of a magazine. Depending on the publication, the food editor may use staff writers to write the articles or hire freelance food writers. Freelance food writers find work by writing letters to editors that outline potential article ideas; they also show samples of their published work to editors. If a freelancer's idea is approved by the editor, the writer writes and submits the article for publication.

Food articles in newspapers and magazines have become much more sophisticated recently. Whereas in the past it was routine to find articles on 25 ways to use up Thanksgiving turkey, nowadays you can read about how people all around the world grow their food, cook, and eat. Newspapers and magazines have discovered that readers long for stories and information on cooking and eating.

A good place to begin a career in food writing is to offer to write a food and cooking column for a local newspaper. Many communities have local weekly papers that are good training grounds for food writers. Once you have a number of clips—copies of your published articles—you can use them like a portfolio when you submit queries to magazines and bigger newspapers.

Becoming a cookbook author is a lot different from getting published in your local paper, and a lot more competitive. Further, publishers are usually looking for authors who are well known, experienced, and expert in their field. To get a book contract, cookbook authors write a book proposal detailing what the book will be like, including a table of contents and a sample chapter. Because the cookbook business is so competitive, most authors are represented by a literary agent. The agent, who usually keeps 15% of the eventual royalty earnings, tries to sell the book idea to publishers. Royalties are usually from 7 to 15% of the book's retail price.

To be a writer, editor, or journalist, you must be able to express ideas clearly and logically, and you should love to write. Creativity, curiosity, a broad range of culinary knowledge, self-motivation, and perseverance also are valuable. Most food writers and authors work as freelancers, so they need to know how to run their own business.

Employment of writers and editors is expected to grow about as fast as the average for all occupations through the year 2012. The outlook for most writing and editing jobs is expected to be competitive, because many people with writing or journalism training are attracted to the occupation.

Employment of salaried writers and editors for newspapers, magazines, and book publishers is expected to increase as demand for these publications grows. Magazines and other periodicals increasingly are developing market niches—including food and cooking. Businesses and organizations are developing newsletters and websites, and more companies are experimenting with publishing materials directly for the Internet. Online publications and services are growing in number and sophistication, spurring the demand for writers and editors, especially those with Web experience. Advertising and public relations agencies, which also are growing, should be another source of new jobs.

Many Chefs working in this field are paid on a freelance or royalty basis. Median annual earnings for salaried writers and authors were $42,790 in 2002. Many Chefs working in food communications are members of the International Association of Culinary Professionals (see Appendix for more information).

# Public Relations

Public relations managers work in hotels, restaurants and restaurant chains, cruise lines, supermarkets, universities, healthcare, and other businesses. The reasons why Chefs and restaurateurs hire public relations managers or agencies are as varied as the types of food they serve. Chefs and restaurateurs use public relations managers to help open a new restaurant, reposition an existing restaurant, raise both media and consumer awareness of the restaurant and/or Chef, or make customers aware of a new menu or service. Public relations managers use every available communication medium to place news and general interest stories about the company in local and broader publications with the aim of attracting customers. For example, public relations managers develop media relationships by keeping restaurant reviewers and food editors informed about special events.

Public relations managers are also involved in drafting press releases. Many radio and television special reports, newspaper stories, and magazine articles start in public relations as press releases. Public relations managers also work to uphold the image of the foodservice after a problem, such as a robbery or food-borne illness, strikes.

Public relations managers also evaluate advertising and promotion programs for compatibility with public relations efforts and serve as the eyes and ears of top management. They observe trends that might ultimately affect the foodservice and make recommendations to enhance the foodservice's's image based on those trends.

Public relations managers may assist foodservice executives in drafting speeches, arranging interviews, and maintaining other forms of public contact; oversee company archives; and respond to information requests. In addition, some handle special events that introduce new products or other activities the foodservice supports in order to gain public attention through the press without advertising directly.

In public relations firms, a beginner might be hired as a research assistant or account coordinator and be promoted to account executive, senior account executive, account manager, and, eventually, vice president. A similar career path is followed in corporate public relations, although the titles may differ. Some experienced public relations specialists start their own consulting firms.

There are no defined standards for entry into a public relations career. A college degree combined with public relations experience, usually gained through an internship, is considered excellent preparation for public relations work. Some entry-level public relations specialists have a college major in public relations, journalism, advertising, or communication. However, many public relations specialists in foodservice have backgrounds and degrees in the culinary/foodservice field, along with an understanding of public relations and marketing, writing and editing skills, and excellent people skills. Creativity, initiative, good judgment, and the ability to express thoughts clearly and simply are essential. Decision making, problem-solving, and research skills are also important. People who choose public relations as a career need an outgoing personality, self-confidence, an understanding of human psychology, an enthusiasm for motivating people, and a passion for the culinary field. They should be competitive, yet able to function as part of a team and open to new ideas.

Some public relations specialists work a standard 35- to 40-hour week, but unpaid overtime is common. Occasionally, they must be at the job or on call around the clock, especially in an emergency or crisis. Public relations offices are busy places; work schedules can be irregular and frequently interrupted. Schedules often must be rearranged so workers can meet deadlines, deliver speeches, attend meetings and community activities, or travel.

Median annual earnings in 2002 were $60,640 for public relations managers. Salary levels vary substantially depending on the level of managerial responsibility, length of service, education, firm size, location, and industry. With frequent openings of new restaurants, as well as the rise of the celebrity Chef, demand for public relations professionals will remain steady.

The Public Relations Society of America accredits public relations specialists who have at least five years of experience in the field and have passed a comprehensive exam. They receive APR status—Accredited in Public Relations. The International Association of Business Communicators also has an accreditation program for professionals in the communication field, including public relations specialists. Those who meet all the requirements of the program earn the Accredited Business Communicator (ABC) designation. Candidates must have at least five years of experience in a communication field and pass a written and oral examination. They also must submit a portfolio of work samples demonstrating involvement in a range of communication projects and a thorough understanding of communication planning.

# Purchasing Managers

Purchasing managers seek to obtain the highest-quality food, beverages, and supplies at the lowest possible purchase cost for their employers. Chefs who especially enjoy the food purchasing part of their job may want to specialize in this area by becoming a purchasing manager for a large hotel or restaurant, hotel or restaurant chain, managed services company, cruise line, or other food-related business.

Purchasing managers determine purchase specifications, choose the suppliers of the product, negotiate the lowest price, and award contracts that ensure that quality, pricing, quantities, and product identification remain consistent. In order to accomplish these tasks successfully, purchasing managers study sales records and inventory levels of current stock, track seasonality and trends, identify foreign and domestic suppliers, and keep abreast of changes affecting both the supply of and demand for needed products and materials.

Purchasing managers evaluate suppliers on the basis of price, quality, service support, availability, reliability, and selection. To assist them in their search for the right suppliers, they review catalogs, industry and company publications, directories, trade journals, and trade shows. Much of this information is now available on the Internet. Purchasing managers research the reputation and history of suppliers and may advertise anticipated purchase actions in order to solicit bids. At meetings, trade shows, conferences, and suppliers' plants and distribution centers they examine products and services, assess production and distribution capabilities, and discuss other technical and business considerations that influence the purchasing decision. Once the purchasing manager has gathered all of the necessary information, he or she places orders and awards contracts with those suppliers who meet the purchaser's needs. Contracts often are for one year and usually stipulate the price or a narrow range of prices, allowing purchasers to reorder as necessary.

## Celebrity Chefs

Celebrity Chefs enjoy something most Chefs never have: a national audience. Take a look at Wolfgang Puck. He cooked for years without an audience until his business took off and grew to include 59 restaurants, six cookbooks, a television show, prepared foods, cookware, and a syndicated news column. Americans' appetite for celebrity Chefs, and especially their cooking shows, has grown tremendously. The Food Network has doubtlessly helped launch many Chefs into stardom. The requirements of celebrity Chefs are huge: they must be very talented in the kitchen, entrepreneurial both inside and especially outside of the restaurant, highly experienced at marketing and public relations, creative, savvy about business, media-trained, and very personable. With such high requirements, it is obvious that you need many years of experience before you can even consider approaching the Food Network or cookbook publisher, or opening more restaurants. In any case, few Chefs ever become celebrity Chefs, and indeed, many Chefs are not interested in pursuing celebrity status. Chefs such as Michael Romano of New York's Union Square Café, would rather run his own restaurant and make wonderful food for a smaller audience.

## EXERCISES

1. Learn about the field of business communications at the website of the International Association of Business Communications (**www.iabc.com**). On their home page, click on "About IABC" in the pull-down box. Which of the job positions discussed there are found in hotels or restaurant chains, managed services companies, or other large foodservices? What does each position do? Do any interest you?

2. Visit the website of Lori Daniel's Two Chefs on a Roll (**www.twochefsonaroll.com**). Click on Resources. What types of resources are mentioned? How would they be useful to a Chef?

3. Many food stylists and food photographers have their own websites, which they use as a portfolio. Find and critique a website for a food stylist or photographer.

4. Find out if there are personal Chefs in your area. Use the phone book or the Internet, or just ask around.

5. Would you like to work in any of the jobs discussed in this chapter? Why or why not?

# Part

4

# INTRODUCTION TO ADVICE FROM INDUSTRY LEADERS

Learning from those who have blazed the trail before us still proves to be one of the best educational approaches besides formal instruction. The following people are some of the most influential and well-respected professionals in our industry today. They are quoted in trade magazines, speak at numerous affairs, set trends—and stop them—helping establish industry direction. As mentors, they have helped many chefs make good decisions that promote success, they have served as a sounding board for important presentations, and much more.

- ◆ *Lee Cockerell, Executive Vice President of Operations, Walt Disney World® Resort*
- ◆ *John Doherty, Executive Chef, Waldorf Astoria Hotel*
- ◆ *Edward G. Leonard, CMC, AAC, Executive Chef of Westchester Country Club, President of the American Culinary Federation, Manager of ACF Team USA 2004 and 2008*
- ◆ *L. Timothy Ryan, EdD, CMC, President, Culinary Institute of America*

Of course, the culinary field boasts many world-class chefs, but we could only pick a few.

To read their journeys will be an inspiration in following their impressive work ethics, dedication, and perseverance. Reading about them will help you create your own legacy. You are at a crucial time in your career to excel from your youthful energy while solidifying your foundation, learning skills that will be invaluable for your future. These industry leaders are not only our colleagues but also advisors from whom we continue to listen and learn. They were born to lead, teach, and mentor. Read on, and you'll know exactly what we are saying.

# A Word from Industry Leaders

## Interview

**LEE COCKERELL, Executive Vice President of Operations, Walt Disney World® Resort**

**Q / What is your present position?**

*A: I am currently Executive Vice President of Operations for the Walt Disney World® Resort in Orlando, Florida. I am responsible for all of the operations, including the four theme parks, the resort hotels, our shopping and dining complex, and our sports and recreation business. I also have responsibility for the operations that support our operations, including security, transportation, engineering and maintenance, textile service, and so on.*

**Q / And you currently have how many employees?**

*A: We currently have 50,000 employees, whom we call Cast Members. Everyone at Disney is a Cast Member, including me. We're the largest single-site employer in the United States, and approximately 36,000 of these 50,000 Cast Members are in operations.*

**Q / That's a tremendous number to be in charge of. How did you first decide to go into cooking, or did you decide to go into management?**

*A: I was one of those young people who had no idea about what I wanted to do when I got out of high school. I asked a friend of mine what his college major was going to be, and he said hotel and restaurant administration. I said, "Okay, I am going to do that too." I'd never even been in a hotel, as growing up on a farm in Oklahoma left no time or money for vacations. I went off to Oklahoma State in 1962 and spent two years there studying hotel and restaurant administration. After my sophomore year, I went off to Lake Tahoe, Nevada, to work at Harvey's Resort and Casino. The first half of the summer I worked in housekeeping doing turndown service in guest rooms. The second half of the summer, I worked in the kitchen. My title was grease man. My job was to roll this cart around and empty the grease from the griddles before they overflowed.*

*I kind of got back to Oklahoma late (accidentally on purpose) and missed going back to school for my junior year. I really did not like school; so here I was, and in those days you either were in school or were soon drafted. I promptly joined the Army, went to Fort Polk, Louisiana, for basic training, and became a cook. In cook's school, I met another soldier. He asked me what I was going to do since we were being discharged. I told him I had no idea. He told me that he was going to Washington, DC, to open the Washington Hilton, this new hotel opening in three weeks, and that I should come along. This was February of 1965, and little did I know that this would be the beginning of a 40+ year career in the hospitality business.*

*I walked into the Washington Hilton on February 26, 1965, and applied for a job. They asked me what I wanted to do, and since I had no idea, I said, "How about a job as a Room Service Waiter?" I had seen this on television and noticed that they made good tips. The lady in the personnel office said that those jobs were all filled but that they needed Banquet Servers. I said fine, and off I went to be a Banquet Waiter. I was fortunate to have one of the Banquet Captains take a liking to me, and he taught me everything. I don't even think I had ever seen a cloth napkin before.*

*I worked as a Banquet Waiter for a couple of years and then had an opportunity to get into a Food and Beverage Control Clerk position, which led to a management training program. This is when I got my first big break. The job paid so little that I had to get a job at night as a waiter in a French restaurant and also another job on weekends for an outside catering company to be able to survive financially. Those were two great experiences as well.*

## Q / Had you graduated already from the university?

*A: No. I didn't graduate from college, and I never went back to finish, so I am probably the worst person in the world to give any advice on the advantages of a college degree. I was lucky, and I would not suggest this strategy to young people today. I do joke sometimes that if I had graduated from college that I would have a really good job today! Despite not graduating, I guess I was lucky that some people noticed my drive, work ethic, and positive attitude. I was Mr. Agreeable. When they asked me to work on New Year's Eve and then be back on New Year's Day morning at 6 AM, I smiled and said, "No problem." Next thing I know, I am promoted. That has been my secret strategy for all of my career.*

*On the other hand, my son received his undergraduate degree from Boston University and years later his MBA from the Crummer Business School at Rollins College. I would have had a fit if he tried to do it my way without a degree. I really was the one who inspired my son to go back and get his MBA. These days, business is far more complicated than it was back in my day. When I started, the electronic calculator had not even been invented. Now with computers, everything can be calculated to the tenth of a cent. A person in business today needs to understand all of finance, marketing, cost management and productivity, industrial engineering, and on and on. I definitely would recommend for everyone to start and finish college; and if you can, go ahead and get that advanced degree as well. It will pay off for you.*

*I worked for Hilton Hotels Corporation for eight years. I held the positions of Banquet Server, Food and Beverage Controller, Assistant Food and Beverage Director, and Food and Beverage Director while working at the Washington Hilton, the old Conrad Hilton in Chicago (now the Chicago Hilton), the Waldorf Astoria, the Tarrytown Hilton, and the Los Angeles Hilton.*

*In 1973, I had the opportunity to join Marriott Hotels. At that time, Marriott had 32 hotels. I spent 17 years with Marriott and saw them grow to over 800 hotels by the time I left in 1990. Today I think they exceed 2,500 hotels or maybe even more. Marriott was great, and this is where I got the best training on how to manage a business. During the 17 years with Marriott, I held the positions of Director of Restaurants, Director of Food and Beverage, Regional Director of Food and Beverage, Area Vice President of Food and Beverage, Vice President of Food and Beverage Planning, and General Manager. I worked for Marriott in Philadelphia, Chicago, Washington, DC, and Springfield, Massachusetts.*

*I loved my food and beverage career, but I must admit that my very favorite position was that of General Manager of a hotel. In that position I had responsibility for everything from food and beverage to marketing, sales, rooms, and housekeeping. I loved taking care of our guests and the associates who worked for me. Once you have been in the food and beverage business, you can do anything. The challenges in the food and beverage area are tougher than any other area of the hospitality business. All areas can be challenging, but food and beverage is fast-paced, with hundreds of decisions to be made daily. This really gets into your blood. I love making decisions at a moment's notice, so running a hotel was an exciting experience for me. I held that position for two and a half years.*

*In 1990, the phone rang one day at my hotel, and it was Disney asking me to interview for a position as the Corporate Director of Food and Beverage and Quality Assurance for the Disneyland Resort in Paris. I went home and asked my wife, Priscilla, what she thought. She immediately said, "Let's go." She said, "Lee, you get to work for Disney, they are going to pay you, and we get to live in Paris. If you don't take this opportunity, you will look back in five years and regret it." I went for the interview. I was offered the job, so I resigned from Marriott the next day and was living in France two months later. It turned out to be the best decision that we ever made. Living in France was exciting and actually exhilarating. I didn't speak French, so I learned quickly what it is like to be illiterate. It was difficult at times; but as they say, "It's the difficult times that make you stronger."*

*I learned a lot about myself on that job. It was the hardest position that I had ever held. When you hear people talk about working 18-hour days, they are usually stretching the truth a bit; but I really did work 18-hour days, from 4:30 AM, finally getting back into bed at 10:30 PM. I must admit that I would not want to work those kinds of hours my whole life, but in a crunch you have to be able to have the stamina and energy to do what is necessary to get the job done. The clock was ticking toward opening on April 12, 1992. After the opening, I was promoted to Vice President of Operations for the six 1,000-room resort hotels. A year later, in 1993, I was promoted to Senior Vice President of Operations for the resort hotels in Orlando at Walt Disney World. A couple of years later, we merged all operations into one operating group, and I was promoted to Senior Vice President for Parks and Resorts; and in 1997, I was promoted to my current position of Executive Vice President of Operations for all of the operations in Orlando.*

*I have now been in Orlando for 11 years and with Disney for 14 years. My wife and I have moved 11 times in my career. We have enjoyed every place we have lived, with no exceptions. I tell people all the time that it really does not matter where you live, as you spend 99% of your time at work or home anyway. So if you love what you do and your home life is good, then it does not really matter where you live. I might make one exception to that. My son and daughter-in-law and my three grandchildren, Julian, Margot, and Tristan Lee, live in Orlando and only one mile from me, so that would cause me to not move for sure.*

*I remember telling my wife when I was with Marriott that we would be moving to Philadelphia and her response was one of alarm when she said, "PHILADELPHIA?!?!" Philadelphia turned out to be one of our favorite places to live, and we made a lot of good friends there who we are still in contact with today, 30 years later. So I can't think of one place that we lived that we did not really enjoy. My wife is a saint. She packed up and moved without hesitation each time I was promoted. Marry a saint if you are going to move a lot.*

*I tell people all of the time that there are three ways that we learn. First is a formal education, then get as many experiences as you can, and last but certainly not least, travel. Those three things make you a well-rounded person. Those three things are very clear to me and important to me. Meeting people in their country or home and getting to know them on a personal basis makes you more tolerant, and you learn that everyone is trying to achieve the same things for themselves and for their families. This is one thing that I know for sure.*

**Q /** **It's interesting that you say that. One of the things that we're trying to establish in this book is a career and education path. This is a very diversified industry, with many career avenues to pursue. If you were going into a culinary program, you would want to get all of your resources together to map out a focused plan. Most aspiring students don't think that far in advance.**

*A:* *My advice to aspiring young Chefs would be a little different. I would approach it in a different way. If you are going to start out in culinary, you will need to gain a lot of hands-on culinary experience and technical skill. Technical knowledge is the ticket that opens the first door to entry into a career. You will gain your credibility for having strong technical skills and knowledge. This means you can cook a great meal. At least with technical knowledge and experience, you can cook one good meal.*

*The second thing to focus on is becoming a great manager. Management is defined as the act of controlling. This simply means that you are well organized, that you have a system in place for planning your day and for following up and for doing the right things in the right order. You know the difference between urgent, vital, important, and limited-value tasks, and you do them in the right order, day in and day out. Being organized is the reason that you can manage a large kitchen and put out hundreds of meals a day and keep your payroll and other costs in line. This ensures that you have the right inventories on hand and that you make a profit. A good system that organizes you ensures that you keep your promises and that your follow-up is excellent and reliable.*

*The next thing you need to be concerned with is keeping up with technical advances so that it will become quickly apparent to you how to apply technology to your business to help with all kinds of business issues, from marketing to cost controls to improving service to improving the quality of your products. Technology will be the answer to many business solutions in the future, so this is one you must pay attention to.*

*Last, but not least, you must be a great leader to be successful. When I talk about leadership, I am talking about the ability to lead and inspire others so that you have followers. Without a team of followers, you will be cooking alone. Learn to do four things with your fellow employees. Make them feel special, treat them as individuals, show respect to everyone (keep your biases to yourself), and train and educate your teams and know their jobs so they can get ahead. When you inspire your teams and watch your own behavior, you can be assured that they will look after your business even when you are not there. Listen to your people, involve them in the decisions, ask their opinion, help them, teach them, and recognize them for their good work. Tell them every day how much you appreciate them. This is the way to build their self-esteem and self-confidence. You will have a loyal and dedicated team if you do this. They will be more than interested in their jobs. They will be committed, and committed is far different from interested. Committed teams accomplish extraordinary things.*

*What happens when you don't understand these four things is this: You are a great Chef, a great technician. You can cook a great meal, but you are not a good manager, so you miss deadlines and your costs are out of control and your inventory system does not work and you are frequently out of ingredients. You don't stay up with technology solutions, so your business is not on the leading edge in the areas of taking care of your guests, employees, and business results. You are not a great leader, so you have turnover and people don't want to work with you because you are abusive and egotistical and your team soon lacks motivation caused by you, their so-called leader in name only, and things go from bad to worse.*

*It's great that the Chef can cook, but the management part is getting everything organized so you have everything to cook. You and your systems must be well organized if you want to stay in business. Lasting leadership is how you get the whole team working together. You have to get the team inspired, feeling good, and wanting to go out every single night to win. If you are a good manager, a great leader, and you are technically competent and pay attention to technical advances, you will do very well.*

*Someone once asked me, "How did you get the job you have now?" I told them that I believe the main reason that I have been so successful is that I have a positive attitude. I wake up every morning and go to work and I stay positive. I advise every leader to be careful what you say and do, as they are watching you and judging you. You are the key to your success. Most leaders underestimate the impact that they personally have on the people and the business. I always smiled and said "yes" when I was asked to do something, from working every single holiday to cleaning out the grease traps. I did not always want to do those assignments, but they never knew that, and it was not long before I got promoted. I got picked because I had the technical knowledge, but most of all I got picked because I had a positive attitude and great relationships with others—with my boss, for sure.*

*So when the job opened for the management training program, I was picked because I had a good relationship with my boss and coworkers. When they asked me to do something, I did it without complaint. That is how the real world works—relationships. When you promote people, you don't have to look in their file. We all know who they are. It is in our head!*

*If you are going to be a great manager and leader, you need to keep balance between your profession and your personal life. You have to make time to be involved in your community. Chefs are asked all the time to be involved in charity and community events. You need to learn to keep all parts of your responsibilities organized. There is enough stress already in the food and beverage business without being disorganized as well.*

*These young adults that are making their choices now in school need a foundation, and it's not so easy to do hands-on training as it used to be. I learned management by working under good managers, and I learned skills on computers by someone teaching me. It's more complicated today.*

*Today, just like back then, you have to count on yourself. You were lucky years ago to have good managers and leaders to learn from. Everyone is not going to be lucky, so they have to count on themselves, as they may not end up with a good mentor to guide them along. I, too, am grateful for the good leaders that I had along the way, and I can tell you I had some pretty bad ones too who I learned a lot from also. My advice is for people to make sure they have the technical skills mastered by going to the right programs or schools and by getting the right experience. Take classes and read a lot about management and leadership and focus on learning about technology. If you have a great leader to learn from, then you are lucky; but most people can make their own luck with hard work, planning, and thinking.*

**Q / What is a typical day for you?**

**A:** *My days are very routine. I get up at 5 AM. I make my wife's coffee so it is ready when she wakes up. I go to Einstein's and have my breakfast, which now is a low-fat yogurt and a cup of coffee with cream and sugar. I read* USA Today, *and I get to my office by 6 AM. For the next couple of hours, until 8 AM, I write my weekly newspaper,* The Main Street Diary, *which goes out to all 50,000 Cast Members on Friday night at 5 PM. I do my email, I plan my day, and I clean up my mail from the day before. My first appointment is at 8 or 8:30. I am driven by my schedule. If something is important, I schedule it, including walking the parks and resorts and other operations. I schedule time to talk with guests and with Cast Members and to do things like teach a time management course once a month. The minute something is important to me, it gets scheduled. Someone taught me long ago to schedule the priorities in your life. That is why my workouts are scheduled appointments in my calendar. I have a full day of meetings, appointments, and walks of our property. I stop working at 5 PM and go work out and stretch at one of our spas until 6:45 and get home around 7 PM. My wife and I have dinner, and we hit the sack pretty early Sunday through Thursday.*

*On Fridays we go to our beach house, which is 90 minutes from Orlando on the Atlantic, and spend the weekend with my son, Daniel, his wife, Valerie, and our three grandchildren. Some people may think that routine is boring, but I have learned that routine is really important. If you want to know what is going on, maintain consistency in your business. We also take our grandchildren to Disney frequently and visit just like any other guest. We wait in line and we do the parks just like a visitor, for me to learn the truth and for the kids to have fun. If you have routine in your life and schedule your priorities, you will find yourself doing what you are supposed to do versus just what you like to do.*

**Q / How do you communicate with your 50,000 Cast Members and get your philosophy to trickle down?**

**A:** *Actually, I worry about my philosophy trickling up and down and all around. I have several strategies on how I do this. When I came to Walt*

*Disney World, in 1993, I worried about this one thing a lot. How was I going to get all of these people to understand what we needed to do? First, I personally write a weekly newspaper for all 50,000 Cast Members. It goes out every Friday at 5 PM. In that paper are numerous columns that I write on leadership expectations, our purpose and role, our vision and how we achieve it, diversity, pre-shift meetings with our teams, general important information, and on and on. This paper is used to not only inform but also to recognize our Cast for the great work they do. We print numerous guest letters that compliment the Cast on the great job they do.*

*I hold monthly meetings with our front-line Cast and our front-line management to listen to them and to explain things to them that are concerns.*

*I make myself available to speak to any group so I have the opportunity to deliver my messages in person. I speak to thousands of people a year.*

*I make myself available to see anyone who wants to see me, as do all of our executives.*

*To deliver my messages, I teach classes on leadership, time management, and how to build commitment on a monthly basis.*

*I visit the operations frequently and talk the Cast about what is important. I inspect the restrooms for cleanliness and check the food. I check the break rooms and cafeterias. I make unannounced visits as well.*

*I pretty much know what is going on, as I talk a lot with front-line Cast who have direct interaction with our guests. After a while, you get a great reputation for this sort of thing, and then the Cast tells you everything. They actually help me do my job better than anyone else because they know why we are trying to do what we do and how we do it.*

*To really know what is going on and to get your messages and expectations embedded, you have to have a clear and routine strategy in place. You need a few simple messages that you deliver over and over and over until you are blue in the face and then deliver them again.*

*Our Cast Members read. They give copies to their friends and neighbors and their children. We even have articles in there on how to raise your children and how to deal with things like getting homework done and how to properly discipline your children, and how to build your relationship with your partner or spouse. We talk about our expectation for being fair and firm. We talk about our expectation to treat everyone with respect no matter what their background, color, race, culture, or sexual orientation. Respect, appreciate, and value everyone is the battle cry around here for how to treat people.*

**Q / So I guess that they know you are going to take action and do something about it. Is that correct?**

*A: That's right. They call me too. I have a confidential voice mail number that I constantly publish. Anyone can leave me a message and whether they leave their name or not, that is up to them. I follow up on every single item. A manager in my office traces every single item until it is resolved one way or the other. This is why I have such credibility. I follow up! When they don't leave their name, I put their message in the Main Street Diary. I say, "This week I got a message about the locker rooms not being clean. Since you did not leave your name, I just wanted you to know that I took care of that and I hope that you are happy. Call me back if they don't stay clean." We just keep pounding away on stuff like this. These are the kinds of things that are important to people. The leaders behave better too, because I tell them if your leader is not behaving, let me know.*

**Q / How do you motivate and educate yourself? You are the leader, the main visionary.**

*A: When I was young, I was very introverted. Actually, I took a speech course in college and I dropped the course the night before I had to give the speech because I was so terrified. I was a quiet, shy little boy. Seventeen years later, I am in an executive position, and I am asked to give a speech to 300 guests of the Chicago Marriott, where I was working as the Director of Food and Beverage. I agreed. What I forgot is that I did not know how to give a speech. I wrote out something on a yellow pad and went out there and made a fool of myself. That fear came rushing back. After that day, I went to get some help from an expert who happened to be Bill Marriott's father-in-law, who taught speech in college. He gave me some of the best advice. He told me, "Lee, always talk about things you are passionate about; use personal examples; and always prepare your own material. Don't let people write speeches for you. Talk about your kids, your mother, your dog, or whatever." So I started doing that, and it works beautifully. What he was telling me was to tell stories and not make speeches. People don't remember speeches but they do remember stories and the people who tell them, therefore they remember the lesson that is taught by the story.*

**Q / Right. You have such a reputation here for speaking that anybody who comes in contact with you usually has a notepad with them and they're taking notes.**

*A: I tell leaders to learn how to be good communicators by watching people who are good at giving speeches in addition to taking classes, and take the time to test out your speaking skills on your staff. This is what I did. Create a few message points that you want to become known for and talk about them all the time. You can tell lots of different stories to make the same points. One day it occurred to me that leaders basically talk for a living. We try to figure out what is going on, and then we figure out what we want to be going on, and then we communicate with our teams to try to get them to do what we want them to do. We communicate in writing, and we communicate by speaking. I think speaking is the best method, as there are far fewer misunderstandings when you speak to people because they can ask questions to clarify.*

*Experience is another way to develop yourself. For example, I had a mentor once who was the Director of Food and Beverage of the Waldorf Astoria. His name was Eugene Scanlan. He later became the General Manager of the Waldorf Astoria. He was the first Executive Chef there after starting as a young apprentice at 17. He was very impressive. He would take me and another young manager for dinner every Monday night to one of the restaurants in the hotel, and he would make sure that we ordered different things, or he would order them for us. He would have us taste each dish, and then he would explain the dish including the ingredients, how it was prepared, and any other history of that dish. He would order different wines and do the same. The first time he ordered raw oysters, I was wishing I was not there. I am from Oklahoma, and I had never eaten a raw oyster. When I looked at it, I was pretty sure I would never like it. I really didn't want to try it; but with Gene's insistence, I ate it, and I liked it. From those early experiences, I learned a lot. First, I learned that it is important to mentor others. This was a real gift that was priceless. To this day, I still encourage others to try things and to get varied experiences. I tell people that there is more to food and beverage than a cheeseburger and a beer, and that there are a lot of other wines besides Cabernet Sauvignon and Chardonnay. This, like all experiences, gets better and better with time. Just think about how much you would miss out on if you did not try things. Just like those oysters that I thought I did not like, it is the same with public speaking. When I first started doing it, I hated it. Now I love it and want to do it as much as possible. In both cases, I had great teachers who showed me the way.*

**Q / Who do you report to?**

*A: I report to the President of Walt Disney World. He has been here since he was 17. He started in a front-line position in Cash Control ringing out the registers at the end of the night. In those days, it was called a Z run. Today, he is the President. It pays early on at least to get with a company that is well-known, I think. I would tell any graduate to go to work for a large well-known quality company when they get out of school and stay five years. In those five years, you will get great training and experience. If at the end of those five years you are not achieving your goals, then move on. A well-known company will open many doors for you for your next move. Don't go jumping around every year from company to company for a few more dollars. You want stability on your resume. After five years, you will still have at least 40 years to work. Get really good at what you do before you start moving around.*

*Here is some excellent advice that my boss gave to someone recently. One of our executives went in and said, "Now that I've become a General Manager, how do I get ready for my next position and promotion? What is your advice?" He said, "I wouldn't worry too much about what you have to do to get the next position. What you should be worrying about is what you have to do now to get that really big job ten years from now." You might have to go back and get your MBA or get certain experiences under your vest over the next ten years. This is a great question, I think. What do you need to do today, this week, or this month that will pay off for 5 to 40 years from now? One thing comes to mind for me and that is to pay attention to your health. And a really big and important one is that if you smoke…stop!*

**Q / Were you always an excellent leader?**

*A:* No, I was not. When I first started my career, I was a great manager because I was so organized that there was no deadline that I could not make. I was a highly disciplined person with a system for keeping on top of things. I was very organized. The problem was that I was so focused on getting things done that I did not focus or pay attention to people. I forgot that it all gets done through people. I know that today, and I use my organizational skills to pay attention to the work and to the people. I am now a great manager and a great leader.

Being aggressive and not being focused on people can get you into a lot of trouble even though you might get short-term results with that style. I got fired once after 90 days on the job. I should never have taken the job in the first place, as it turned out that the place was going bankrupt, which they forgot to tell me in the interview. My wife had told me not to take the job in the first place, but of course, I was pretty aggressive and knew everything, so I took it. This was a good lesson in listening to others. I went home and told my wife, Priscilla, that I got fired. She said, "Good, I hate living here." Later on in my career, I got passed over for a big promotion, and it turned out to be because of my aggressive style which was to get things done at the expense of people sometimes. I was more focused on getting things done than I was on developing and inspiring people to get things done. This was in 1985. I really knew then that I had to take a long, hard look at my management and leadership style. I started studying leadership. I went to seminars. I read a lot about leadership. I thought a lot about my behaviors and how they affected people. I became totally aware of myself and the impact that I had on people. I made a lot of changes. I learned how to listen, how to show respect to others, how to involve others in the decisions, how to build others' self-confidence and self-esteem. I am a lot better today than I was then, and I continue to get better. I even became a better leader for my family and friends as well.

I can guarantee you that in the first half of my career, there were thousands of people who couldn't stand me, and now in the second half of my career, there are thousands who have great deal of respect for my leadership. I just wish I could get them all together for a weekend to clean up my reputation from the early days! Today, I have a great amount of respect and admiration for the people I lead, and I make sure that I let them know that. In the old days, I thought that people had to listen to me. I have learned that it is the other way around if I want to get great results. I now know that I serve them and not that they serve me.

Being in a place where you are learning is really important too. I go out to the operations a lot. One day I asked one of our cooks how long he had been in one of our restaurants, and he told me he had been there since it opened. I asked him why he was still there, and he told me that he stayed because the

*Chef taught him something new every day. That is a great lesson for all leaders. Develop your people. Get them ready so they can move up to the next level and have a better life.*

*I am doing a lot of research right now on the subject of commitment. How do leaders inspire commitment from their teams? I know for sure that the following things are important: (1) Make your employees feel special. (2) Treat your employees as individuals. (3) Treat everyone with respect. (4) Develop your employees, teach them, and know their jobs. Do these things, and you will have a committed team who will do anything for you, and they will go all the way. Being interested in your job and being committed are two different things.*

*I think a lot about my responsibility to create an environment where people are happy. Happy people don't quit, and they live longer. My main responsibility at Walt Disney World is to create a happy, healthy environment where every single Cast Member can achieve whatever they are capable of. Get people into jobs that they can be happy in. Many unhappy people are just in the wrong positions.*

**Q / What do you see as trends in the industry?**

*A: The industry will always be changing. I think we've got to have an industry where people can have a balanced life. The days of working 14 and 15 hours a day and six and seven days each week are over. I make sure that our people get two days off a week, and we expect a ten-hour workday on average. That's enough. Go home and have a life. We really insist that our people get off to do what they have to do. Go see your son or daughter in a school program or to that teacher's meeting or whatever. Don't miss these things and then have regrets some day. This is one reason that we have such low turnover.*

**Q / In your organization there must be so many job opportunities: teachers, Chefs, Sous Chefs, and so on.**

*A: Yes, there are, and add to that purchasing, test kitchens, restaurant managers, and on and on. We have 7,000 leadership positions at Walt Disney World. There are lots of opportunities here.*

**Q / In the test kitchens, would Chefs develop new menus? Is that a whole division?**

*A: We have our Chefs work together to create new things. We have a team just working on coffee ideas. We have a team just working on desserts and others that just think about children's menus, and others on healthy eating, which is a big craze again right now. We have people just focused on wines. We make over 1,500 wedding cakes a year. You need a lot of culinary and artistic talent to get all of this done.*

**Q / What is the volume of food and beverage?**

*A: This year, we did just about $1 billion in food and beverage sales at Walt Disney World® Resort in Orlando, with over 100,000,000 food and beverage transactions* *from full service to quick service to buffets to carts and snack bars. We have hot dog carts that do $5,000 a day. I wish I owned that cart. Walt Disney World is an exciting and magical place to work, and I just love working here. This is a place where we really do make dreams come true and a place where everyone who works and plays here finds a real sense of joy and inspiration.*

# Interview

**JOHN DOHERTY, Executive Chef, Walforf Astoria Hotel**

**Q / What volume of business do you handle?**

**A:** *The Waldorf Astoria is a $55 million food sales operations hotel. About 70% of that is banquet driven. Furthermore, we support two restaurants. Oscar's is a three-meal casual American brasserie which serves anywhere from 800 to 1,200 people a day. It has its own kitchen, Chef, and culinary brigade. The Bull and Bear is a steakhouse that has its own self-contained kitchen as well. Another area of food and beverage is room service that has its own à la carte kitchen for 1,400 guest rooms. We have 130 culinary staff, 7 of which are Chefs, 13 Sous Chefs and then the remainder are Chef de Partie, Commis, Tourant, and about 10 or 12 utility personnel.*

**Q / Is a Chef a member of the hotel's Executive Committee?**

**A:** *Yes, in this company. That is very important because an Executive Chef today is first and foremost a businessperson who represents the culinary department. You need to learn how to conduct yourself as a professional businessperson. In the eyes of the board of directors of a company and the corporate office, or if it's a small company the owner or General Manager, your role is to keep that business financially sound. So, controlling your expenses, payroll, food cost, and providing food that is going to make the business grow as well as provide a profit is your first and foremost objective. If you can do that, then you will have the freedom with your staff to buy fun china, display pieces, and interesting venues for your buffet. Without those resources, you can't improve your product. So, you have really got to be focused on the bottom line. The old philosophy of I'm here to be a Chef and I take care of the food and the Food and Beverage Director takes care of everything else are long gone. That is a clear road for disaster for any Chef.*

**Q / When you first decided that you wanted to do this, what road did you take to get to your present position?**

**A:** *The story started when I was 14 or 15, pumping gas on these long gas lines in the 1970s. It was a freezing cold winter, and I was* *miserable. My friend said there was a busboy opening where he worked and I said, "If it's warm, I'm there." So I went and talked to the boss and was hired on the spot. First day at work, they sent me to the kitchen to wash dishes. I didn't care, it was warm. That lasted about two weeks, and then they kept pulling me to help them with the food prep, which I didn't mind. It was fun. Back then it didn't matter what I did, I was making money. Next I got to do cooking. It was a real mom-and-pop restaurant. They did everything from scratch. Only later on in my career did I really appreciate what they did, the pride they had. I noticed right away the pride and the care that they took in preparing food for their customers. It was inspiring, and I really loved it. The owner saw the interest that I took in their business. I noticed right away the pleasure I got in preparing food, taking the care, making people happy, and getting the positive feedback right away. I found the passion in this part of the business.*

*Next I got a job in another restaurant that paid more money. It was a very busy restaurant. This is where I really learned to work hard and meet the challenge of a fast business, yet still take pride and care in everything that I did. My next-door neighbor was a professor at Nassau Community College and told me there was a school that you could attend to actually do this for a living. So I looked into it and I learned about the Culinary Institute of America (CIA) and that was it. I was a junior in high school and that was when my decision was made.*

*At that point I started going to a technical high school in the afternoon. At the tech school, the Chef enrolled me in statewide competitions and doing the hotel show circuit. I really had a good jump start in my career. By the time I got to the CIA I had already been cooking for probably close to three years and had done food shows and competitions and even won some awards. I was well prepared, which made getting through school easier. I don't remember why or how, but somebody gave me the book about the French Pavilion. That inspired me to only want to work at the greatest places. One of the Chefs at school in my first semester asked me what I wanted to do, and I said someday when I am 30 or 40 years old I hope to be good enough to work at the Waldorf. He said you get down there this weekend and you tell the Chef you will do anything. Tell them you will peel onions or sweep the floors and he'll give you a job. So, that is what I did. I came down to the city, bewildered as a 19-year-old kid can be, and looked up at the Waldorf and went and held my breath to find the Chef. I told the Executive Chef, Arno Schmidt, that I had this externship program and I only want to do it here, and I'll do anything, whatever it takes. Believe it or not, he gave me a chance at it.*

*So I came here at 19 years old between my first and second years at school and worked in all the different kitchens. It was a crazy, crazy place. It was a monster. The amount of covers that we did, we did parties up to 5,000 people, it was a crazy, exciting place but I loved the energy. I loved the potential that it had. Every day the Chef would come out and do a special party for somebody in The Towers. It was very special food, and that is what I wanted to be part of. I did my externship and went back to school. I had it all set up that when I graduated I would come back to work here. Then, I graduated on a Friday and came to work here that Monday. That was 26 years ago.*

*My resume is on the back of my business card, and there is still some room.*

### Q / What kind of foundation does a Chef need nowadays?

*A: I think if somebody wants to be a Chef, the foundation of that career is cooking. The saddest thing in our business is people who go through culinary school, graduate, and think that they are now qualified to be a Sous Chef. They get out of school and take these jobs as Sous Chefs, and spend a lot of time managing processes. That is what Sous Chefs do besides cook: manage processes, anticipate problems, and work the problems out of the operation. The downfall is they never get enough time to develop their skills and palate. They never learn how to be a great cook. If their passion and their dream are to become a great Chef, if you want to be an authority and an expert in what you do, you have to build your foundation.*

*School is an introduction to concepts. You still have to think and remember what you have learned. It is not instinctive. So for that knowledge to become instinctive, you have to go out and practice it and develop those skills so you don't even think about it. It is the same with your palate—it takes years to develop your palate. The more depth a person gives their education by cooking, the greater they can be. I use the analogy of a skyscraper, and if your education is the foundation, the deeper it goes, the higher you can build the building.*

*A Chef also needs to take advantage of opportunities and make a difference. What are opportunities? Opportunities are where the operation is less than what it should be. It could be an opportunity to fix something, for example. I tell cooks to make a difference. That might mean walking down the hallway and picking up a piece of paper on the floor. Anytime you see the garbage overflowing, you will pick up the phone and call the steward to get it. Anytime you see a light bulb out, you are going to call engineering. Silly little things like that. Why wait for somebody else to do it? Make a difference, and that means taking the initiative. The same thing goes with food. You see an opportunity to make the food better, make a suggestion. Don't come with complaints. I don't like when people come with complaints. Come with an observation and solution and be willing to get involved to correct it. I've got so many things going on I don't need another problem on my shoulders every day, although those are important things. Come with solutions, be prepared.*

### Q / Are schools preparing students well for the workplace?

*A: Yes, they offer many more courses, and the facilities are ten times better than when you and I went there. The one thing that hasn't changed is people. Again, people need to take their careers into their own hands. That's a problem today. Too many people have expectations that they need a mentor or that they are owed something. If you are going to wait for somebody to push you along in your career, you are never going to reach your goal. So set your goals, set your targets, and make a plan for what you've got to do to get where you want to be because nobody is going to give it to you.*

**Q / What do you love most about your job every day? Now it's 26 years. Obviously you are an incredible self-motivator and driving force.**

*A: I love coming here every single day. I bring work on the train. I bring it home. I'm obsessed with it. I love the challenge. I love my team. I love the excitement that this place brings. I love the constant need to be creative and to change and develop myself. So we are constantly reestablishing our goals. It's an awesome challenge.*

*I told my Chefs last January that by June I want our customers to eat and ask the waiter, "Is there a new Chef here?" I want dramatic, bold change, and that doesn't come easy. You have to forget what normally works, and you have to try new things and be around people who are willing to take the risk. It's effort. You can do anything. You just have to be willing to work harder. And you have to surround yourself with people who are willing to work harder also.*

**Q / What are things that maybe you don't like so much that are challenges on a daily, weekly basis?**

*A: In any large operation, it is difficult to make changes because you have to change behavior. The bigger the operation, the more you have to change people's behavior. To create a culture where people buy in and commit to the change is very difficult to do. In a small operation there are fewer people to communicate with, so it's easier to watch what everyone does, to follow up and hold people accountable. In a big operation, that is really tough. Sometimes I wonder how much faster and better I would be able to make this kitchen if it was half the size. Then on the other hand, I don't know if I would find it as stimulating, interesting, and diverse because what we do here is different every day.*

**Q / What do you see for the industry at this point and in the future?**

*A: Trends come and go, so I've always been reluctant to jump on any trend. But you look at trends and you try to take something from that trend that you could infuse into your cuisine to make it popular to the mainstream. And that is what we have to do in a larger place like this. So that keeps you fresh and new, but you have to be very careful. American cooking is going to be less about a theme and more about an individual style of the Chef—whatever his or her personal blend is. And I think that is going to continue because the Asian, the Latin, the French, Mediterranean, and American regional influences aren't going away, so you are going to find some places that do one or the other and then you are going to find a lot of Chefs that combine two, because that works. And the concepts get a little cloudy, but people seem to be willing to deal with that as long as the food is great.*

**Q /** Do you have anything you would want to tell a young, aspiring culinarian?

*A: Dream about what is going to make you happy and satisfied and then you have to plan the steps that are going to take you there. Because you are not going to get there unless you have the knowledge and skills to do that job. Life is all about making decisions, thoughtful decisions. Understand that every decision you make has a consequence and potential for greatness, so try to make a decision that is going to get you where you want to be. Along the way, you need to treat people with the same respect that you want to be treated with. Honesty and straightforwardness are valuable in business today. If you are confident, make the right decisions, work hard, and do a good job, you will never ever get sucked into a political situation. Politics and the workplace are driven by incompetence. It's people covering their butts and trying to blame other people for things that have gone wrong. Make the right decision, give your best effort, do the best you can. You won't always win, so when you don't, it's okay to say here's where we went wrong, here's what we have learned from it, here's what we are going to do different starting tomorrow. Also, be a team player, and by that support everybody around you and treat everybody with respect. Work hard and everything will fall into place. I promise.*

# Interview

**EDWARD G. LEONARD, CMC, AAC,** Executive Chef at Westchester Country Club, President of the American Culinary Federation, Manager of the 2004 and 2008 Culinary Olympics Team

**Q /** In your job as Executive Chef at the Westchester Country Club, who do you report to?

*A: I report directly to the General Manager of the club, and informally to the Executive Director. We have a staff of 58 who work in culinary, stewarding, food and beverage operations.*

**Q /** What is a typical day at work?

*A: I work anywhere from 9- to 14-hour days. My day starts off by making all the rounds to the kitchens, talking to the Chefs to see what is going on for the day. There are meetings, such as F&B meetings or staff meetings, depending on the day of the week. Meetings might include reviews of the business plans, budgets and forecasts, basically communicating overall production of the staffs and understanding what every department is faced with.*

**Q / I know your club has a large membership. What kind of secondary management do you have in the kitchen below your position?**

*A: The Executive Sous Chef, Banquet Chef, Banquet Sous, Restaurant Chef, Restaurant Sous, Pastry Chef, and Beach Club Chef all report to me and are all in different places, so it keeps me traveling around the club grounds.*

**Q / What is the number-one way that you get your message across to the line employees?**

*A: There was a lot of the challenge in the beginning. It really took a lot of communication and constant reinforcement. The culture, the philosophy, what we call culinary pride is visible all over the kitchen. In the beginning it was not so much informing supervisors but carrying the message to all and having staff meetings on all levels.*

*Once objectives, policies, and standards are in place, they are communicated to the group of supervising Chefs. The menu-changing cycles are difficult because the problem is you have to really rely on your supervising people to carry the message to the staff and constantly reinforce the standards you have tried to develop. They are going to have to go through the dishes with all the cooks, constantly reiterate the instructions, and ensure that they are carried out each time that dish is produced. It requires a lot of constant interaction with all levels because it's always important to ensure that everyone still understands the mission and the goals.*

**Q / How many hours do you work every week?**

*A: It varies, but probably minimum is 55. It can go as high as 70+. There are opportunities to be successful in this field, but none come without hard work. The hours you need to put in to pursue your goals, and keep it at a superior level, is a lot. There is just no shortcut, there's no easy road to make lots of money and get a lot of titles. There is a discipline of learning the many aspects of the business and working your way up. Find great Chefs to work with, learn our craft, take time to learn how to become businesspeople, and all the while try to progress to that next level. It's a real climb that needs a five- to ten-year game plan of where you are going to go and how you want to get there. It's time and effort that are well spent.*

**Q / What was the career track to your position?**

*A: It's been a very diverse, odd career track in terms of a Chef. I've spent a lot of time in the corporate world and a lot of effort learning ancillary skills. From a young age, I worked my way up. I worked in kitchens in restaurants, country clubs, and hotels for about six years. Then I went into operations, food and beverage management, and sales and marketing, mostly to earn more money and learn new things. I didn't realize at that time how much it really helped me. Even in this job, though it's a culinary position, it is a real management position, and all my experiences have helped me greatly in driving what I have at the club. Understanding the importance of mission statements, philosophy, culture, manuals, numbers, operations, marketing, all these disciplines and attributes have helped me. The more diverse your skills are, the more successful you will be in whatever endeavor you take on.*

**Q / What was it that made you decide to pursue a culinary career?**

*A: I think the interest in food and of course my family. My grandmother, who I was close to, cooked all the time. She was from Italy, and my family loved to eat and enjoy food; it was just so much a focus of life. Everyone was present at Sunday dinners at 2 PM. I really just always had the interest in me. I also liked art, music, and sports. But I guess when I wanted to make a living, it came to down to food and cooking were the things I liked.*

**Q / What would you have done differently in your career?**

*A: That is a hard question. I've always been one to move forward. I am very fortunate that I'm just extremely self-motivated, always pushing for excellence. I'm always been my harshest critic. I think if you get promoted too early in your career, especially in a kitchen, you may wind up with no one to mentor you. Early in your career, you just need to spend as much time as possible learning, because in our craft there is always something new, something different, or something we have yet to master. Even today, at 44, with whatever people say I've accomplished, there is not a week that goes by when I don't pick something up from somebody. When you judge a competition, you learn something, such as a new dish, or something that is not so good. But it's constant learning. You share with colleagues, you network.*

*But going back, I don't regret anything. There were times when I was going into management, for example, and maybe I didn't understand how a skill, such as public speaking, would pay off to a Chef in the end. However, today I'm very thankful that it actually did happen, and all of those experiences were really valuable for my career.*

**Q / Tell me about your philosophy of management and leadership.**

*A: I've seen lots of good managers. But managers do not necessarily make leaders. It's very important to create a culture, to let people know why they are there, to get people to buy into it. Leadership is that ability to take people and bring them to places that they would never have been without your influence, your assistance, and your help. Leadership means having presence, being able to walk into a room with charisma being evident. I think leaders are people that excel above that management level. When you decide to lead the team, you have to lead them in the direction of good for all, and that direction has to be the mission that the organization needs to achieve. Giving opportunities is another aspect of great management. That is a heavy responsibility, because you are the one who is there to generate success.*

*It is also important to realize that everyone has a life and should be able to live it. You must give people the opportunity to achieve a balance of life. There is something in your employees' lives besides work. That is something that has taken me time to learn, and something that I don't always do for myself. But I definitely do with my staff because if they are happy at what they are doing outside of work, then they are happy when they are in the kitchen working.*

*Excellence is not something staff can achieve all the time. But I think it is up to you as a leader to instill the challenge of achieving excellence in your staff. If you don't try to exceed customers' expectations, keep raising the bar of what you are doing, then you can never improve yourself or the organization. So, I created the culture of yesterday's meal is good, but how are we going to make today's meal better. What else can we do to take something that everybody enjoyed and make it just that much better or special and create an experience for customers? Everybody looks for a wow factor today, whether it be through your food, ambiance, service, or a combination of them. I think that is always what keeps people coming back: creating an exceptional experience. Most importantly, though, it is generated from the work of the team you have formed and basically making sure everybody is on that same page and working toward a common goal. If you have done your job as a leader, then everybody buys into it and believes in it.*

### Q / What do you like most about your job?

*A: It would be that I have the freedom and the respect from management which allows me the ability to do anything that I see fit, from a great hamburger to classical cuisine, from Italian to buffet work to ice carvings. I like the diversity. I like the challenge. I like the multitasking. I love the variety of challenges. If I went into a job that was the same every day, it would be hard for me to cope. I like the ability to go in every week or every day and not know what's going to happen. That just really keeps energizing me. There are some events you always have, like the holidays. I'm challenged to make the holiday events better than we did last year.*

*One of the best things is at this stage of my career, I am fortunate to get to work with a lot of students. I think there comes a point in one's career regardless of what you have done personally when you need to give back. You're successful; your business has been good to you; you have an obligation to give back. I am lucky to be shaping young people and encouraging them. It is also a great pleasure to be able to guide young cooks to the type of success where you can stand back and let them have opportunity to shine in the light of it. I think what I like most is I have the ability to affect a lot of people right now and to give people opportunities that I unfortunately didn't have when I was coming up in this business.*

### Q / What do you like the least?

*A: As much as I love what I do, it does consume me at times. That does take time away from things that are important to you. My hobbies tend to be my craft, so there are things that I wanted to do throughout my life that I still quite haven't found the time to do. There is time with my family that I miss when I get home at night late and they have had dinner already. That is a tough thing to come to terms with sometimes. It's just the sacrifice that you make in this business. When everybody is enjoying themselves, such as during the holidays, you are there providing for people's enjoyment. So you don't tend to have that joy as much, and that is a sacrifice that you make. That's tough, but I do sit back now and take time for myself a little more often.*

**Q / What advice would you give a student who is reading this book?**

*A: Although many students are attracted to fine dining, this segment may be less profitable and the hardest to work in. Other opportunities in the field are tremendous, from research and development to supermarkets to corporate environments. I hope students will sit back, read this book, take the advice, and say, "Wow, I need to learn. I need to have passion. I need to really love this business." This is not the industry where students can go because they don't want to study or learn a lot of facts and details. That's wrong. You need a great education and a love for food to be successful. It's a myth that you don't need to go to school in the restaurant industry. I think, hopefully, they would come out of it saying, "I can be successful and have a life." But if you really want to be super-successful and be the Master Chef or the star Restaurant Chef, that is a huge commitment, and you're going to have to sacrifice. It is going to take a lot out of you, and maybe that balance won't be there for a period of time. But for most students who aspire perhaps to be Executive Chef one day, they will have to study hard and work hard.*

## Interview

**L. TIMOTHY RYAN, EdD, CMC, President, The Culinary Institute of America**

**Q / Your current position is President of the CIA, which sounds like a big responsibility as well as a lot of fun. Can you give me a brief description of what your job encompasses?**

*A: Well, it is a lot of fun. The CIA has been in existence since 1946. We have alumni all over the United States and all over the world, and 2,400 degree-seeking students. We also educate about 10,000 continuing education students a year in each professional culinary field, and we do that here at Hyde Park and our campus at Napa Valley, which is called Greystone.*

**Q / Can you give me an idea of where your career started and what was the path you chose to stay on task?**

*A: I was born and raised in Pittsburgh, Pennsylvania, so that's where I started. By chance, when I was 12,*

*I started to work in a local restaurant as a dishwasher. The way that worked out was some kids in my neighborhood had jobs washing dishes. They were older than me, and they wanted to go to a ballgame on a Saturday night. They knew that they couldn't go if they didn't have somebody to cover their shift. So they came to me and convinced me to do it even though I told them I didn't know what to do. They said the Chef was cool and you got paid ten bucks or something like that. So I did it, and actually it was neat, and the Chef was cool. It seems that many of us who are Chefs started out washing dishes. That's a real common denominator.*

*I came from a poor family, and we were lucky if we went to a restaurant once a year, so I wasn't exposed to restaurants and Chefs. I mean this was the farthest thing from my mind.*

*Here I was in this kitchen. For a 12-year-old boy, I got to run this cool machine and squirt stuff. Through the stacks where the shelves were, I had a complete view of the kitchen. There were these folks in white jackets, checkered pants, hats, and knives moving around; it was like a ballet. The Chef was a big, tall, handsome guy who was Italian. You could see that everybody respected this guy. He was very charismatic. It was fun washing dishes and doing all those things. I got to watch all this stuff, be exposed to things I had never seen before. At the end of the night, he took this huge wad of bills out of his pocket, more money than I had ever seen at one time. He peeled off a ten-dollar bill and gave it to me. He told me I did a good job and asked me if I wanted to have a steak dinner, which was another big deal. Then he asked me if I would come back. You bet I did; that is how it all started.*

*Prior to that, I thought I wanted to be a lawyer, but I didn't know any lawyers. My family didn't know any lawyers. We were poor folks from the inner city of Pittsburgh. My only exposure to a lawyer was from television. I kept coming back to this restaurant. I was exposed to a different kind of role model. He was respected, talented, and rich. He lived in a mansion and owned a couple of restaurants. So now here was a person that I knew and was working for, and could see, listen, watch, and talk to, so I switched my mind. Okay, I didn't want to be a lawyer anymore because that was just something on TV. That was great by my family, so all that worked out well for me and I continued to wash dishes and pots and loved it.*

*Shortly thereafter, I was watching the cooks and knew I could do that. That's what I really wanted to do, and I asked the Chef if I could cook. And he said, "No, you don't want to do that." I said, "Yeah, I do." "No, I don't think so," said the Chef. "You stay in school and get an education. What did you want to be before this?" he asked. "Well, I thought I wanted to be a lawyer." "You be a lawyer." I was really sort of crestfallen. He said this is a tough industry and you may not understand that, when everybody else is playing, that's when you work. "I'm not sure you're really ready for a lifestyle like that.*

*Why don't you be a lawyer?" the Chef said. So, like most kids, as soon as you hear "no," you want it ten times more.*

*So I kept pestering him, and he finally let me start to do some basic things. In reflecting back, those were some of the fondest memories of my times in the kitchen because I knew nothing, so everything was a new adventure. Many Chefs can relate to when the callus on your fingers where the knife hits gets bloody and sore, and how proud you were of that callus developing because it meant you were starting to become more professional. I would write things down in my little notebook. When I learned how to make hollandaise sauce or all the different things that I learned how to make, I loved it. He was a great guy and great mentor; he really did things the right way and instilled some lessons that I keep to this day about how to cook, how to take care of the customer, and how to work. I always draw back on those things.*

*At the same time, as I went into high school, I thought I also did have a natural sort of affinity for the work. I had been inclined to go to college, and it was an expectation to go to college. So really, not knowing any better, I said I'll just go to college for that. I walked to the local Carnegie library, which was about three blocks from my house, and I started to look for colleges that I could go to learn how to be a Chef. And there was only one—located in New Haven, Connecticut, and that was the Culinary Institute of America. So I remember that very vividly discovering in book after book, in reading about the CIA, that this place existed for me.*

**Q / And that was 1974–1975?**

*A: I was here from 1975 to 1977. Then I went out in the restaurant business and got to travel the world, spent time in France and Switzerland. I went back to the restaurant business in Pittsburgh, my hometown. That's where I met Ferdinand Metz, my predecessor at the CIA. Incidentally, I'd never intended to go back to my hometown, but two months into my studies, my father passed away. I had two younger sisters, so I was compelled to come back to my hometown. Then I just ended up staying there. I had already envisioned that I would go to work in New York, but that didn't work out. That was just my destiny. Sometimes that's the way life works out. I was in the restaurant business and doing very, very well, and made a nice reputation for myself in Pittsburgh, and that's where I met Ferdinand Metz.*

**Q / Where was he in Pittsburgh?**

*A: He was in charge of research and development at H. J. Heinz Company, which has their world headquarters in Pittsburgh. I met Ferdinand and then, shortly thereafter, he became president of the school. He had been to the restaurant several times. I was the first person that he really hired because he was looking to do some different things at the school. He brought me here to help develop and open the American Bounty Restaurant 20 years ago.*

**Q / Twenty years ago, there was no American cooking.**

*A: Right. If you talked about American food, people said that is hot dogs and hamburgers. So, the American Bounty restaurant concept was a revolutionary idea.*

**Q / We've always had American food and the regionality of America; it just needed to be brought into a restaurant.**

*A: That's why I came to the CIA. I really didn't want to teach. That wasn't part of my plan. Chefs of my generation, I think in particular, were dominated by the French. The French were the best Chefs in the world, the most famous Chefs in the world. So I had a French restaurant, that was my inclination. I often tell our students here today that Paul Bocuse was, in our business, the equivalent of Elvis. He was the Beatles. Chefs of my age aspired to be Bocuse.*

*But when Ferdinand came along, I wanted to talk about this American restaurant. I did some research on it, but I wasn't that excited about it. I thought he was a cool guy, and I certainly had never envisioned that I would come back to my alma mater, though I loved it. Particularly at this point in my career, it wasn't where I was at. I wanted to be the Bocuse of America, just as Dean Ferring, Bradley Ogden, and Jeremy Towers did, and that was in the back of my mind. Ferdinand was very talented and well respected, so I figured I'd come here and do a little stint for two to three years and then be back out on my uncharted course.*

*But that's not the way that it worked out. The reason why is because I really found what I believe to be a higher calling once I got here. As much as I love being a Chef in the restaurant business, as a faculty member, I feel that I'm changing people's lives. And I'm going to make a far bigger impact here, even if I had become the American equivalent of Paul Bocuse. I'll be touching people's lives in a real and important way, helping them fulfill their dreams and helping them shape the profession and change the industry. That's what kept me here for all these years, from my faculty position, to about six administrative jobs, each with different adventures and different responsibilities. I did well, and that is how I ended up as President.*

**Q / I want to talk to you about education, which is incredibly important in this day and age. You said it yourself: In 1977 you went out, you worked, got a little education, and went back out in the field. In 18 months you trained to be a Chef. That's not the case anymore. The foundation of education is as crucial as the foundation of culinary skills. Because this particular book is for new culinarians starting out, I would like your thoughts on this. It was very cool to be a dishwasher and work 100 hours a week to learn. Basically those days are over, and the educational part of this industry is so important.**

*A: I think, and I know you agree with this, that you learn a lot at the dish table in an establishment. I can still, sitting here right now, feel what it's like to stand at my dish machine, smell what it was like to be there, and remember the interactions I saw between people when they were good and when they totally blew it. I remember many occasions when I saw things going on in jobs, even when I was a cook. I said if I ever have a chance to be the boss, I would never do that. And so those are some of the most valuable lessons. This is an industry like many professions where you still have to pay your dues. You can't rocket to the top until you learn all those important lessons, earn your stripes, and know what it's like to be a dishwasher. Then, when you're supervising dishwashers in the future, they know you know what it's like. And that's important in our industry.*

*So I still believe in hands-on. But one of the things that I think is important for anybody who wants to get into this business, or any line of work, is to have a tremendous thirst for knowledge and desire to constantly get better. So that thirst for knowledge led me in later years to continue to formalize my education and go back and get a bachelor's degree and then an MBA and finally to get a doctoral degree. But, along the way, I always read to stay informed. That's the way I approach things. When I picked up golf, I read every book, trying to learn as much as I can. So...my inclination was go to a library and see how much I could learn about being a Chef, about cooking and the industry. I had the desire for knowledge coupled with the desire to constantly get better. Even as a dishwasher, one of the things that made me stand out is I didn't just mindlessly do the job. I was trying to think about how I could do this better and faster. I would go to the Chef and say I think if we moved these shelves over here, it'll be better for the waitstaff because we won't drop as many plates on the floor. I remember him saying, "Nobody ever asked that before. I think you're right. Let's move them." I remember how empowering that was to me. Hey, they took my idea. I think having a thirst for knowledge coupled with a desire to constantly improve can propel you to the top of whatever part of the industry you choose.*

*I do think that formal education is becoming increasingly important. I shouldn't say formal education. Let me clarify it: education beyond the education of our craft—beyond the culinary part. The world is so much more complex. Business is so much more complex. We are dealing with legal issues, financial issues, union issues, human resource issues, and marketing issues. The dish table and stove and culinary classes in and of themselves don't necessarily prepare you for that. You have to have those skills, but you need to know something about finance, marketing, operations management, and so on. The easiest way to do that, of course, is through practical experience, but if you have schooling before the practical experience, you can avoid a lot of mistakes. Today, the school of hard knocks doesn't exist in the way that it used to, because money is too tight, business is too competitive, and people aren't going to pay you to learn on the job and make all these mistakes the way that we once could. We have to provide for that kind of education. It's not just business education either that's important. The components of traditional liberal arts education are important, as well, because it gives you a broader perspective of the world. Chefs can't be isolated and focused only on food. You become a better person, a better leader, and I think a better Chef when you know something about art, literature, and languages, for example.*

**Q / Languages are huge.**

*A:* *Right. If you're a Chef today, you need at least a rudimentary understanding of Spanish, the first language in the kitchen. You know one of the issues in the industry now is the downsizing. Not all hotels, for instance, have the layers of management as in years past. A Chef today must be an astute business person, not only managing costs but an influential part of budgets and forecasts. It is difficult to learn on the job. Number one, the work ethic doesn't allow the 12- to 15-hour days anymore. Number two, the Chef doesn't have time to take you under their wing to explain the whole financial process, which really is an academic part of education.*

**Q / What were the investments you made in the last ten years that really advanced your career?**

*A:* *Well, with respect to communication, in high school I took two courses on public speaking. I don't remember the tangible lessons that we learned, but what I do remember is that they forced me to get up in front of a group and make a speech. Once I did that, I realized I was actually pretty good at it and stopped being afraid. From the time I was a high school student, I felt that I could get up and communicate with people. I've continued to build on it. I made huge investments of my time, money, and effort to better educate myself. There are two thirsts that I find are within me: professional and educational. On the professional side, I was involved in many competitions as well as the culinary Olympics. When I was a student I did not envision myself doing that. They showed us an Olympic film, and I thought it was exciting. I've been fortunate through my career to meet people along the way who encouraged me. I was too naïve at the time to disagree.*

*For example, I was in Pittsburgh and doing well as a Chef, and there was a man named Roland Schaeffer who was on the 1980 Olympic team. Roland said, "You should try out for the Olympic team." I said, "You got to be kidding. I've never competed in a show." He said, "There's one coming up in Pittsburgh. Why don't you—even if you haven't competed? What's great in the Olympics is the quality of Chefs and you're a good Chef, so why don't you try?" I said, "Roland, I really don't know how." He said, "Well, you do know, and if you need tips, give me a call." I thought about it. Roland set up a little demo, which was maybe 40 minutes, showing how to aspic food for a show. Very naïvely I said, okay, I can do that with no fear.*

*With no fear, I competed in the food show fully expecting that I would get a gold medal and I did. So my naïveté was fortified and supported. On the basis of my two competitions, I agreed to try out for the Olympic team and basically was off and running on what was a great adventure for me. I constantly try to stress to the students here who are interested in competitions like the Olympics this old Zen proverb, "It's the journey that's the reward."*

*I know you're a seasoned competitor, Lisa, and you have all kinds of medals and diplomas, 99% of which you don't know where they are. The first couple you get are very rewarding. But then they sort of get lost along the way and, on reflection, it's everything you learn and the people that you meet along the way. So, if I look at the Olympic experience, there are lifelong relationships that I have developed as well as I learned so much about how to be a Chef. I learned to be innovative, which is one of the challenges of the Olympic team. They set new industry standards.*

*The very top Chefs, the big names in the country and the people who are famous and are making money, they distinguish themselves by coming up with original ideas and, by and large, everybody else copies them. So being part of the Olympic effort taught me how to be creative in a sort of pressure-filled environment. We had practices each month and, if you came to practice without good ideas or with some copied version of somebody else's stuff, your teammates let you know it. I think that that was a good experience and huge investment in time and effort. Then I prepared for the Master Chef test, which is a huge investment but a tremendous benefit.*

*On the other side, my educational side, there were two people who encouraged me to pursue education. One was Ferdinand. Very early on, I said to Ferdinand, "I'd like to be President some day. What do you think? When you're done, how about me? I could do it." That's not exactly how the conversation went, but he said, if you want to do that, you better go back to school because, in the future, the President must have an MBA; Ferdinand has an MBA. He said the president must have a master's and you will probably need to go beyond that. I said, "Well, I only have an associate's degree, I don't even have a bachelor's degree." He told me to start right away plugging at it. He also went to night school to get his degrees. So I agreed.*

*There was another gentleman whom you may know, Herman Zaccarelli. He was in charge of continuing education at Purdue University, and very well known in our industry. He was a big supporter of mine and really encouraged me to pursue higher education. He's retired and living in Florida now. He was really a very inspirational person for me. I was working here and went to the University of New Haven in Connecticut to get my bachelor's degree. Maybe took two classes a week. So I would drive down for two hours after work and sit in class for three hours or whatever it was, drive two hours back. I was on the Olympic team as well. I was doing all these other things. I wasn't married, so that made it a little bit easier. I was a single person just focusing on my career. But I just kept going step after step and I eventually, after years, checked off my bachelor's degree and then said, okay, well, I guess I need to work on an MBA and did that. And then I looked at a doctoral degree. And I'm glad to have finished that.*

**Q /** **The 1988 Olympics was a national highlight with a phenomenal team. Winning gold medals in both hot and cold food is an incredible accomplishment while getting your Master Chef certification, which is very difficult. Then continuing your education, bachelor's, master's, and doctoral degree at a very young age is again quite amazing. In recapping these accomplishments, so future leaders can see how one can obtain so much through a career, how do you motivate and encourage your students?**

*A: Let me tell you just one other story, and it has to do with another point that really impressed me and shaped my future. My predecessor was a great mentor to me, and at one time, we had very much a sort of Yoda-Luke Skywalker relationship, then that sort of grew into more of a partnership over the years. But, when Ferdinand first came to the CIA, he called me, and I was in the restaurant kitchen in Pittsburgh. He said, "I want to talk to you about coming up here." And I was so stunned even that he called me. He had just come back from the 1980 Olympics. I said, "I know you just got back from the Olympics. That's so great. You guys did so well. Congratulations!" This was before he said I want to talk to you about coming up to the school because I was just trying to buy some time and catch my breath for the pure fact he even called me. I was being so complimentary about the Olympics that he sort of nicely cut me off after a little bit and he said, "Yes, we're happy too, but that's in the past and I want to talk to you about the future." I mean they literally had just gotten back and had done better than any other American team had up to that point. It struck me that he was not even stopping to rest on his laurels for one minute. But you always have to be looking forward, looking ahead at what your next challenge is. Individuals—people fundamentally require challenges. Without a challenge, we shrivel up and die. That's something I believe in.*

**Q /** **Young adults today are more savvy. Like you say, you went to restaurants once a year. Children today go to restaurants once a week, 90% of them. They watch cooking shows. They're online. They're moving and shaking. They listen to music, have the TV going, work online, and do their homework all at the same time. What are your work ethics here at the school to prepare students for the real-world profession and the need to produce leaders, not just cooks?**

*A: Well, as you said, the students today are different than when we went to school. We're thinking about that all the time here because we have to approach it in different ways. The students will shape the industry in a different way than we did. I think that one of the things they'll force us to do as an industry is to become a little bit more user-friendly in terms of the workforce and to take a different approach to hours. Our generation worked 18 straight hours, no problem. That's not a way to have a life. So they're going to be creative enough to reform things, which will ultimately be better for the industry.*

*It will attract more people that are smart and creative. That's the way that we're going to progress.*

*Then how are we trying to educate the leaders? One is through higher levels of education, just like we talked about; it's not only about cooking. If you're going to learn to golf, you have to hit golf balls. If you're going to learn to cook, you have to cook food. The only way that happens is in the kitchen. You can't do it out of a book, though academic studies are important. You can't do it watching videos. You have to take a knife in hand and roll up your sleeves. That's the way you build those skills.*

*In addition, we educate leaders by conveying excellence. We're trying to instill a spirit of excellence, to constantly think how to make things better and how to be a professional.*

**Q / You've been a big fan of professional organizations.**

*A: Absolutely. The American Culinary Federation (ACF) was a big part of my development. If you look at leadership development, I was a public relations chair for the ACF for many years, and then I became Vice President, President, and Chairman of the Board. I learned a lot of things about interaction, communication, and leadership. I still promote organizations like ACF or the Research Chefs or the National Restaurant Association. Those are great times when leaders are getting together. In formal sessions, you learn a lot, and in the informal sessions, just talking to your colleagues is one of the most powerful things that you can do. I still continue to support all the professional associations; I think they're critical.*

**Q / When I went to culinary school, I was going to be a Chef. Years later I realized that I didn't want to continue to be a Chef; I wanted to get into management. I wanted to run the operation, directing the entire team and preparing financials. This industry is so wide open. Can you talk about what you envision in nontraditional career paths such as research and development, manufacturing, business and industry, healthcare, or teaching? Have you seen this industry change in the diversity of available professions?**

*A: Absolutely. I think that there are more opportunities for young people in this business now than ever before. There's a marketing adage that says, "Over time, all markets fragment." If you take a look at our profession, when you and I went to school, we thought of being a Chef. And to us that meant being a person in a singular, independent restaurant. This is still an option today, but there is a whole world of new openings. As the world becomes more sophisticated and Chefs become more sophisticated, they see more opportunities. So, when we were younger, you looked to a Paul Bocuse or Andre Soltner in America. They were icons that had single restaurants. Our generation of Chefs said I can do well in a single restaurant, but there's all these other business opportunities. Why don't I have ten restaurants, and I'll be much more financially successful. So the top Chefs in the country nowadays have multi-*

*ple restaurants. One example would be, of course, Thomas Keller. He's had the singular restaurant, the French Laundry. Now he's going to open his second restaurant, Per Se, in New York. And Charlie Trotter did the same. There's too many opportunities out there, and Chefs are becoming businesspeople just like anybody else. Chefs who worked in restaurant kitchens somehow got involved in research and development and said, I can have a completely different kind of lifestyle now. So I think our profession is responding to all the opportunities out there, probably taking a more savvy approach and a more businesslike approach to those opportunities than we did in the past.*

# Part
## 5

# IINTRODUCTION TO MANAGING YOUR CAREER

You have skills that employers want, but those skills won't get you a job if no one knows you have them. Good resumes, cover letters, applications, and a job portfolio all broadcast your abilities. They tell employers how your qualifications match a job's responsibilities. If these paper preliminaries are constructed well, you have a better chance of landing interviews—and, eventually, a job.

Technology has added a new twist to preparing resumes and cover letters and sending them to prospective employers. The availability of personal computers and laser printers has raised employers' expectations of the visual quality of resumes and cover letters applicants produce. The widespread use of the Internet to post jobs and resumes has also created situations where there are simply millions of resumes on the Internet, and a single job opening might generate hundreds or even thousands of responses. Busy reviewers often spend as little as 30 seconds deciding whether a resume deserves consideration. In some companies, if your resume is not formatted properly for computer scanning, it may never reach a human reviewer. Chapter 14 offers advice on how to write up your resume to help you pass the 30-second test and win interviews. Chapter 15 helps you put together a job search portfolio that will highlight your accomplishments and make you stand out from other applicants.

Before you can interview, you must learn how to locate and contact prospective employers. Which of the following techniques do you think is most likely to get you a job?

1. The classified advertisements in the newspapers

2. The Internet

3. Networking contacts

The correct answer is #3. Networking contacts are much more likely than other sources to connect you with your future employer. Chapter 16 helps you use a variety of techniques to find potential employers and write appropriate cover letters to send with your resume so you get your foot in the door.

Chapter 17 discusses the steps involved in interviewing, including choosing what to wear and bring, anticipating questions you will likely face, closing the interview, and following up. If you prepare for the interview properly, as this chapter discusses in detail, you will certainly not be as nervous when the time comes, and you'll do a better job of projecting your knowledge, skills, and abilities.

The final chapter, chapter 18, talks about your career. A career is not simply a series of jobs but rather is a series of progressive achievements. This chapter discusses how you can set goals, work with others, involve mentors, be an active member in professional organizations, become certified, and participate in lifelong learning to ensure progressive achievements in a long, successful, and enjoyable career.

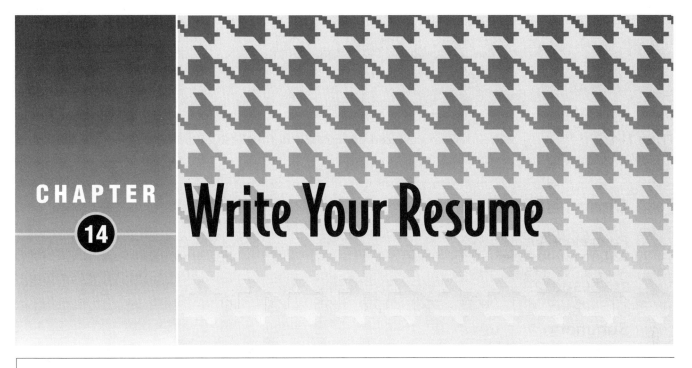

CHAPTER
14

# Write Your Resume

## Introduction

**THE RESUME IS ONE OF THE MOST CRITICAL STEPS** in securing a job. Remember that your resume precedes the interview and is the only impression you make on your potential employer before you are (hopefully) asked to make a personal impression in an interview. Most resumes are glanced at for less than a minute—less time than you might wait for a red light to turn green. Resumes that are wordy or hard to read end up in the trash, and many resumes wind up being filed away forever. Only a small percentage of resumes ever make it to the interview step.

But employers still ask for resumes, and a good resume provides a competitive edge. Your resume tells potential employers what you have accomplished already and what you can do for them now. Look at the resume as an advertising tool; it sells your talents and skills to an employer, much as a 60-second commercial sells to a consumer. Good resumes can awaken an employer's interest in you and get you what you want—an interview.

## Types of Resumes

Resumes fall into one of three categories: chronological, functional, or combination (combines characteristics of both the chronological and functional). The type you choose should emphasize your strengths and deemphasize your weaknesses. most resumes these days are combination resumes, as you will see shortly.

The *chronological resume* (Figure 14-1) lists the jobs you've had by date of employment, starting with your most recent job and working backward. The education section lists your education in reverse chronological order as well. Use the chronological resume if:

- ◆ You have recent and continuous work history in the field you are looking for a job in.

**Figure 14-1**
Chronological Resume

- ◆ You have progressed up a clearly defined career ladder and are looking for advancement.

---

# Cheryl Richardson

*Permanent Address:*
92 Longwood Road
Aurora, NY 11593
315-555-1212

cherylrich@yahoo.com

*Current Address until June:*
233 University Avenue
Ithaca, NY 12830
315-555-1213

## Summary

Dean's List college student in culinary arts, recently promoted to Line Cook at nationally known Moosehead Restaurant.

## Work Experience

7/04–Present **Line Cook** at Moosehead Restaurant, Ithaca, NY
Work at sauté or grill station for lunch or dinner meals in a well-known restaurant featuring healthful natural foods cuisine. Perform mise en place and food preparation. Follow safe and sanitary food procedures.
○ Test and evaluate new recipes.
○ Won Employee of the Month (June 2005).

10/02–6/04 **Preparation Cook** at Moosewood Restaurant, Ithaca, NY
Performed all preparation tasks in kitchen emphasizing scratch cooking and vegetarian dishes. Completed all duties in timely fashion while maintaining sanitation standards.
○ Received "Excellent" performance evaluations.

Summers 2001 and 2002 **Assistant Cook** at Lenape Summer Camp, Seneca Falls, NY
Under Head Cook's direction, did basic food preparation tasks, cooking, and baking. Assisted in purchasing, receiving, and inventory management.

## Education and Certification

May 2006 Bachelor of Professional Studies in Culinary Arts, Olympia University, Ithaca, NY

Dean's List every semester (Anticipated)

Treasurer, Culinary Club (sophomore year)

ServSafe® Food Protection Manager, #2364656 (National Restaurant Association Educational Foundation)

Employers especially like to see a clearly defined career ladder in your listing of jobs; it lets them know what you can do right now. Do not use this type of resume if you are just starting out, trying to switch fields, or have large gaps in employment—then it would be better to use the functional resume.

**Figure 14-2**
Functional Resume

The *functional resume* (Figure 14-2) also includes a listing of your work experience and education, but in a brief form toward the end of the resume. Most of the functional resume is a summary of your skills and accomplishments, such as

---

## Tim Fitzpatrick

3626 Chestnut Drive, Salina, CA 84529     408-392-8942     tfitz@aol.com

*Seeking an entry-level Cook position in a restaurant.*

College student in culinary arts with diverse cooking and foodservice experience, including food preparation and supervising.

Culinary Arts:
- Experienced with kitchen food preparation and cooking equipment.
- Competent in basic food preparation techniques, including cutting.
- Use standardized recipes.
- Follow portion control guidelines.
- Plate and garnish foods.

Sanitation
- ServSafe® certified.
- Follow appropriate cleaning and sanitation procedures.

Supervision
- Supervised five employees.
- Scheduled, trained, motivated, and coached employees.
- Solved problems.

---

### Employment

**Cold Food Preparation**, Bay Community College.   September 2004 to present.
Part-time.

**Head Waiter**, The Tides Retirement Community.   June 2003 to August 2004.
Part-time.

**Waiter**, The Tides Retirement Community.   June 2002 to June 2003.
Part-time.

### Education

Associate in Occupational Studies in Culinary Arts anticipated May 2005. Bay Community College.

specific culinary skills you've used or menus you've developed and served. Use the functional resume if:

- You are applying for a job that is quite different from your current or past job.
- You have little to no work experience in this field.
- You are reentering the job market after a break.

The functional resume emphasizes what you can do and deemphasizes where you have worked. Many skills, such as management skills, are transferable between industries, and this type of resume especially helps people who are switching to the culinary field or just starting out after college.

Many employers look on functional resumes with some level of distrust. While they can see what sorts of skills and abilities you have, they don't know where you learned them. This is a good reason to consider the next type of resume.

The *combination resume* (Figure 14-3) combines features from both the chronological and functional resumes into a type of resume that is increasingly popular. Basically, you showcase your skills and achievements at the beginning of the resume, typically in a section entitled *Profile* or *Summary*. Then you go on to describe your jobs and education in reverse chronological order. It's a format that almost any jobseeker can customize to meet his or her needs.

**Figure 14-3**
Combination Resume

---

# Richard Plumb, C.E.C., A.A.C.

211 West Greenwich Avenue
Greenwich, CT 07041
203-437-9365 (h) 203-530-8821 (c)
brewchef@yahoo.com

**Profile**

- Experienced Executive Chef and Director of Operations. Have operated multiple restaurants accommodating over 500 guests.
- Developed kitchen and menus for new brewery restaurants.
- Excel in developing successful menus and recipes.
- Proven team-building and motivational skills have kept staff turnover below 40%.

**Experience**

**Director of Operations/Corporate Executive Chef**
Boston Hops, Inc., New York, NY          May 2000–present
Responsible for menu development, kitchen/bar design, opening plan and execution, training, and staff hiring for three new brewery restaurants.

- Redesigned kitchen.
- Upgraded menus.
- Developed corporate buying policies, recipes, restaurant standards, and training manuals.

**Executive Chef/Back of House Director of Operations**

Greenwich Regency, Greenwich, CT   May 1996–May 2000

Responsible for 32 Cooks and 6 Sous Chefs in a $9.5 million food and beverage operation. Also supervised stewarding, purchasing, and receiving.

- Five-year average of 29% food cost and 30% labor cost.
- Employee retention improved 75%.
- Operation featured in numerous publications.

**Chef de Cuisine**

Pebble Creek Café, Purchase, NY        June 1993–March 1996

Instrumental in kitchen and restaurant design of American regional restaurant. Responsible for all costs for front and back of the house. Developed menus, monthly marketing tools, and advertising strategies.

- Increased quarterly sales 25%.
- Demonstrated project planning and design skills.

**Education**

A.O.S. in Culinary Arts, April 1991
New Jersey Culinary Institute

Nutritional Cuisine, January 2000, New Jersey Culinary Institute, 20-hour course

**Certification**

Certified Executive Chef, January 2000
American Culinary Federation

**Associations**

Active Member, American Culinary Federation, since 1991

The Chefs Association of Westchester and Lower Connecticut since 1991, President from 1994 to 1995

**Awards**

1998 Chef of the Year, The Chef's Association of Westchester and Lower Connecticut

Delaware Valley Chefs Association Culinary Competition, ACF Silver Medal, 1997

U.S. Team Member, International Ice Carving Competition, Gold Medal, 1996

Southern New Jersey Chefs Association Culinary Competition, First Prize, Poultry Platter, 1995

# The Ingredients of a Great Resume

A great resume sells a potential employer the idea that you are the person to do the job. Your resume will do this most effectively if you remember that it is not just a job description of your current and past jobs. For a resume to be great, you need to:

- Choose and highlight the parts of your background that position you for the type of job you are currently seeking.
- Discuss what you did in other jobs, but especially how well you did it.
- Include measurable achievements and accomplishments.

Following are guidelines for what to include and what not to include on your resume.

## WHAT YOU MUST INCLUDE

Most professional resume writers agree that you must include these sections in your resume.

- Contact information (including a businesslike email address)
- Profile (short summary of qualifications)
- Professional experience
- Education
- Professional licenses/certifications (such as ServSafe®)
- Professional affiliations (such as membership in the American Culinary Federation)

Additional sections that present information such as computer skills and awards are also appropriate.

## WHAT YOU MIGHT INCLUDE

You might include a job objective, a short statement of the type of job you are looking for. It is important that the job objective be concise and not too broad—for example, "Job Objective: Sous Chef in Club Setting." Some applicants like to use an objective; others don't. The information stated in your objective will be stated in your cover letter, so it is not absolutely essential that it be on your resume. However, if you are not sending your resume in for a specific job opening, it's a good idea to include a job objective because the employer is not immediately associating your resume with a specific opening.

Place your job objective below the contact information on the resume and check that it is appropriate each time you send your resume out. You want your stated job objective to closely match the job you are applying for.

## WHAT TO OMIT

Don't put any of these on your resume:

- ◆ Reference information (just state that a list of references is available)
- ◆ Availability
- ◆ Salary history
- ◆ Diversity issues
- ◆ Photographs

It is customary to give out your reference list only at an interview or after you have been interviewed for a job. Availability is also a subject that can be addressed in an interview. You don't want to advertise that you are available immediately—it makes you look desperate! Salary is yet another issue that should be discussed later. As described in the chapter on interviewing, it is best not to discuss salary with the employer until you are offered the job. Once you receive an offer, you are in a much better position to negotiate a good salary.

# Write Your Resume

The type of resume discussed in detail here is a combination resume, which begins with a profile in which you highlight your qualifications and accomplishments. Then it moves on to a chronological review of your professional (work) experience, education and certifications, professional affiliations, and other information you want to include. Use the Resume Worksheet (Figure 14-4, also on the CD-ROM) to help organize the information for your resume.

**Figure 14-4**
Resume Worksheet

---

**CONTACT INFORMATION**

Name: _____

Home Address: _____

_____

School Address (if applicable): _____

_____

Telephone Numbers: _____

_____

Email Address: _____

Other Information: _____

**PROFILE**

_____

_____

_____

_____

_____

_____

_____

**PROFESSIONAL EXPERIENCE**

Dates Employed (month/year): _____

Employer's Name: _____

Employer's Address (City/State): _____

Job Title: _____

Responsibility #1 _____

    Duties and Accomplishments

    _____

    _____

    _____

    _____

Responsibility #2 _____

    Duties and Accomplishments

    _____

    _____

    _____

    _____

Responsibility #3 _____

    Duties and Accomplishments

    _____

    _____

_____

_____

Responsibility #4 _____

    Duties and Accomplishments

_____

_____

_____

_____

**EDUCATION**

School: _____

Major(s): _____

Minor: _____

Date Graduated/Dates Attended: _____ Degree Seeking/Granted: _____

Grade Point Average: _____

Academic Honors _____

_____

_____

_____

Scholarships: _____

Co-ops or Internships (Where, When, Description)

_____

_____

_____

Extracurricular Activities: _____

_____

_____

_____

Offices Held: _____

_____

International Travel (Where, When, Description): _____

_____

_____

_____

Special Projects/Team Projects (When and Description): _____

_____

_____

Relevant Coursework (only for current students and recent graduates): _____

_____

_____

_____

_____

_____

High School (if going to include): _____

Continuing Education: _____

_____

_____

_____

_____

_____

Certifications (certificate number and expiration date when applicable): _____

_____

_____

_____

_____

**PROFESSIONAL AFFILIATIONS**

_____

_____

_____

_____

**OPTIONAL SECTIONS**

Volunteer Work:

_____

_____

Computer Skills:

_____

_____

Foreign Language Skills:

_____

Awards/Honors:

_____

_____

Publications:

_____

_____

Presentations:

_____

_____

## CONTACT INFORMATION

At the top of every resume is your contact information, including your mailing address, telephone numbers, and email address. It is acceptable to use the postal abbreviation for your state instead of writing out the name of your state. For example, use CT for Connecticut. Don't use any abbreviations for your street address (such as Ave. for Avenue) or city (such as NYC for New York City).

When typing out your phone number(s), be sure to include your area code and designate which number is which, as shown in the following example.

(H) 272-356-7890
(C) 272-367-5237

You may put parentheses around the area code, but don't put 1 before the area code. Make sure you have a reliable answering service for every phone number you put on your resume, including cell phones. Of course, once you send your resume to potential employers, you must frequently check for voicemail messages.

If you are a student still in college, it is best to give both your college and home addresses and telephone numbers and to note when to use each address. For example, you might state next to your college address something like "Contact through May."

Email represents yet another way to communicate with employers. You should definitely list an email address, one that sounds professional. Don't type partyguy@aol.com on your resume and then wonder why you aren't getting any phone calls. Job hunting requires a suitably professional email address. Your Internet provider may allow you to pick several email addresses, so choose one with a neutral feel. It's quite common for job hunters to reserve one email address for the resume. If you want a new email address, check out the free email accounts available from companies such as hotmail.com.

## PROFILE

Your profile section appears right below your contact information so the employer can quickly get an idea of who you are, what you can do, and how you can contribute. This section can be titled Profile, any of the following names, or any appropriate name you can think of.

- Career Profile
- Professional Profile
- Summary
- Qualifications
- Summary of Qualifications
- Areas of Expertise (or Proficiency)
- Key Strengths
- Core Competencies
- Professional Highlights
- Achievements or Accomplishments
- Highlights of Skills and Experience
- Highlights of Qualifications

This section can take the form of a bulleted list, a paragraph, or both. Whichever format you choose, make this section brief and focused. Highlight your experience, accomplishments, and skills. As you write a rough draft of this section, make sure it answers this question: "If I had only 30 seconds to get someone to hire me, what would I say?"

When you mention your skills, be sure they are directly related to the type of position you want. Also, stating "hardworking employee" is not nearly as strong as evidence such as "Promoted from preparation cook to line cook within three months because of excellent knife skills and work ethic."

Here is a bulleted profile for a highly experienced Certified Master Chef.

### PROFILE

- Successfully completed Certified Master Chef test.
- Over 20 years' experience in quality food preparation.
- Thorough understanding of all facets and styles of foodservice.
- Well versed in many ethnic and international cuisines.
- Able to produce quality results while adhering to well-planned budgets.
- Over 8 years' experience in multi-unit management.
- Excellent human resource management skills, maintaining a departmental employee retention average of 3.5 years.
- Self-motivated quality- and cost-oriented manager.
- Highly trained in nutritionally conscious cuisine.

Of course, this profile is pretty long because of the chef's extensive and noteworthy culinary career. Yours will most likely have fewer bullets. Note that the most significant achievements are noted first. The format of this profile could be changed by combining the first three bullets into a short paragraph and then bulleting the remaining points.

A profile for someone coming out of college with some work experience in the industry might look like this.

### PROFILE

Hardworking and reliable culinary student distinguished by:

- Over two years' experience as a preparation cook promoted to line cook at La Brasserie.
- Silver Medal earned in ACF-sanctioned hot food competition, category K.
- President's Honor Roll every semester.
- Strong interpersonal and organizational skills.

Make sure that some of the points you make are measurable achievements, like earning a silver medal at a culinary competition, saving the department $25,000 a year in labor, and cutting staff turnover by 25%.

You may want to write up your profile after you have completed the work experience and education sections. Once you have those sections ready, it will be easier for you to see which skills, achievements, and experience you want to highlight in the profile.

## PROFESSIONAL EXPERIENCE

After the profile, your next section will probably be professional experience, although in some cases it may be education. A good rule of thumb is to put the stronger section first. For example, if you are seeking a job as a culinary educator, the amount of formal education you have is important, so you may want to highlight your degrees at the top of the resume. For most culinary positions, put your work experience first unless you have almost no experience.

You don't have to call this section Professional Experience. Other possible names include the following.

◆ Professional Background
◆ Employment History
◆ Work History
◆ Work Experience
◆ Experience
◆ Career Track
◆ Employment Chronicle
◆ Career History
◆ Career Path

To start writing up your work experience, use the Resume Worksheet to write down information about your jobs. Start with your current or most recent job, and then work backward. The worksheet will help you decide what to include on the resume.

When writing up your job duties, think in terms of the broad responsibilities you had and the specific tasks or duties you performed for each responsibility. For example, on your resume you could state the broad responsibility first, and then present a bullet list of important duties and notable accomplishments. Don't just discuss what you did; also include *how well* you did it. Employers want to see measurable achievements. It helps them see how you can contribute to their organization's bottom line. Here's an example.

EXECUTIVE CHEF                                                    June 2000–July 2004
Big Oak Café, Troy, New York

Supervised and coordinated the food purchasing and production for kitchen producing 1,000 meals/day.

◆ Purchased over $1 million of food and supplies yearly.
◆ Saved $25,000 in the first year after improving bid system and updating purchase specifications.
◆ Developed and instituted regular seasonal menu changes.
◆ Reduced kitchen labor cost by 5%.
◆ Quality of food consistently rated "good" or higher.
◆ Conducted formal monthly training sessions and daily coaching of employees.

Note three points in this example. First, there are no complete sentences; each statement is a phrase. Second, each phrase begins with a specific, descriptive verb. For example, instead of a general verb such as *manage,* use precise verbs such as *organize* or *direct.* Table 14-1 lists action verbs you can use when preparing your resume. Try to avoid phrases that begin with "Responsible for"; instead, find an appropriate verb. Third, once you have climbed the career ladder, it is assumed you can cook. So talk about how many people you supervised, the volume of the business, and how you managed costs. If your experience is mostly cooking, be careful of repetitive wording when describing your jobs. For example, don't keep listing "sautéed fish and chicken" for each job.

## Table 14-1 Verbs for Resumes

### Communication Skills

| | | | | |
|---|---|---|---|---|
| arranged | composed | edited | motivated | publicized |
| addressed | conferred | explained | negotiated | published |
| authored | corresponded | formulated | persuaded | wrote |
| clarified | drafted | informed | presented | |

### Creative Skills

| | | | | |
|---|---|---|---|---|
| conceptualized | designed | fashioned | illustrated | originated |
| created | established | focused | invented | performed |

### Culinary Skills

| | | | | | |
|---|---|---|---|---|---|
| arranged | converted | filleted | judged | prepared | simmered |
| assembled | cooked | finished | measured | produced | specified |
| baked | cooled | flavored | microwaved | purchased | steamed |
| boiled | cut | formulated | pan-broiled | roasted | stored |
| braised | deep-fried | garnished | pan-fried | sautéd | thickened |
| broiled | designed | griddled | performed | scaled | used |
| calculated | determined | grilled | planned | seasoned | |
| chose | dressed | identified | poached | set up | |

### Financial Skills

| | | | | |
|---|---|---|---|---|
| administered | computed | formulated | purchased | sold |
| analyzed | contracted | increased | recommended | trimmed |
| audited | cut | marketed | reconciled | |
| balanced | decreased | planned | recorded | |
| budgeted | eliminated | projected | reduced | |
| calculated | forecast | provided | saved | |

## Table 14-1 Verbs for Resumes (continued)

### Human Resource Skills

| | | | | |
|---|---|---|---|---|
| coached | encouraged | instructed | oriented | specified |
| counseled | evaluated | interviewed | placed | staffed |
| delegated | facilitated | mediated | promoted | streamlined |
| developed | guided | moderated | recruited | taught |
| empowered | helped | motivated | represented | trained |
| enabled | hired | negotiated | screened | |

### Management Skills

| | | | | |
|---|---|---|---|---|
| accepted | created | finished | optimized | restored |
| accomplished | defined | focused | organized | restructured |
| achieved | delivered | founded | originated | revamped |
| adapted | demonstrated | formulated | overhauled | revitalized |
| administered | designed | generated | oversaw | saved |
| advanced | developed | guided | performed | scheduled |
| advised | devised | headed | persuaded | solved |
| allocated | diagnosed | identified | planned | spearheaded |
| analyzed | directed | implemented | prepared | streamlined |
| appraised | diversified | improved | presented | structured |
| approved | eliminated | increased | presided | summarized |
| assigned | engineered | innovated | prioritized | supervised |
| assisted | enlisted | inspected | processed | surveyed |
| chaired | established | installed | produced | traveled |
| clarified | evaluated | instituted | provided | trimmed |
| conducted | examined | introduced | regulated | upgraded |
| consolidated | executed | launched | remodeled | |
| contributed | expanded | led | repaired | |
| controlled | expedited | maintained | represented | |
| coordinated | facilitated | monitored | resolved | |

### Marketing and Sales Skills

| | | | | |
|---|---|---|---|---|
| compiled | distributed | generated | maintained | obtained |
| consolidated | expedited | increased | marketed | stimulated |

Another way to format your work experience is to start with a short paragraph listing your responsibilities and duties, and then have a bulleted list of your accomplishments. Here is how that approach looks:

EXECUTIVE CHEF                                      June 2000–July 2004
Big Oak Café, Troy, New York
Supervised and coordinated the food purchasing and production for kitchen producing 1,000 meals/day. Purchased over $1 million of food and supplies yearly. Developed and instituted regular seasonal menu changes. Conducted formal monthly training sessions and daily coaching of employees.

*Performance Highlights*
- Saved $25,000/year after improving bid system and updating purchase specifications.
- Reduced kitchen labor cost by 5%.
- Quality of food consistently rated "good" or higher.

When thinking of your accomplishments and achievements, ask yourself if you ever did the following.

1. Save your employer money—If so, how much?
2. Increase sales—If so, how much?
3. Increase profitability—If so, how much?
4. Bring in new business—If so, how much?
5. Increase employee retention—If so, how much?
6. Decrease payroll costs, including overtime—If so, how much?
7. Increase guest satisfaction—If so, how much?
8. Increase profitability—If so, how much?
9. Decrease or keep food cost constant—If so, how much?
10. Increase check average—If so, how much?
11. Reduce purchasing costs—If so, how much?
12. Update and improve policies and procedures?
13. Initiate and implement new menus or programs?
14. Implement new hardware, software, or other systems?
15. Improve productivity?
16. Improve communications?
17. Design new training programs?
18. Introduce new standards?
19. Streamline operations, functions, or support activities?
20. Realign staffing to meet business demand and/or decrease costs?
21. Receive a prize/honor/award from an employer, school, or professional organization?

22. Manage special projects, such as kitchen renovation or purchasing new equipment?

23. Develop unique skills or qualifications?

24. Have public speaking experience?

25. Have culinary industry certifications?

Quantify your achievement whenever possible, as in "Increased check average 5%."

## EDUCATION AND CERTIFICATIONS

Next, discuss your education and certifications. Conceptually divide this section into three parts:

1. College

2. Continuing education (or lifelong learning)

3. Certifications

If this section is long, you can certainly separate it into two or three sections.

As long as you are in college or have graduated from college, you probably do not need to include high school information. If you went to a particularly prestigious high school or one with a well-known culinary program you were in, you might include the name of the school and program and the year you graduated.

With regard to your college education, the following items are the bare minimum you must put on your resume.

◆ Type of degree received, your major, and date of graduation—always list the degree before the name of the college or university at which you earned it. If you have graduated, you could use this format:

> Bachelor of Professional Studies in Culinary Arts, 2003
> Culinary University, Denver, Colorado

If you are still in school, give the month and year when you anticipate completing your degree. For example:

> Associate in Occupational Studies in Culinary Arts anticipated May 2006
> Culinary University, Denver, Colorado

If you are not that close to finishing your degree, you could say this:

> Currently pursuing an associate degree in Occupational Studies in
>    Culinary Arts
> Culinary University, Denver, Colorado

If you minored at college in an area related to culinary, mention that as well. If your college major was unrelated to the culinary field (such as German or history), mention your degree but don't specify your major.

◆ Names of colleges and universities you've attended—if you transferred from a community college, for example, to a four-year college and earned your degree, it is not absolutely essential to mention the community college. However, if it might work to your benefit to mention the community college, as when the community college's culinary program is well known, include it on your resume. You can also include your cumulative average if it is good —meaning at least over 3.0 if your school uses a standard 4.0 scale. List your cumulative average like this: 3.0/4.0.

Of course, you can include many other aspects of your college education on your resume.

◆ **Academic honors**—Note academic honors such as Dean's List, awards, honor societies, and scholarships.

◆ **Internships**—Mention where you completed your internships; note the time frame and what you did.

◆ **Activities**—Many college students don't have much work experience, so listing involvement in school or extracurricular activities is important. Employers look for this because such involvement shows initiative. If you were involved in a culinary club or association, especially if you held an office, include this information on your resume. Holding an office shows leadership. Include volunteer activities.

◆ **International study**—Include where you studied, when, and a brief statement of what you did.

◆ **Special projects/Team projects**—If you don't have much work experience, you may want to briefly describe a special college project, perhaps a team project, if it is related to the position you are applying for. For example, you may have worked on a project involved in catering events or culinary competitions.

◆ **Courses taken**—Listing four to eight relevant courses may benefit you if you are a recent graduate and don't have much work experience.

After your college section, mention relevant continuing education courses you have taken. These could include classes provided by an employer, workshops or seminars attended at industry-related conferences, continuing education courses taken to maintain American Culinary Federation certification, and formal education courses such as computer classes taken online or in the classroom. Specify the year in which you took the training. If the training was particularly lengthy, you can also add the number of hours or days it required. Don't forget to include computer courses you have taken.

You can also include certifications, such as Certified Culinarian, in this section, or you may want to list them in a separate section. Specify the certification you have, the certifying organization, and when you received the certification.

## PROFESSIONAL AFFILIATIONS

Your memberships in appropriate professional associations show your enthusiasm and dedication to your career. Membership is also important for keeping up in the field and networking with colleagues.

## ADDITIONAL SECTIONS

- **Computer skills**—Every job requires computer skills. List the software programs you can use with at least basic proficiency.
- **Foreign language skills**—If you are fluent in a language other than English, especially Spanish, do mention it on your resume. If you are not fluent but can read, write, or speak well, include this information too. Just make sure you write down, for example, "Speak Spanish" or "Read French."
- **Volunteer work**—Chefs frequently do volunteer work with food banks and other organizations. Mention relevant volunteer work you have performed, along with the name of the organization and the year.
- **Awards/Honors**—List awards and honors you received, from employer awards to medals won at culinary competitions. Give the name and year and describe the award/honor, if necessary.
- **Military service**—Mention the branch of service in which you served, your highest rank, your dates of service, decorations or awards, and special skills or training you received that could further your career.
- **Publications**—If you have published an article in an industry magazine, a book, or any other relevant material, list the title and publication date.
- **Presentations**—List presentations you made at professional meetings and in other professional settings.

## REFERENCES

Resumes usually do not list names of references. Most resumes close with the statement "References available on request."

## ROUGH DRAFT

Now that you have completed the Resume Worksheet, you are ready to make a rough draft of your resume. For your rough draft, just concern yourself with the information you want to include and how you want to say it. Don't worry about laying it out on your fancy resume paper yet. At this stage, you just want to decide what to say and what to leave out. Use Figure 14-5, which is also on the CD-ROM, to start working on your rough draft.

Contact Information _____

Profile (or other name) _____

Professional Experience (or other name) _____

Education and Certifications (or other name) _____

Professional Affiliations (or other name) _____

Additional Sections (Volunteer Work, Computer Skills, Foreign Language Skills, Awards/Honors, Publications, Presentations)

**Figure 14-5**
Resume Draft

Don't plan to tackle this project in one night. You will need a number of work sessions to get a rough draft that you can type up. Take time to edit your resume at every step. Ask friends, teachers, and family for ideas and feedback.

# More Resume Guidelines

Once you have prepared a rough draft, it's time to take a look at these points to consider as you type it.

## LENGTH

The length of your resume usually depends on the amount of work experience you have. Although you may have been told that your resume shouldn't exceed one page, if you have carefully chosen relevant material that requires two pages, that's fine. Resumes are frequently two and sometimes three pages long. If you have over ten years of experience, a two-page resume is common. Make each page a full page. If your last page has just a few lines on it, compress your information to remove the excess page.

## FONTS AND FORMATTING

Use the following guidelines to format your resume.

- As for any business document, allow 1-inch margins on the sides, top, and bottom of each page.
- Double-space between sections and entries. Single-space paragraphs and bulleted lists. Be consistent with your spacing.
- Paragraphs should be short—five or six sentences at most. Break longer paragraphs into two or more.
- Pick out a font that is up-to-date and crisp.

  - Arial
  - Bookman
  - Century Schoolbook
  - Franklin Gothic
  - Garamond
  - Palatino
  - Tahoma

Times New Roman is also an acceptable font, but because it is used frequently, it is less distinctive than those noted here. The serif fonts—those where the letters have small lines extending from them, usually at the top and bottom—are often easier to read than sans-serif fonts. Examples of serif fonts include Bookman, Century Schoolbook, Garamond, and Palatino. Arial, Franklin Gothic, and Tahoma are sans serif fonts. Avoid fonts such as Courier that give the same amount of space for each letter, even though some letters are wider than others.

◆ In most cases, your best font size will be 10, 11, or 12, although headings and your name should be taller. For example, if you use 12-point Arial for the body text, try 14-point Arial for section headings and 16- to 18-point Arial for your name at the top of the resume.

◆ Don't clutter your resume with too much text. Lots of white space makes your resume easier to read. White space is the space on a page not occupied by text or pictures.

◆ Use underlining sparingly, if at all. Instead of underlining, try boldface, which is often better at getting attention. Further, it's easier to read boldface type than underlined text. Boldface works well for section titles and job titles. Don't use boldface to attract attention to a word or phrase in a sentence.

◆ Don't type words with all capital letters; this is very difficult to read. Instead, capitalize the first letter (if appropriate) and then switch to lower case.

◆ When you make a list, use a bullet (round, square, or diamond-shaped) or a tiny box instead of a hyphen. Use the same bullet style for each section or for the entire resume.

◆ Use a horizontal line to separate your contact information from the rest of the resume. The line helps organize the contents of the resume. You might set off each section with horizontal lines, as in Figure 14-1.

◆ To give your resume a consistent flow, maintain the same style from beginning to end. Every section should have the same design elements. For example, if your education heading is bold and centered, every section heading should be bold and centered.

## FORMATS

Most resumes are set in either one or two columns. The resumes in Figures 14-2, 14-6, and 14-7 use one column, while the resumes in Figures 14-1 and 14-3 use two columns. The one-column format allow you to fit a little more information on a page because more space is available (but you still must leave plenty of white space). You can certainly type up your resume in both formats and then decide which looks best. You can even combine both formats by using one column for your contact, objective, and profile sections and then switching to a two-column format for the rest of the resume.

Let's take a look at the five resumes in this chapter to develop a better idea of ways to format a resume.

# Cheryl Richardson

*Permanent Address:*
92 Longwood Road
Aurora, NY 11593
315-593-8270

cherylrich@yahoo.com

*Current Address until June:*
233 University Avenue
Ithaca, NY 12830
315-229-5987

## *Summary*

Dean's List college student in culinary arts, recently promoted to Line Cook at nationally known Moosehead Restaurant.

## *Work Experience*

**Line Cook** at Moosehead Restaurant, Ithaca, NY                 July 2004–Present
Work at sauté or grill station for lunch or dinner meals in a well-known restaurant featuring healthful natural foods cuisine. Perform mise en place and food preparation. Follow safe and sanitary food procedures.
- Test and evaluate new recipes.
- Employee of the Month (June 2005).

**Preparation Cook** at Moosewood Restaurant, Ithaca, NY        October 2002–June 2004
Performed all preparation tasks in kitchen emphasizing scratch cooking and vegetarian dishes. Completed all duties in timely fashion while maintaining sanitation standards.
- Received "Excellent" performance evaluations.

**Assistant Cook** at Lenape Summer Camp, Seneca Falls, NY        Summers 2001 and 2002
Under Head Cook's direction, did basic food preparation tasks, cooking, and baking. Assisted in purchasing, receiving, and inventory management.

## *Education and Certification*

Bachelor of Professional Studies in Culinary Arts                 Anticipated May 2006
Olympia University, Ithaca, NY

Dean's List every semester

Treasurer, Culinary Club (sophomore year)

ServSafe® Food Protection Manager, #2364656 (National Restaurant Association)

**Figure 14-6**   One-Column Resume Sample

- Figure 14-1. The body of this resume is set in 12-point Arial, the section heads in 14-point bold, and the person's name in 16-point bold. In this use of the two-column format, the dates of employment and college are in the left column, and the job and education information are in the right column. The horizontal line under each section heading, along with the appropriate use of white space, make the headings stand out and result in an easy-to-read resume.

- Figure 14-2. This one-column resume uses 12-point New Century Schoolbook as the body font. The section names and contact information are set in 14-point type and the person's name in 16-point. The name, job titles, and section names are bolded. Because this is a functional resume, the middle has horizontal lines that emphasize the person's skills. Italics are used in this section for the top line.

- Figure 14-3. This two-column resume uses a *T* set of lines to make it look appealing. The font is Palatino, with 12-point type for the body, 16-point for the section heads, and 18-point for the person's name. The name, section heads, and job titles are bolded. The first column contains the section names and the second column the dates and information. If you like how this format looks but have more than one address to list, you can start your vertical line just below the contact section.

- Figure 14-6. This one-column resume shows the section heads in italic bold and centered with horizontal lines above and below for emphasis. The body font is 12-point Garamond, and the name font is 18-point. Section heads, job titles, and the person's name are bold. The years of employment are kept to the right.

- Figure 14-7. This resume is typical of someone with a lot of experience, expertise, and involvement in the culinary profession. The font is Franklin Gothic (12-point body text, 14-point section heads, and 16-point name). Section names, job titles, and employer names are bolded. The section headings appear to the left and have a horizontal line coming out to add emphasis and clarity. The body text is tabbed in to make the section headings more prominent.

The formats of these resumes can also be found on the CD-ROM.

## KEYWORDS

Keywords are nouns or noun phrases that state job titles, skills, duties, and accomplishments (see Table 14-2). Some employers scan resumes into a database. Keywords help the employer identify applicants who may be able to fill a specific position. This is described in length in a moment. For now, you want to use appropriate keywords when possible in your resume. Another source of keywords is job advertisements.

# Brad Barnes, C.M.C., C.C.A., A.A.C.

213 Davis Avenue, Christianson, NY 10735
203.555.0150
BandBsolutions@aol.com

## Qualifications

Culinary Skills

- Very strong experience in quality food preparation.
- Thorough understanding of all facets and styles of service.
- Well versed in many ethnic and international cuisines.
- Able to produce quality results while adhering to budget.
- Committed to upholding the highest standards of operation in the professional kitchen.
- Highly trained in nutritionally conscious cuisine.

Management Skills

- Self-motivated, quality- and cost-directed manager.
- Solid experience in multi-unit management.
- Excellent human resource management skills, maintaining a departmental employee retention average of 3.5 years.
- Skilled in sanitary management of food preparation facilities.
- Experienced in public speaking, presentations, and seminars.
- Developed, wrote, and presented educational videos.

## Professional Experience

**Chef/Owner, B & B Solutions**                              2001–present
Partner in Food and Beverage Management firm currently operating food and beverages operations in two Manhattan properties: the Embassy Suites Hotel in Battery Park City and the Hilton Times Square. Food and Beverages is a freestanding entity and is required to be totally self-sustaining while providing 24-hour room service, an employee cafeteria, and many other hotel services.
- Report on profitability, quality, and operations to our client.
- Reversed the operations from substantial losses to break even.

**Corporate Executive Chef, ITB Restaurant Group**          1992–2001
Oversaw profitability, training, menu development, and staffing of kitchens in three restaurants while acting as executive chef for the flagship operation, 64 Greenwich Avenue.

**Figure 14-7**   Resume of Very Experienced Chef

**64 Greenwich Avenue Restaurant**, 125 seats/$2.4 million annual sales
Responsible for design of the kitchen as well as the purchase of all
equipment. Developed all menus. Developed profit and loss prospectus for
opening food sales.

- Increased profitability of food sales by 10% since the opening
  through a customer-driven sales-oriented approach to menu
  development as well as a concentrated effort to retain employees
  and increase productivity.
- Practiced an aggressive approach to purchasing by constantly
  researching new resources while maintaining a good business
  relationship with purveyors.
- Initiated our banquet/catering division in order to expand sales as
  well as make better use of available staff and facility.
- Banquet/catering has grown to 35% of annual sales at a higher
  profitability than à la carte service.
- Maintained a constant learning atmosphere in the kitchen through
  promotion from within and the rotation of culinary school externs
  in the facility.

**The Black Bass Grille**, 65 seats/$1.4 million annual sales
**The Black Goose Grille**, 120 seats/$2 million annual sales
Set tone and style of menus and worked with the Chef to produce profitable,
customer-driven menus that stayed within our philosophy and food standards.
Wrote and implemented all front-of-the-house training procedures.

- Assured profitability of each kitchen through guidance in food cost
  control, staffing, and time management.
- Produced all graphics for seasonal menus.

**Executive Chef, The Black Bass Grille**                    1989–1992
Hired to change the style of food and service from a tavern-style pub to a white-
tablecloth casual dining restaurant.

- Raised check average from $20 to $37.
- Increased yearly sales from $780,000 to $1.4 million.
- Analyzed lunch business, which showed a history of poor customer
  counts, then recommended closing for that meal period, saving the
  company about $16,000 annually.
- Purchased new equipment per budget to facilitate new style of
  service.

**Executive Chef, Greenwich Island Catering**, $1.8 million annual sales 1987–1989
Supervised all food production and event logistics, including staffing and
equipment setup. Maintained three daily corporate accounts.

**Executive Chef, The Brass Register at Four Squares**, 225 seats, 240 banquet seats, $1.6 million annual sales                                                                    1980–1985
Worked as Sous Chef and then Executive Chef. Started catering and banquet service.

## Education and Certifications

A.O.S. in Culinary Arts, Culinary Institute of America, 1987

Nutritional Cuisine course, Culinary Institute of America, 1995

Certified Master Chef, American Culinary Federation

Certified Culinary Administrator, American Culinary Federation

Certified ACF International Judge

Certified ServSafe® Food Protection Manager

Certified TIPS Alcohol Service Trainer

## Professional Organizations

Member, American Culinary Federation

Member, American Academy of Chefs

Member, World Association of Master Chefs

## Honors and Awards

President's Medal from the American Culinary Federation

Coach and Design Director for American Culinary Federation Team USA, 2004 and 2000

Hermann Rusch Humanitarian Award for Contributions to 9/11 Relief Effort

Two Gold Medals, IKA/HOGA Culinary Olympics, Frankfurt, Germany

"Chef of the Year," The Chefs Association of Westchester and Lower Connecticut

## Table 14-2 Culinary and Management Keywords

### Culinary Keywords

| | | |
|---|---|---|
| Back-of-the-house operation | Food service management | Mise en place |
| Banquet operations | Front-of-the-house operations | Multi-unit operations |
| Banquet sales | Garnish | Profit and loss responsibility |
| Budget administration | Guest relations | Portion control |
| Catering operations | Guest satisfaction | Presentation |
| Club management | Information technology | Product positioning |
| Corporate dining room | Inventory control | Project design |
| Customer retention | Labor cost controls | Project management |
| Customer service | Leadership | Purchasing |
| Employee training | Marketing | Sales |
| Food and beverage operations | Menu planning | Service management |
| Food cost controls | Menu pricing | |

### Management Keywords

| | | |
|---|---|---|
| Benchmarking | Leadership | Problem solving |
| Communication | Leadership development | Profit and loss management |
| Consensus building | Long-range planning | Quality improvement |
| Corporate culture | Multi-unit operations management | Sales management |
| Corporate mission | New business development | Team-building |
| Decision making | Organizational development | |

## VOICE AND TENSE

Even though you never say *I* on a resume, the subject of each phrase is indeed *I*. Be sure your verbs agree with the first person. Use the past tense of verbs when talking about past jobs and events. Use the present tense when describing what you do in your current job.

## SPELLING, PUNCTUATION, AND GRAMMAR

When in doubt, use a good dictionary and a style guide. Use the dictionary to determine when certain words are hyphenated or capitalized. Also:

◆ Capitalize job titles, department name, company name, and towns/cities. Capitalize the first word of each bulleted item.

- Do not use abbreviations. Spell out abbreviations and acronyms, unless they are certifications that follow your name. For example, in Ron Smith, CPC, Certified Pastry Culinarian does not need to be spelled out.
- It is common practice to spell out numbers one through nine and then write the numbers 10 and above as numerals.
- Use colons and semicolons correctly, as well as apostrophes. Remember that *it's* means "it is," and the form *its'* does not exist in English.
- Put one space between a period and the first letter of the next sentence.
- Put a comma between a job title, the company name, and the location.
- Always put a comma between the name of a town or city and the state.

## HONESTY

This guideline is simple: Be honest. Don't even try to be dishonest. The culinary world is really quite small, and you don't want to get a reputation for twisting facts. Even if you get something past an employer who hires you, many contracts include a clause that says dishonesty in the hiring process can result in job termination later.

## PAPER

As you can guess, white or conservative colors such as ivory and light gray are best for resumes.

If you use watermarked paper, be sure to print your resume on the correct side of the paper. Hold up a piece of watermarked paper to the light; the correct side is facing you if you can read the watermark. Be sure the paper you use is at least 20# weight and is suitable for your printer.

# Edit and Proofread

Once you have typed up a resume, it's time again to edit and proofread. The most common mistakes are simple typographical and spelling errors. Computer spellcheckers do not catch correctly spelled words used incorrectly—*of* for *on*, for example, or *their* for *there*. You want your resume to stand out, but not for the wrong reasons. Avoid mistakes: Have several people proofread your resume before you send it anywhere.

Use Table 14-3, Resume Checklist, to make sure you have a polished product.

## Table 14-3 Resume Checklist

_____ 1. Is your resume easy to read?

_____ 2. Is your resume attractive?

_____ 3. Is there enough white space? Is each section distinct?

_____ 4. Have you kept every paragraph under five lines?

_____ 5. Is your contact information all correct?

_____ 6. Are your qualifications at the top of the resume easy to scan? Do they make you an attractive candidate? Does the list include at least one substantial accomplishment?

_____ 7. Does your resume highlight relevant education and work experience?

_____ 8. Does your work experience include measurable accomplishments?

_____ 9. Did you use action verbs when describing past work experiences?

_____ 10. Have you omitted references to salary and reasons for leaving jobs?

_____ 11. Is your highest educational attainment shown first?

_____ 12. Have you included relevant continuing education?

_____ 13. Have you included certifications you have, such as sanitation?

_____ 14. Did you mention special work-related skills?

_____ 15. If you are still in college, did you mention college activities and clubs you were involved in and offices you held?

_____ 16. Have you proofread your resume and allowed at least one other person to edit and proofread as well?

_____ 17. Can someone quickly glance at your resume and see the most important points?

## Scannable Resumes

Many large companies, and a growing number of small ones, use computers to sort the hundreds of resumes they receive. These companies scan paper resumes into a computer database. When managers need to fill a position, they program the computer with keywords that describe the qualifications they want in a candidate. The computer then searches its database for resumes that include those keywords. The resumes with the most matches are forwarded to the managers.

Before you submit your resume to a company, call them to find out if it scans. If it does, be sure your resume's design is computer-friendly. Resumes that will be scanned should contain no graphics or formatting that a computer might misinterpret. Follow these steps to increase a scanner's ability to read your resume:

- Use nontextured white or very light paper with black letters.
- Choose a plain, well-known font such as Helvetica, Arial, or Times New Roman.
- Use a 12-point font for all body text and 14-point for all headings.

- Do not use underlines or italics, and do not use asterisks or parentheses. Modern systems can understand bold, but older systems might not. You can still distinguish headings by using capital letters.
- Use a one-column format.
- Avoid boxes, graphics, columns, and horizontal or vertical lines.
- Put your name on its own line at the top of each page. Also, give each piece of your contact information (address, phone number, email address) its own line.
- Use round, solid bullets.
- Do not staple or fold your resume.

Figure 14-8 contains an example of a scannable resume.

**Figure 14-8**
Scannable Resume

---

Cheryl Richardson

92 Longwood Road
Aurora, NY 11593
315-593-8270
cherylrich@yahoo.com

Summary

Dean's List college student in culinary arts, recently promoted
to Line Cook at nationally known Moosehead Restaurant.

Work Experience

7/04-Present    Line Cook at Moosehead Restaurant, Ithaca, NY
Work at sauté or grill station for lunch or dinner meals
in a well-known restaurant featuring healthful natural foods
cuisine. Perform mise en place and food preparation. Follow safe
and sanitary food procedures. Test and evaluate new recipes. Won
Employee of the Month (June 2005).

10/02-6/04 Preparation Cook at Moosewood Restaurant, Ithaca, NY
Performed all preparation tasks in kitchen emphasizing scratch
cooking and vegetarian dishes. Completed all duties in timely
fashion while maintaining sanitation standards. Received
"Excellent" performance evaluations.

Summers, 2001 and 2002.    Assistant Cook at Lenape Summer Camp,
Seneca Falls, NY. Under Head Cook's direction, did basic food
preparation tasks, cooking, and baking. Assisted in purchasing,
receiving, and inventory management.

Education and Certification

Bachelor of Professional Studies in Culinary Arts Anticipated
May 2006, Olympia University, Ithaca, NY

Dean's List every semester

Treasurer, Culinary Club (sophomore year)

ServSafe® Food Protection Manager, #2364656 (National Restaurant
Association)

# Everything You Need to Know about References

Before making a hiring decision, most employers want to speak with people who know you well. You should find three to five people who agree to recommend you to potential employers.

Choosing references can be difficult, especially for people with little work experience. But you may have more options than you think. The people you ask to be references should be familiar with your abilities. Supervisors from paid or unpaid jobs, teachers, advisors, coaches, and coworkers are all good choices. Select the most willing, articulate people you can. Always ask permission of the people you want to speak for you before including their name on your reference list.

After choosing and contacting references, type a list with the name, address, telephone number, and email address of each one, and briefly describe his or her relationship to you. Bring copies of this list, along with your resume, to interviews.

When people agree to be references, help them help you. Send them a copy of your resume or application to remind them of your important accomplishments. Tell them what kinds of jobs you are applying for so they know what types of questions to expect.

# EXERCISES

1. Learn more about resumes at the monster.com website:
   **http://resume.monster.com/resume_samples/**

2. Use the Resume Worksheet (Figure 14-4, and on the CD-ROM) to write up the information for your resume. You will probably not include everything you write on this worksheet on the resume itself, so just be complete.

3. After you have gathered the information for your Resume Worksheet, write your first draft using Form 14-5 (on the CD-ROM). Be sure to use action verbs from Table 14-1 and keywords from Table 14-2.

4. Type your rough draft in at least two different formats. The CD-ROM contains five formats. Which looks best?

5. To evaluate your resume, use Table 14-3 or go to monster.com and click on "Quizzes." Then click on "Resume Readiness Quiz."

# Put Together a Job Search Portfolio

## Introduction

**FOR YEARS, CERTAIN PROFESSIONALS,** such as artists and writers, have used portfolios in their job searches to showcase their abilities and qualities. If you have ever seen an artist's portfolio, you know that it contains an organized collection of his or her work based on artistic style, growth, abilities, and aspirations. By the time you finish looking at the collection, you have an excellent idea of who the artist is and what he or she does well.

Portfolios aren't just for artists and writers. Professional portfolios can be used by anyone who has a career to:

◆ Get a new job.
◆ Get a raise.
◆ Get a promotion.
◆ Help direct and keep records of lifelong learning.

In addition, you may be asked to maintain a learning portfolio in college. A learning portfolio, which tends to be quite long, documents learning in specific areas, such as sanitation. A job search portfolio focuses on your work-related skills, abilities, and qualities that are necessary to do a job. This type of portfolio is much shorter, from 10 to 30 pages in length.

Think of a job search portfolio as a catalog of your skills, abilities, and qualities. It is not simply a resume. A portfolio goes beyond your resume to demonstrate what is on your resume and be a visual representation of your strengths. Artifacts such as menus and photographs of plated meals are a visual way to demonstrate your accomplishments. Just as archeologists use artifacts to reconstruct a civilization, a portfolio reconstructs your career.

Why bother with a job search portfolio? Assembling a portfolio has many benefits:

◆ It helps you assess your learning and work experience and compare it to an employer's need for skilled, capable employees.
◆ It helps you prepare for interviews.
◆ It gives you a competitive edge when applying for a job because it showcases and convinces others of your skills, abilities, and qualities, and it demonstrates the results of your work.
◆ It helps you communicate clearly.
◆ It provides evidence of your potential.

Creating your own job search portfolio is a fairly simple process, as we now show.

## Choose How to Organize

You can organize your job search portfolio in a number of ways. Here are some possibilities:

◆ **By Resume Category**—Probably the most popular and easy-to-use method, the resume-based portfolio provides artifacts for each of your resume categories. See Table 15-1 for a list of potential resume categories.

| Table 15-1 Potential Categories to Use in a Resume-Based Portfolio | | |
| --- | --- | --- |
| Professional Experience | Computer Skills | Publications |
| Education | Foreign Language Skills | Presentations |
| Certifications | Community | Military Service |
| Continuing Education | Service/Volunteer Work | |
| Professional Affiliations | Awards/Honors | |

◆ **By Date (Chronological)**—With a chronological format, a portfolio is normally divided into the years you were in college, the years you worked for ABC Restaurant, the years for worked for XYZ Catering, and so on. This portfolio is essentially divided up by jobs you have had.
◆ **By Skill (Functional)**—A functional portfolio is organized according to your skills. For example, an experienced cook may include the following skills categories: cooking, garnishing, catering, sanitation, and technology.
◆ **By Theme (Thematic)**—A thematic portfolio is divided just as it sounds—by theme. For example, a chef working in a catering company may organize the portfolio into sections based on types of catering affairs: weddings, corporate affairs, and so on.

So which one do you pick? Whichever format will encompass the artifacts you need to exhibit to get the job you want. Think about the requirements for the job you are looking for, and then choose a format that helps prove you are the best match for the job.

Whichever format you choose, you should include a resume and reference list at the front of your portfolio. You may also want to include a work philosophy or philosophy of cooking statement. This is a description of the guiding principles that drive you and your cooking, including your philosophy of foods and cooking, your work ethic, management philosophy, and so on. Your cooking philosophy may be, in brief, to emphasize local, organic foods in simple meals, or to blend traditional with contemporary cooking. In a healthcare setting, your cooking philosophy may be to provide home-style, attractive meals that patients enjoy. In any case, your work philosophy should be about one paragraph long. It could even be put into a bulleted list of three to five points.

## Collect the Contents

Next, you need to collect and select artifacts for the portfolio. To do this:

- Think about *what* you do, the *skills* involved, as well as *how* you do it. For example, if one of your job duties is to cook for banquets, you could select a banquet menu and photographs to show your cooking and food presentation skills.
- Collect artifacts that match the requirements of the job you are looking for.
- Choose items that are your best examples and show mastery. You don't want to include average or mediocre results.
- Try to find examples of your work in which you were the only or at least the major contributor.
- If you are still in college, or only out a short time, don't forget that you have developed work-related skills while playing team sports, performing volunteer work, doing hobbies, and going to school. For example, being a member of a sports team requires discipline, motivation, teamwork, and energy. Performing volunteer work shows dedication and can develop any number of skills. Doing hobbies shows motivation and skills. Working on team projects at school requires teamwork, problem-solving skills, and people skills. It's fine to include these activities and skills in your portfolio.

Here are examples of artifacts you might select for your portfolio. They are divided into resume categories.

- **Professional Experience**—This section is likely the heart of your portfolio. Table 15-2 shows categories of skills and gives you more ideas for artifacts. You may want to subdivide Professional Experience into skill groups if you have more work experience. Table 15-3 gives examples of personal qualities you may want to highlight in your portfolio.

**Table 15-2 Possible Skills Categories**

| | |
|---|---|
| Communication Skills | Management Skills |
| Creative Skills | Marketing and Sales Skills |
| Culinary Skills | Supervisory Skills |
| Financial Skills | Technology Skills |
| Human Resource Management Skills | Sanitation and HACCP Skills |

**Table 15-3 Personal Qualities to Highlight in Portfolio**

| | | |
|---|---|---|
| Accurate | Flexible | Punctual |
| Adaptable | Hardworking | Reflective |
| Careful | Honest | Reliable |
| Cheerful | Imaginative | Resourceful |
| Confident | Innovative | Self-starter |
| Cooperative | Logical | Sensible |
| Courteous | Loyal | Sensitive |
| Creative | Motivated | Steady |
| Efficient | Open-minded | Tactful |
| Energetic | Patient | Trustworthy |
| Enthusiastic | Persistent | |
| Ethical | Practical | |

- Photographs of plated foods, buffet tables, etc.
- Samples of menus
- Samples of recipes
- Customer survey results
- Performance evaluations
- Documentation of accomplishments: increases in sales, decrease in costs, etc. (Can use bar graphs, pie charts, or other graphics.)
- Major projects completed
- Training materials developed
- Title page of report written
- Newspaper/magazine clipping describing event you contributed to
- Thank-you letters from customers
- Menu clip-ons or other artifacts of marketing/sales skills
- Memos, reports, letters that show communication skills

- Education
  - Copy of your college diploma(s)
  - College transcript
  - College course descriptions
  - Copy of scholarship letters
  - Copy of awards or honors
  - Copy of honor society memberships
  - Photographs or other artifacts of internships
  - Photographs or other artifacts of extracurricular activities
  - Listing of leaders of an organization in which you held an office
  - Photograph or other artifacts of service project participation
  - Title page or other artifacts of relevant course projects such as a business plan
- Certifications
  - Copy of certificates
  - Description of what you are doing to recertify
- Continuing Education
  - Chronological list of workshops, seminars, courses attended during last five years
  - Certificates (if available) from workshops, seminars, courses
  - Chronological list of trade and industry shows attended in last five years
- Professional Affiliations
  - List of organizations to which you belong, the year you joined, offices you held, and boards or committees on which you served
  - Copies of current membership cards
  - Artifacts of leadership positions held
- Computer Skills
  - Copies of menus, brochures, etc., you created
  - List of computer software you are proficient in
- Foreign Language Skills
  - List of college courses you have taken in a foreign language
  - Statement of which languages you are proficient in, and whether you are proficient at reading, writing, or speaking each one
- Community Service/Volunteer Work
  - Photographs, flyers, menus, or other artifacts of volunteer work
  - Awards/Honors
  - Copy of certificate or photo of you accepting award/honor
- Publications
  - Title page of published articles
  - List of publications, including date published and publisher's name
  - Front cover of publications
- Presentations
  - List of presentations, including date, topic, and place
  - Sample of visual aids, handouts, etc.

◆ Military Service
   ◆ Copy of honorable discharge showing years of service
   ◆ Verification of military education
   ◆ Certificates verifying competence in various areas
   ◆ Commendations and awards
   ◆ Photographs of accomplishments

When collecting artifacts from work, include only items you clearly own or have permission to include. You do not want to reveal proprietary information about your current or past employers.

If you want to develop a functional, or skills-based, portfolio, use Table 15-2 as a starting point to develop your own skill categories.

## Get Supplies

To put together a job search portfolio, you will need supplies.

1. A slim three-ring loose-leaf binder with inside pockets: A zippered binder is a good idea if anything could accidentally fall out of the binder. A viewbinder, which has a place to insert a front cover you create, may also be a good idea.

2. Sheet protectors or insertable plastic pockets: Sheet protectors are great for storing many of your artifacts. They also make it easy to add and remove pages from the binder.

3. Tab dividers: Make a tab to identify each category of your binder. Because sheet protectors are larger than 8½-x-11-inch paper, buy extra-wide tab dividers so the tabs can be easily seen and read.

4. Plastic photo sheet holders: Plastic sheets that hold several photographs are helpful when you want to display more than one or two photographs on a page.

You will also need heavy (24#) bright white paper to mount many of your artifacts and card stock for your titles and captions.

## Put the Portfolio Together

Before you assemble your portfolio, remember that more is not better. A job search portfolio is a professional portfolio, not an overstuffed photo album. The following steps will guide you through the process of finishing your portfolio. You will also find them on the CD-ROM as the Portfolio Worksheet.

1. Decide which type of portfolio, or which mix of portfolio types, will best serve your purposes. There is no one correct way to organize a portfolio, so choose what works for you.

2. Come up with a tentative list of categories to include in your portfolio.

3. From the artifacts you have collected, pick out the best sample(s) for each category. Keep in mind that it's better to have a few good samples than lots of mediocre or redundant ones, and that *an interviewer can absorb only six to eight samples.* If you see you have too much for one category, you can either split it up or be pickier about what you select. If you have too little, don't worry. One or two artifacts are probably enough in areas not directly related to work. You may also be able to combine two categories, such as certifications and professional affiliations.

4. Decide if you want your first tab to be for your resume and reference list. If you prefer, you can put these documents in the pocket of the loose-leaf binder. Decide if you want to include a work philosophy in your portfolio. If so, put it at the front.

5. Create an index tab for each category.

6. Prepare the artifacts to go into the binder. For each item, such as a photograph or a menu you created, you need to write a title and caption. After you think of a concise title for each artifact, work on the caption. Your caption can be in the form of a short paragraph or a bulleted list. When writing your caption, consider these questions:

   ◆ What is being shown? What were the results?
   ◆ How did you accomplish this task? What skills did you use? What personal qualities, such as persistence and flexibility, were important?
   ◆ Why were you doing this? Why is it important?
   ◆ When did this happen?
   ◆ Who worked with you on this?
   ◆ If anything presented is confidential, do you have permission to use it?

   See Figures 15-1 and 15-2 for examples of titles and captions.

7. Now you can start laying out your pages. A portfolio contains diverse artifacts and will not look professional unless you unify its presentation by being consistent in your layout. As with your resume, allow 1-inch margins on the sides, top, and bottom of each page. Pick a simple typeface that is easy to read. Don't clutter your pages; lots of white space makes them easier to read. White space is the space on a page that is not occupied by text or pictures. Avoid headlines in all capital letters. It's actually easier to read headlines in lowercase with the first letter of each word capitalized.

8. If your artifact consists of one or two photos, you can tape them on the paper after you have printed the headline and caption. Place the headline at the top of the paper and the captions along the sides of the pictures or at the bottom (see Figures 15-1 and 15-2). Whichever style you choose, be consistent. If your artifact is a menu you want to put in a sheet protector, print your title and caption on card stock, cut it out, and let it float on top

# College Menu Project

## Starters

### WINGS & THINGS
A pound of our own buffalo wings served with celery sticks and choice of ranch or blue cheese dressing.

### CHICKEN FINGERS
You get 5 chicken fingers with choice of ranch dressing or barbecue sauce served on a bed of lettuce.

### FRESH VEGGIE PLATE
Celery sticks, carrot sticks, cucumbers, cauliflower, and tomatoes. Choice of ranch or blue cheese dressing.

### ONION STRAWS
Served with carrot and celery sticks, choice of ranch, French fry sauce, or blue cheese dressing.

## The French Fry Factory

### FRESH MADE FRIES
Try our own fresh-made fries with your choice of two sauces—French Fry Sauce or Lethal Weapon Sauce.

### FRESH MADE FRIES WITH CHOICE OF THREE TOPPINGS
Melted cheddar cheese - Chili con carne and shredded cheese topped with sliced green onions - Green chili Verde with shredded cheddar cheese and sliced green onions, salsa and sour cream.

## Dinners

### Eagle's Nest Specialty: THE COMMANDER'S CUT
A 32 oz. Porterhouse steak; charbroiled to your taste.

16 OZ. T-BONE

10 OZ. RIB EYE

6 OZ. NEW YORK STEAK AND 3 SHRIMP

8 OZ. GRILLED HALIBUT STEAK

SHRIMP DINNER (5 piece)

8 OZ. BONELESS GRILLED CHICKEN BREAST

SOUTHERN FRIED CHICKEN (4 pieces)

All dinners served with choice of French fries, baked potato with sour cream and whipped butter, salad with choice of dressing, vegetable of the day, roll and butter.

## Sandwiches

| | |
|---|---|
| ONE-THIRD-POUND CHEESEBURGER | CHICKEN AND SWISS |
| MALIBU CHICKEN | ALASKAN HALIBUT BURGER |

All sandwiches are served with French fries, lettuce, tomato, red onion, and pickle.

This menu was developed in Fall 2003 as part of a menu project that required menu development and food costing skills. After determining food costs, menu prices were developed. This menu was designed for us on a military base, specifically for an officers' dining facility. Each student did their own work.

**Figure 15-1** Sample Portfolio Page with One Caption

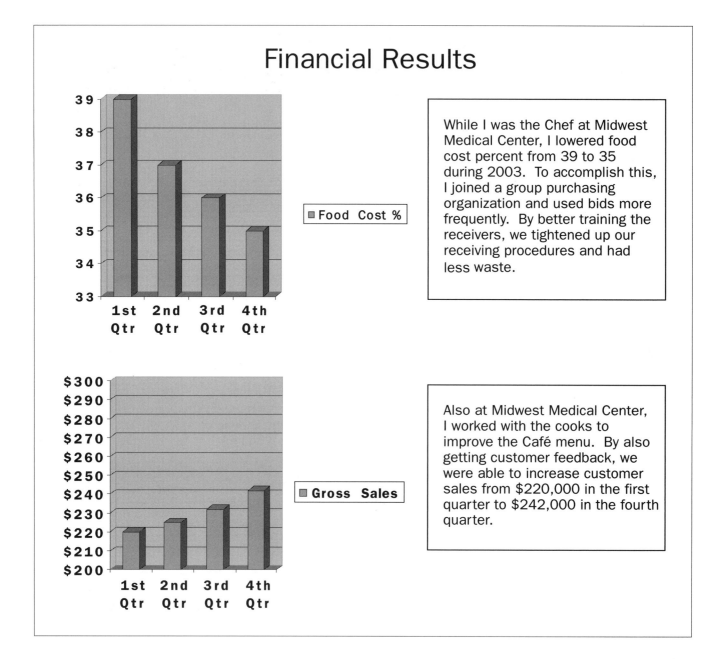

**Figure 15-2**
Sample Portfolio Page
With Two Captions

of the menu in the sheet protector. If you use photo sheet holders, use card stock to identify the photos and put it with the photo if room permits. You can always put the titles and captions on a separate page if needed. You may need to use reduced-size copies or photographs of large and bulky three-dimensional objects (medals, etc.) to fit them into the binder.

9. Within each category, you can place your artifacts in chronological order or put your best ones first. Do whichever will work better in an interview.

10. Type up a table of contents for your portfolio. Place it up front in the loose-leaf binder.

# Evaluate Your Portfolio

Once you complete your job search portfolio, ask a friend or teacher to review and critique it with you. Here are possible review criteria:

◆ Is the job search portfolio neat and organized?
◆ Is the portfolio labeled properly?
◆ Is it easy to use?
◆ Does your portfolio market your talents?
◆ Does your portfolio highlight what you need to secure a job?
◆ Are the titles concise?
◆ Are the captions engaging and polished?
◆ Do your artifacts support your knowledge, skills, abilities, and qualities?
◆ Have you avoided using jargon, acronyms, and abbreviations people won't understand?
◆ Have you proofread it for spelling and grammatical errors?

# Use a Portfolio in an Interview

Before you go into an interview, make sure you have the statement "Portfolio Available" at the bottom of your resume. With luck, the interviewer has read that and remembers the information.

During the interview, look for appropriate opportunities to show parts of your portfolio. For instance, when you are asked about your experiences making food attractive, tell the interviewer that you have a portfolio. Turn to the appropriate section in your portfolio and show the interviewer a page or two. If you are seated at a table with the interviewer, you can place your portfolio in front of him or her. Otherwise, you may want to remove the page from the binder and hand it to the interviewer. Artifacts have only a few lines of text at most, so you need to chat about them, especially in relation to the interview question. Here are additional guidelines:

◆ Don't narrate your portfolio page by page during the interview. Instead, draw on it only as the interviewer's questions come up. If the interviewer asks for an overview of your portfolio, briefly describe how it expresses your work philosophy and how it is organized. This will enable the interviewer to know enough to question you further.
◆ You can change the contents of your portfolio based on the job for which you are applying. For example, if you are a Chef in a restaurant and you are looking for a Catering Chef position, you may want to add more quality artifacts to the Catering Skills category.
◆ Watch to see how interested the interviewer is in your portfolio. Try to gauge the level of interest. If the interviewer does not show a lot of interest, don't draw on the portfolio more than once or twice.

◆ Know that you will have interviews in which you do not use your portfolio at all. Some interviews are mainly to determine if you fit in well with the other players. A portfolio is more useful when the interviewer is still trying to ascertain if you have the knowledge, skills, abilities, and qualities to do the job.

## Keep Your Portfolio Up to Date

Once you have gone to the trouble of assembling a job search portfolio, spend the time to keep it up to date. Review your portfolio two or three times a year. Remove anything that is out of date, and add new artifacts (with titles and captions) to the appropriate categories. Your objective is to keep your portfolio as responsive to future needs as possible.

## EXERCISES

1. Go to the following website, which contains links to many excellent resources on portfolios. Read one of the resources and write a paragraph on what you learned. http://www.acinet.org/acinet/library.asp?category=2.1

2. Collect artifacts for a job search portfolio. Next, follow the steps to putting together your portfolio, using the CD-ROM.

3. Have a friend or teacher review and critique your portfolio. Make changes as needed.

4. Evaluate your portfolio using the Evaluation of Portfolio found on the CD-ROM.

5. In a mock interview, use your portfolio to support your candidacy.

# Beyond the Want Ads: How to Locate and Contact Prospective Employers

## Introduction

**ALTHOUGH THE JOB ADVERTISEMENTS** in the Sunday newspaper seem to offer lots of positions, trying to find one only by wading through the want ads does not often work—especially as your salary requirements rise. This is because most jobs are never advertised. Research indicates that one of the most effective ways of finding out about jobs is by getting leads from people you know, a technique called *networking*. Many more positions are filled every day through word of mouth than through printed advertisements. Another effective way to find a job is to identify potential employers yourself and call them or knock on their door. This chapter discusses a variety of ways to find jobs. As you read, keep in mind that the more methods you use to find a job, the higher your odds of getting a job offer.

## General Guidelines

You may find it frightening to look for a job because it means calling strangers, being interviewed, taking risks. While the experience may pull you out of your comfort zone, it is clearly an opportunity to meet new people, grow, and become more confident. The worst that can happen is that someone may not want to help you in your job search, may not want to interview you, or may give the job to someone else. The best that can happen is that you will find a job you love.

Looking for a job requires time and organization. Be prepared to spend *at least* 20 hours each week using the methods discussed in this chapter to find out where job openings exist, get your foot in the door, and be interviewed. You will be busy emailing, faxing, snail mailing, and delivering your resume to lots of potential employers (many of whom will never call you). Set aside certain times every day to follow up leads, make phone calls, check for job openings, and so on.

To be successful with your job hunt, you must be organized. One way to organize is to write up an index card for each job you applied for. Write notes on everything about the job and the company on the card. Keep the index cards in alphabetical order by the phone. If you get a phone call from one of the companies, all your information is right there. You can also keep track of resumes you send out, phone calls, and other communications by maintaining a job log. The CD-ROM contains a job log that you can use to review progress in your job search.

Here are more suggestions:

◆ Use a variety of strategies to find a job. Don't rely just on the Internet or just on the want ads. The more strategies you use and the more time you put into your search, the more likely you are to find a job.

◆ Don't get discouraged if you don't get a job within a month. Short job searches can take up to four months.

◆ Learn about the companies you have applied to or want to work for.

◆ Make an effort to contact and establish a relationship with the hiring manager. Although you may think you are being a nuisance, you're not. Your initiative and persistence are positive traits.

◆ Always follow up after you have sent someone your resume. If you email your resume to someone, mail a paper copy with a cover letter to the person that same day. Then, within a week, if you have not been contacted, call the person to check if your resume arrived and if you can set up an interview. Reiterate your strong interest in the job and the company.

◆ Be persistent. If the restaurant said last month that there were no openings, go back and find out if the situation has changed. If Human Resources said you should call back in a week, make the call.

◆ Think beyond the employers you are currently considering to additional employment possibilities. Sometimes jobseekers are convinced they will get a job from the handful of employers they are seriously looking at. Don't put your job search on hold while you wait. You are wasting valuable time that could be spent looking for other jobs.

◆ Don't limit your search to a certain dining segment such as fine dining. Any type of foodservice is a great place to develop preparation skills. Mastering the quick pace of grill, sauté, and pantry can be of value in many jobs with fabulous menus and talented Chefs. A banquet cook working in an upscale catering house can apply the precious skills he or she is learning in many other settings.

- The more people you talk to on the phone and visit in person each day, the quicker you will find a job.
- For smaller restaurants, it's much better to go in person rather than use the telephone. If the restaurant or company is big enough to have a Human Resources department, then it is appropriate to call. In either case, avoid calling at times when the staff are likely to be busy (mealtimes) or otherwise unavailable.
- Be prepared to tell employers why you are the best choice for the job and what makes you different from other candidates.
- Always write thank-you notes after interviews (see Chapter 17).
- Always treat everyone courteously during your job search. Even if no jobs are available, someone may still be able to refer you elsewhere, so you always need to make a good impression.

The final guideline is to accept rejection. It's a normal part of the job search process. Don't take it personally. It doesn't mean you won't find a job. You will; it is just going to take more work.

# Search the Hidden Job Market

The hidden job market includes jobs that are not advertised in the Sunday paper, posted on the Internet, or otherwise made public. Networking is by far the best way to find jobs on the hidden job market. Another way to uncover job openings is to target and contact specific employers.

## NETWORKING

Networking is not only helpful in finding jobs but crucial for advancing in your career. What is networking? Networking is investing regular time and effort to create and maintain career-related contacts. It is building relationships and sharing information. What does networking look like?

- Networking involves talking with people at work and outside of work to gather information, ask for advice, or learn about something new in the culinary field.
- Besides networking face-to-face, you can also network on the phone, via email and electronic bulletin boards, and by letter; however, talking in person is most effective.
- Networking is a two-way street. You get useful information and contacts, and in turn, you act as a resource and contact. Networking involves give and take: Expect to give as much as you get.
- You don't need special skills or training to network. If you can cultivate a business or personal relationship, you can network.

Who should be in your network? Here are some groups to consider:

◆ Current and former coworkers
◆ Classmates and alumni of schools and universities you attended
◆ Members of professional organizations to which you belong
◆ Relatives, friends, and neighbors
◆ Members of social, recreational, religious, volunteer, and other groups to which you belong
◆ Members of community groups and activities

Your network may include people you have never met. You may hear about them through a mutual contact, or you may get their name from a directory or other resource.

Professional networking groups may also be of help. For a listing of networking and support groups by geographical region, use the helpful Riley Guide (www.rileyguide.com/support.html).

## *Making a Contact List*

Networking can help lead you to unadvertised jobs. To use networking to full advantage, it is important to compile a list of your contacts. Some people keep a list of contacts in a computer database, while others prefer using index cards or another written format for each contact. Some information to maintain on each of your contacts is shown in Table 16-1. As the table suggests, you should rate how well you know each contact and how much each can help in your job search.

---

**Table 16-1 Important Contact Information for Networking**

Name of person

Title

Employer

Contact information

Email address

How do you know this contact?

How well do you know this contact? (You may want to rate this on a three-point scale such as 1: Know well, such as friend or family; 2: Know somewhat, such as colleagues; 3: Don't know well or don't know at all.)

Discussion notes/referrals

Follow-up

How valuable for job search (You may want to rate this on a three-point scale such as: 1: Well connected, with good information and help; 2: Moderately connected, can provide some information and help; 3: Not likely to be of much help.)

Comments

## *Communicating with Your Contacts*

Before discussing how to use your contacts to find a job, it's important to understand the basic guidelines for networking.

- Seek advice, information, or feedback rather than solutions or jobs.
- Avoid asking for information you can get through personal research.
- Ask for information your contact would be comfortable giving to you.
- Know what you want, be clear about your objectives, and clearly state them.
- Let your contacts know how much you appreciate their help by sending them a thank-you note or letting them know how you used the information they gave you. Also, offer to help people who have helped you.

When you are looking for a job, drawing on your contact list for help is a logical choice. Not that you are going to call people to ask them directly for a job—what you want is to find out who has job openings and the names of additional contacts that may lead to other job openings. In the process, you will gain even more networking contacts.

Most people you call will be happy to help you, but they don't have much time, so make your point quickly and directly. To do this, you need to develop a script, or a sales pitch, ahead of time. Your script should include the following elements.

- Your name
- How you know this person or who referred you
- Your occupation
- Your current situation
- What type of job you're looking for and what you have to offer
- A request for names of people to contact regarding job openings, a request for the contact to get in touch with you if he or she hears of a job opening, or a request for advice

Try to fit your information into a 20- to 30-second message. For example, an actual script may look like this.

My name is John Lucas. Ted Alvarez gave me your name. Do you have a minute or two to talk? [After hearing "yes."] I've been the Chef at the Makefield Country Club for the past three years, and sales have grown 20%. I would like to get some hotel cooking experience, and Ted told me you might know some people I could contact for a hotel Chef position.

Once you have written a good script that is neither too long nor too short, you need to practice it so it comes across clearly and confidently. You also want it to sound spontaneous and genuine, not memorized.

When you are well prepared, these calls will be much easier than you anticipated. You have nothing to lose by calling; if you don't make the call, you'll never find out if there was good information or a job lead at the other end. If you do call and make it easy for the contact to help, you will most likely be successful. At the worst, you'll a bit uncomfortable. Each call you make is good practice for the next.

Of course, you can also use email and letters to contact people. Again, you need to include all the elements just discussed, being sure to get to the point in an upbeat manner. Letters usually work best with contacts you know fairly well.

In addition to using your contact list, be sure to attend association meetings, conferences, annual meetings, and so on. These are all excellent places to network with fellow professionals. People are approachable at meetings, so use your script to find out about job openings and obtain more contacts. Also, be sure to bring lots of business cards and copies of your resume so you can distribute them if asked. Keep a record of all the contacts you make, what the results were, and any follow-up that is needed. This will help you organize your time and monitor your progress.

## TARGETING EMPLOYERS

Many employers want you to find *them*. This technique is a little different from networking in that you identify employers for whom you would like to work, and then contact them to see if any jobs are, or will be, available.

Don't confuse this technique with mass mailings of resumes to zillions of potential employers. You want to conduct a targeted mailing, not a mass mailing. First, research a number of potential employers, choose which ones to contact, and write an individualized letter to a specific person at each. As with other techniques, follow up with phone calls.

The best place to start learning about prospective employers is often the Internet. Many employers have websites that are full of information, although they usually just share the good stuff! Look for sections on websites entitled About Us, Pressroom, or News Releases. These sections contain background information and the company's most recent news. In addition, publicly held companies publish annual reports that tell you just about everything you need to know. The president's message in the annual report can say a lot about the mission and future of a company. The annual report is often on the website.

Other good places to learn about employers include networking contacts, trade magazines, and association publications. Local business journals also have information about local businesses including news, names, and upcoming events. For a list of business journals for many U.S. cities, go to www.bizjournals.com/journals.html. Also, be sure to look in your local Yellow Pages.

After you have compiled a list of interesting employers, you need to find the person who does the hiring. Although it is easy to simply send your resume to Human Resources, that does not always get results. Based on the type of position you want (for example, Banquet Chef), find out who the supervisor is (for example, the Executive Chef) and send him or her your resume. To get the Executive Chef's name, ask an appropriate networking contact or simply call the operation and say, "I'm writing a letter to the Executive Chef and I would like the spelling of the name. Can you help me?" The section on cover letters later in this chapter will help you send an appropriate letter with your resume.

# Search the Public Market

The public job market includes jobs that are advertised in some way to potential applicants. This section discusses college placement offices, job fairs, classified advertising, recruitment and executive search firms, and employment agencies. Employer websites also often list job openings. This topic is discussed in the section on using the Internet in your job search.

## COLLEGE PLACEMENT OFFICE

The college placement office provides a number of services to students. Many offer workshops on writing resumes and taking interviews, for example. Go to your college placement office often to check on workshops and other resources as well as to become friendly with the counselors, who can be helpful in your job search. Check often on when recruiters are coming and sign up for interviews.

## JOB FAIRS

Job fairs, like interviews, are face-to-face meetings between jobseekers and employers. They are among the easiest places to find good job leads. Every employer attending is there to find good candidates for open positions.

At a job fair, jobseekers gather information about a company to help them decide if they want to apply for a job. Recruiters staff booths and answer questions, distribute brochures, accept resumes, and conduct mini-interviews. Here are tips to help you make the most of a job fair:

- Dress as you would for an interview (see the next chapter).
- Bring a briefcase, expandable folder, or bag to organize materials such as brochures, application forms, and business cards.
- Bring copies of your resume.
- When you arrive, take a quick walk through the fair. Time is limited, and booths are probably crowded. Plan a route to see the companies you really want to visit. Save visits with the best prospects until you've first warmed up with a few other employers.
- If you are with a friend, don't appear to be inseparable when visiting booths. A professional image is easier to maintain if you speak to employers alone.
- Be professional during the mini-interview. Introduce yourself, shake hands, answer questions, and ask questions. Find out about the company, the types of jobs available for people with your education, and hiring procedures. Don't leave the booth without getting the recruiter's business card.
- After each mini-interview, it is often a good idea to take a few notes so you don't confuse what each recruiter told you.

## CLASSIFIED ADS: WHERE TO LOOK

You are certainly at least a little familiar with the want ads, a feature of newspapers across the country. Small hometown papers, major metropolitan dailies, and national newspapers have classified sections that include "Help Wanted" advertisements. Hometown papers tend to offer local jobs, often with small companies, while the bigger metropolitan dailies offer jobs over a much larger geographical area; they may even offer national or international jobs. Some newspapers post their classified advertising online.

Another source of want ads is trade magazines such as *Nation's Restaurant News* and association publications. Look in the back of these publications for classified advertising, including want ads. You may already subscribe to some of these publications or have access to them at a college library.

When using classified ads, keep these tips in mind:

◆ When you start seriously looking at want ads, scan all of them to see which categories and position titles you need to target. Don't limit yourself to looking under *C* for cooking and Chef jobs. Employers often list cooking or Chef jobs under the industry name: restaurant, hotel, hospital, etc. Get an idea early on where to look for relevant ads.

◆ Read the ads daily. Sunday editions tend to have the largest classified section.

◆ When you are deciding which ads to respond to, try to select only those where you meet all or most of the qualifications and you are really interested in the job.

◆ Be careful about responding to blind ads—that is, ads that do not mention the name of the employer. Blind ads may be used by employers who don't want anyone to know that a position is, or will be, vacant. Sometimes, though, blind ads are used by companies that have a poor reputation. Treat blind ads carefully; if you are working, the employer might be yours!

◆ Answer ads promptly, within a few days, because openings may be filled even before the ad stops appearing in the paper.

◆ Keep a record of all advertisements to which you respond. You might simply staple the ad to a copy of your cover letter.

## RECRUITMENT AND EXECUTIVE SEARCH FIRMS

Recruitment firms, also called search firms or headhunters, work for client companies to find qualified candidates to fill specific positions. In return for finding a candidate who gets hired, the recruiting firm is paid a fee by the client. Recruiters work in many industries, including the foodservice and restaurant arenas. Many recruiting firms specialize in a specific industry. Several Internet websites offer directories of recruiting firms; use these to locate only those companies that recruit for your job.

Executive search firms, as their name indicates, focus on recruiting for top management positions. Both recruitment and executive search firms can give you access to jobs you might not otherwise hear about.

If you want to see if a recruitment or executive search firm has jobs you might be interested in, check the website to see if you can register and if they have any positions posted that could be a match for you. If so, send an email with your resume to the recruiter and then follow up with a telephone call several days later. Recruiters usually know their client companies well and are good at matching candidates to jobs. Other ways to get in touch with a recruiter are to ask someone to introduce you (the preferred way) or to send a strong cover letter with your resume. Keep in mind that the recruitment firm is much more approachable than the executive search firm because the recruitment firm normally is filling midlevel positions or lower, whereas the executive search firm is only filling top management positions.

If you are contacted by a recruiter and you pass the initial phone screening, the next step is normally to meet with him or her. Recruiters interview you carefully to see if your background and personality match the needs of any of their clients. Be prepared to fill out application forms and to have your references checked carefully.

If a recruiter has a position that would be suitable for you, he or she should be your guide throughout the interview process. Good recruiters prepare you for your interviews and provide coaching and feedback. They also help to arrange appointments and travel schedules.

## EMPLOYMENT AGENCIES

Employment agencies can be public, meaning they are run by the government, or private.

### *Government Organizations*

In line with the U.S. Department of Labor's vision for helping jobseekers, CareerOneStop (www.careeronestop.org) is a collection of electronic tools that are available because of a federal-state partnership. Each of these tools offers help to jobseekers.

1. America's Job Bank: This is the biggest and busiest job market in cyberspace. Jobseekers can search for job openings and post their resume where thousands of employers search every day.

2. America's Career InfoNet: You can get millions of employer contacts, wage and employment trends, and state-by-state labor market information here.

3. America's Service Locator: This service directs you to one-stop career centers in your area that offer job postings, information about local employers, help preparing for job interviews, posting of your resume, and many other services.

### *Private Employment Agencies*

Private employment agencies can be helpful, but they are in business to make money. Most operate on a commission basis, with the fee dependent on a percentage of the salary paid to a successful applicant. You or the hiring company will pay the fee. Find out the exact cost and who is responsible for paying associated fees before using any service.

Although employment agencies can help you save time and contact employers who otherwise might be difficult to locate, the costs may outweigh the benefits if you are responsible for the fee. Contacting employers directly often generates the same type of leads that private employment agencies provide.

# The Internet

The Internet can yield tons of information relevant to your job search. Most of what the Internet has to offer can be categorized into one of these five areas.

1. Advice and counseling on resume writing, interviewing, and other career and job search topics
2. Information on employers and salaries
3. Networking
4. Resume postings
5. Job postings

Job-hunting using *only* the Internet is likely to be a waste of time. The Internet is just one way to search for a job. Use it wisely, or you will be spending a lot of time job searching online with little in return.

Table 16-2 lists some of the more prominent online job sites. Most of these sites offer career advice, resume postings, and job postings. Big job sites, such as Monster, may offer some type of networking program as well. Table 16-3 lists websites that specialize in culinary and foodservice jobs as well as professional organizations that post jobs. The websites noted are all current as of this writing but may have changed by the time of your search.

## ADVANTAGES AND DISADVANTAGES

Using the Internet has advantages and disadvantages. One advantage is that you can work at home (as long as you have a computer with Internet access) at any time you choose. Once you have posted your resume on job sites that offer search-while-you-sleep capabilities, the job site actually contacts you (by email), or the employer, or both, that a match is found. Using appropriate keywords in your resume (as discussed in chapter 14) is crucial to finding matches. Job sites can also often provide salary information—which, however, may be inaccurate—

## Table 16-2    General Job Boards

| Monster Boards    www.monster.com |
| --- |
| To search jobs, choose a Keyword, Location, and Job Category. Highlight Restaurant and Foodservice for Job Category. Use any of these keywords: Chef, Sous Chef, Chef/Manager, Executive Sous Chef, Chef/Kitchen Manager, Executive Chef, Corporate Chef, Food Production Manager. You can also post your resume and get job search advice and information. |
| Flipdog    www.flipdog.com |
| Flipdog is a Monster company and offers most of the same features. |
| America's Job Bank    www.careeronestop.org |
| Sponsored by the U.S. Department of Labor, this site allows you to search through a database of over one million jobs, create and post your resume, and do a job search in your local area for Food and Lodging jobs. |
| Career Builder    www.Careerbuilder.com        Yahoo!    hotjobs.yahoo.com |
| These sites offer places to find jobs, post your resume, and check out advice and resources for finding a job. |
| U.S. Office of Personnel Management    www.usajobs.opm.gov |
| This site offers a place to find jobs, post your resume, and check out advice and resources for finding a job with the federal government. Click on Search Jobs. Enter Cook Supervisor, Chef, Cook. |

## Table 16-3    Websites for Culinary Jobs

| | |
| --- | --- |
| American Culinary Federation | www.acfchefs.org |
| Chef2Chef Culinary Portal | www.chef2chef.net |
| Chef Jobs | www.chefjobs.com |
| Chef Jobs Network | www.chefjobsnetwork.com |
| Foodservice.com | www.foodservice.com |
| National Restaurant Association | www.restaura |
| Star Chefs | www.starchefs. |
| Restaurant Report | www.restaurantr |

and employer websites contain much useful information, including job postings.

Yet another advantage of the Internet is that websites dedicated to helping people network have been growing. For example, Ryze.com and Linkedin.com help you build a professional network with fellow members. Networking sites not only help you contact people you wouldn't ordinarily meet but also many members are hiring managers from national companies who are looking for a better way to meet qualified candidates than posting a job and receiving thousands of emails.

Although the Internet may sound like a jobseeker's dream, it has its downside. Because so many job hunters are using the Internet, you have a lot more competition when you respond to a job posting. Also, job postings are sometimes out of date, incomplete, or simply too plentiful to search through without a big expenditure of time. Using the Internet is also impersonal. It may be high-tech, but it's also low-touch.

## OVERCOMING THE SHORTCOMINGS OF THE INTERNET

So how do you overcome the shortcomings of the Internet? First, be sure to use other job search methods, especially networking (both in person and online). Second, show initiative and persistence by taking a couple of steps beyond just emailing your resume to an employer. After you email your resume to a potential employer, mail a paper copy with a cover letter to the person that same day. Then, within a week, if you have not been contacted, call the person to check if your resume arrived and if you can set up an interview. By this time, you should have done some homework on the company so you can sound interested and knowledgeable on the telephone. Don't forget that job candidates stand out when they make the extra effort to contact and establish a relationship with the hiring manager. Although you may think you are being a nuisance, you're not. You are being graded by how much you show initiative and persistence.

## ELECTRONIC RESUMES

You need two versions of your resume when using the Internet. Both versions contain the same words; they just use different formatting. One version is your original word-processed version. Employers sometimes ask that you attach your word-processed resume to your email. The second version is referred to as an ASCII resume or a plain text resume. If you're clueless about ASCII, don't worry. ASCII is just a form of computer file that is easily understood by many kinds of computer programs. In short, it's a stripped-down version of your resume file that doesn't contain special formatting or symbols. It can be used to paste your resume into specified fields on job board resume builders or online job applications. It can also be pasted directly into an email message. To create an ASCII resume, follow these four steps.

1. **Reset your margins.** Each email software program has its own length of lines that is acceptable. To make your resume easy to read on any software program, your resume should have no more than 65 to 70 characters and spaces per line. Set your page margins at 1 inch for the left margin and 1.75 inches for the right margin.

2. **Convert your resume to an ASCII file.** Word-processing software, like Microsoft Word, can easily create an ASCII file. Just click Save As (using a different file name) and select Text Only with Line Breaks. If you are an XP user, select Plain Text, and when the File Conversion window appears, click Insert Line Breaks under Options. Then click OK. Your file should now have a.txt extension (the part of the file name after the period).

3. **Clean up your resume.** Make the following changes to your resume:

   - Change bullets to asterisks or hyphens (dashes).
   - Get rid of columns or tables. (This might be easier to do before step 2.)
   - Delete any references to "Page 2" and multiple appearances of your name.
   - For emphasis, use all capital letters rather than boldface or italics, but do so sparingly. Words with all caps are harder to read.
   - Use the space bar, not the tab key, to space text.
   - Rearrange text as needed. Review every line for extra spaces, words in wrong places, and so on.
   - Space between sections to make them stand out.
   - Check for misspellings and grammatical errors. Misspelled keywords will be skipped over by the scanner.

4. **Check it out.** Once you're finished with the conversion and cleanup of your resume, open it using WordPad, Windows' simple word-processing program. Print your resume and then review it. If you've done your job right, it should look good. Also, paste it into an email to yourself or a friend so you can check it out one more time.

## RESPONDING TO JOB POSTINGS

Now that you have a plain text resume to use on the Internet, here are tips for responding to job postings online.

- Put the job code, job title, or number in the subject line of your email.
- Address your cover letter directly to the recruiter.
- Unless you are asked to attach your resume, put your ASCII-formatted resume in the body of the email message. This will get your information in front of the recruiter's eyes. When you attach a resume, it takes more time for the recruiter to get to; also, the file may be rejected by email systems due to virus concerns. However, you can also attach your resume, giving the recruiter a choice.

Some job sites sort resumes by date of submission, with the most recent resumes up front. Renew your resume every two to three weeks to keep it fresh.

## Following Up on Contacts

Job candidates stand out when they make the extra effort to contact and establish a relationship with the hiring manager. Although you may think you are being a nuisance, you're not. This is how you can make an impression *and* stand out from the other candidates.

## Filling Out Applications

Many jobs require applicants to complete an application instead of, or sometimes in addition to, submitting a resume. Application forms make it easier for employers to evaluate and compare a group of applicants because the forms ask the same questions. In many cases, it is harder to compare resumes. If an employer asks you to fill out an application, do so graciously. Don't bother offering your resume in place of the application. If an employer uses application forms, you must fill one out to be considered.

When given an application form, read it over completely before you begin. Use your resume to help you fill in the necessary information. Write neatly in black or blue ink. Answer every question on the application. Write "none" or "not applicable" if a question does not apply to you.

Although applications do not offer the same flexibility as a resume, you can still find ways to highlight your best qualifications. For example, you can use strong action verbs to describe your job duties and accomplishments. If you do not have paid experience, you can list volunteer job titles.

Applications often ask for your salary history, and your application may be considered incomplete without it. If you are unsure of the exact numbers, write in an approximation, Usually, approximations are acceptable.

If possible, make a copy of your completed application. If you go back for an interview, take the copy with you.

## Writing Cover Letters

Every resume you send, fax, or email must have its own cover letter. Sending a resume without a cover letter is like starting an interview without shaking hands. The purpose of the cover letter is not simply to say what job you want and repeat what is in your resume. The best cover letter:

- conveys your enthusiasm and energy.
- sparks the employer's interest.
- creates an impression of competence.
- positions you above the competition.

Ultimately, you want your cover letter and resume to generate enough excitement to get you called in for an interview.

So how do you write a great cover letter? It's not hard. Just check out these tips:

- Every cover letter should have a professional appearance. Use a block or modified block format that fits on one page, looks neat, and contains no errors. In the block format (see Figure 16-1), all text starts at the left-hand margin, except if you want to put your name and contact information centered at the top, as on your resume. In the modified block format (see Figure 16-2), indent the first line of each paragraph five spaces and place the date, "Sincerely," and your name in the middle of the page. Use the same stationery your resume is printed on for your cover letters.

    Whenever possible, send your letter to a specific person rather than to an office. Consider how differently you respond to a letter addressed to "Occupant" than one addressed to you. If you do not know whom to address, call the employer and ask who is hiring for the position. Check that the name you use is spelled correctly and the title is accurate. Pay close attention to the correct use of Mr. or Ms. Use a colon after the name in the salutation, not a comma, as follows.

    Dear Mr. Smith:

- If you are responding to a want ad, you can skip the salutation line and go right to the opening paragraph. If you absolutely can't get a name, use "Dear Sir/Madam:" as your salutation.
- The first paragraph, called the opening paragraph, should tell the employer which job you are applying for and the connection you have to the company. If someone the employer knows suggested you apply, mention that recommendation. If you are responding to an advertisement, refer to it and the source that published it. You can also put a position reference line between the address and the salutation, as in Figure 16-2.

    Your knowledge of the company might give you another opportunity to connect yourself to the job. You could briefly cite a recent success or refer to its excellent reputation for catered events, for example. You might also want to state why you would like to work for the employer. Don't go overboard; save the specifics for the interview.

- In the next paragraph, the main paragraph, you highlight your knowledge, skills, abilities, accomplishments, and successes that relate directly to the position for which you are applying. The idea is that the cover letter should *complement* your resume, not just repeat it. One way to do this is to summarize your most relevant credentials using a bulleted format. Leave no doubt in the reader's mind that you can contribute to the success of the operation.
- In your final or closing paragraph, thank the reader for his or her time, request an interview, and repeat your home phone number. The closing is your chance to show commitment to the job.

Figures 16-1 and 16-2 show sample cover letters.

# Heather Plumb, C.W.P.C.
211 West Greenwich Avenue
Greenwich, CT 07041
203-437-9365 (h) 203-530-8821 (c)
brewchef@yahoo.com

May 6, 2005

Mr. Ted Carlisle, CEC
Executive Chef
Wamtuxet Inn
10 Shore Drive
Madison, CT 08483

Dear Mr. Carlisle:

As a Certified Working Pastry Chef, I am looking for a position in a larger operation. Marybeth Gilmore gave me your name because she said you plan to expand your bakery operation soon. I have read many fine reviews of the food at the Wamtuxet Inn and feel I could contribute to its fine reputation.

My ten years of employment in the pastry field show increasing responsibility, dedication, and solid accomplishments, such as the following.

- Developed and executed a new menu for 100-seat bakery operation.
- Increased sales 30% by selling bakery items to area restaurants.
- Reduced employee turnover from 75% to less than 25%.
- Developed new quality-control standards.

Thank you for your time reviewing my enclosed resume, which can only briefly highlight my qualifications. I look forward to an opportunity to meet with you to discuss how my interests and qualifications can best meet your needs. I will call next week to schedule a convenient time for an interview. In the meantime, please feel free to call me at 201-437-9365.

Sincerely,


Heather Plumb, CWPC
Enclosure

**Figure 16-1**
Cover Letter Using Block
Format with Centered
Name and Address

E-mail cover letters are much briefer than typed letters to ensure ease of readability. Addresses are not needed, but be sure to use a salutation. The position you are applying for should be typed into the subject line. As in Figure 16-3, include a bulleted list of reasons you should be considered for the job, along with

211 West Greenwich Avenue
Greenwich, CT07041
203-437-9365 (h) 203-530-8821 (c)
brewchef@yahoo.com

May 6, 2005

Mr. Ted Carlisle, CEC
Executive Chef
Wamtuxet Inn
10 Shore Drive
Madison, CT 08483

Re: Pastry Chef Position

Dear Mr. Carlisle:

As a Certified Working Pastry Chef, I am looking for a position in a larger operation. Marybeth Gilmore gave me your name because she said you are looking for a new Pastry Chef for your expanded operation. I have read many fine reviews of the food at the Wamtuxet Inn and feel I could contribute to its fine reputation.

My ten years of employment in the pastry field show increasing responsibility, dedication, and solid accomplishments, such as the following.

- Developed and executed a new menu for 100-seat bakery operation.
- Increased sales 30% by selling bakery items to area restaurants.
- Reduced employee turnover from 75% to less than 25%.
- Developed new quality-control standards.

Thank you for your time reviewing my enclosed resume, which can only briefly highlight my qualifications. I look forward to an opportunity to meet with you to discuss how my interests and qualifications can best meet your needs. I will call next week to schedule a convenient time for an interview. In the meantime, please feel free to call me at 201-437-9365.

Sincerely,

Heather Plumb, CWPC

Enclosure

**Figure 16-2**
Cover Letter Using
Modified Block Format

a request for an interview and any other required information. Unless you have specific instructions on how to email your resume, paste it below the cover letter in text format and also attach it as a Word document. Microsoft Word is the industry's standard word-processing program.

---

Subject: Pastry Chef Position

Dear Mr. Carlisle:

My strong qualifications for the available pastry chef position and a referral from Marybeth Gilmore have prompted me to contact you. In addition to being a Certified Working Pastry Chef, I have over ten years of experience in the pastry field and have:

- Developed and executed a new menu for 100-seat bakery operation
- Increased sales 30% by selling bakery items to area restaurants.
- Reduced employee turnover from 75% to less than 25%.
- Developed new quality-control standards.

I would like to meet with you to discuss how I could contribute to the fine reputation of the Wamtuxet Inn. Thank you.

Heather Plumb, CWPC
********************************

My resume is pasted below in text format, and I have attached a Word copy if you prefer to download it.

---

**Figure 16-3** Email Cover Letter

## EXERCISES

1. Write up at least 12 cards on networking contacts. Rate how well you know each contact as well as how much each can help in your job search.

2. Write a short script to use when calling someone about a job. Be sure to mention the contact who referred you.

3. Find out what services your college placement office provides.

4. Find out if there are any job fairs in your local area over the next eight weeks. Use the Internet, classified advertisements, and other sources.

5. Attend a job fair and speak to at least two recruiters. What is different about talking to a recruiter compared to having a formal interview?

6. Attend a meeting or conference of a professional organization such as the local chapter of the American Culinary Federation and make at least two networking contacts.

7. Generate a plain text (ASCII) resume and bring it to class to compare with another student's.

8. Find an advertisement for a job that interests you. Write a cover letter to the employer. Bring it to class and discuss it with another student.

9. Write a cover letter to be emailed to respond to the job advertisement in exercise 8.

# CHAPTER
## 17

# Three-Step Interviewing

## Introduction

**LANDING AN INTERVIEW IS IMPORTANT.** It's like getting your foot in the door. This chapter will help you ace the interview process, get a job offer, and negotiate the details of the job offer. Interviewing is more than going in and answering somebody's questions in an intelligent manner. In an interview, you are selling yourself —your skills, abilities, accomplishments, personality, and more. The length of this chapter alone shows that there's a *lot* more to interviewing, so start reading! You don't want the interview door to slam shut and leave you without a job offer.

So what is that interviewer thinking about? He or she is looking to see if:

♦ You are qualified to do the job (experience, knowledge, skills, and abilities).
♦ You would fit in with the company and the people with whom you would be working.
♦ You are hardworking, persistent, and passionate about your career.

An interviewer is always looking for a candidate who is a good fit with the job, the supervisor, and the company. Much like every restaurant has its own ambiance, each company has its own atmosphere, and you may very well prefer working for one company over another simply because it matches your style better. An interviewer is also always looking for candidates who talk about what they can do for the employer, not what the employer can do for them.

Most interviews are either with a gatekeeper or the person with the authority to hire you (often the person who will be your boss if you get the job). Gatekeepers are people in human resources departments, employment agencies, or executive search firms who interview you to determine if you should go on to the next interviewing step. They do not determine if you get the job but rather if you should stay in the running. Gatekeeper interviews are also called screening interviews. If you interview satisfactorily with the human resources representative of a managed services company, for example, you will be invited to meet directly with the supervisor of the operation where you would be working. Similarly, if an employment agency or executive search firm interviewer thinks you are a good candidate, he or she will send you to interview directly with the client—your potential employer. In many cases, especially with smaller employers, you interview with the hiring manager from the start. This type of interview is called a selection interview.

Screening interviews may be done by phone. These interviews are becoming more popular because they avoid the time and money expenses associated with face-to-face interviews. If you get through the phone interview, you will be invited for a face-to-face selection interview.

Even without a screening interview, you may be interviewed several times before a hiring decision is made. With each successive interview, you can expect more technical questions and a closer consideration of how you will fit in. For instance, in your first interview you meet the person who will be your supervisor if you are hired. The interview reveals that you are qualified to do the job. Next, you are invited back to be interviewed by other people on the team. The emphasis in the second interview is not so much on screening you out as on how you will fit in. In addition, the interview covers how you will contribute to the company and be a valuable employee.

Beyond the screening and selection interviews is the confirmation interview. In this third type of interview, the person who will be your supervisor introduces you to his or her boss, usually as a matter of courtesy for approval. In most cases, the superior approves the selection and the supervisor makes the offer. During the confirmation interview, it is important to establish a good rapport with the superior. The issue is not whether or not you are qualified for the job; the superior wants firsthand assurance that you are a great choice. This is not the time to sell yourself too hard. Just be likable and emphasize how you are productive and can meet goals.

Interviews involve three steps. The first step includes everything you need to do *before* the interview, such as learning about the employer and deciding what to wear. The second step is the interview itself. The final step is what you do after the interview, including sending a thank-you note, evaluating your presentation, and following up.

## Before the Interview

### 1. LEARN ABOUT THE PROSPECTIVE EMPLOYER

Knowing about the employer before you go in for an interview has many positive benefits.

◆ It increases your confidence.

◆ The interviewer looks more favorably on candidates who took the time to research the employer than on candidates who didn't. You appear more knowledgeable, serious, and committed.

◆ It will be easier for you to initiate and follow a conversation about the employer.

◆ It will be easier for you to determine how your knowledge, skills, and abilities can benefit the employer.

The Internet is the best place to begin researching prospective employers. Many employers have websites that are full of information—although they usually just share the good stuff! Look for sections on websites entitled About Us, Pressroom, or News Releases. These sections offer background information and the company's most recent news. Try to find the information noted in Figure 17-1, Interview Form (on page 336). Here are additional resources:

◆ Publicly held companies publish annual reports that tell you just about everything you need to know about a company before an interview. To get a copy of the most recent one, check the employer's website, call the shareholder relations department, or ask a stockbroker. The president's message in the annual report may say a lot about the mission and future of a company.

◆ Other resources for business and financial information on employers include the following.
  ◆ www.ceoexpress.com. This site has links to lots of newspapers and business periodicals. It also has several search engines.
  ◆ www.hoovers.com. This site has business information on most American companies.
  ◆ www.sec.com. This is the site of the Securities and Exchange Commission and has financial information on all public companies.

◆ Your college placement office may have information about the employer. This resource normally maintains files on employers who visit the campus to conduct interviews.

◆ People who work for the employer, or used to, may be good resources. Some of your college teachers may be able to give you the name of such a person. If the person says negative things about the employer, keep in mind that you might have a totally different experience.

◆ Current and past issues of industry periodicals such as *Nation's Restaurant News* may be helpful. You can search the contents of the following industry periodicals at www.findarticles.com: *Food Management, Hotels, Hotel and Motel Management, Nation's Restaurant News,* and *Restaurant Hospitality.*

◆ The employee who schedules your interview may be able to mail you descriptive literature such as company brochures, an annual report, and employee newsletters.

## 2. PREPARE YOUR QUESTIONS

At some point during the interview, usually toward the end, you will have an opportunity to ask your own questions. This is your chance to find out more about the employer, the job, and who you would be working with. After all, you may have to decide if you want to work there.

It is important to ask questions during your interview. By asking good questions, you show the interviewer you are interested, smart, and confident. Your questions enable the interviewer to see a little more of who you are as well as establish a rapport. Even if the interviewer answers all of your questions in the course of the interview, you should ask at least one when he or she turns the interview over to you. Table 17-1 lists many questions you might ask during your interview.

An interview is not the time or place to inquire about salary or benefits. You don't want to seem more interested in financial rewards than in contributing to the company. If asked about salary requirements, try to convey flexibility. The best time to discuss salary and benefits is after you are offered the job. At that point, you are no longer the seller; you are the buyer, and you have more leverage.

### Table 17-1 Great Questions to Ask the Interviewer

What do you think is the most important contribution the company wants from its employees?

What is the company's mission? (If you found the mission statement on their website, ask the interviewer to discuss it.)

What are the goals of the company for the next five years?

How would you characterize the company's culture? What are its values?

Do you have a job description for this position I can look at?

Why is the position being filled?

What would be my day-to-day responsibilities?

Do you have an organizational chart I can look at? How is the kitchen organized?

What specific skills and abilities are you looking for?

How does this position contribute to company goals?

If I am hired, what will be my first assignment?

What are this job's biggest challenges?

What do you want the person who gets this job to achieve?

What is the budget for this area?

What is my spending authority?

Which committees would I take part in?

How would I be evaluated in this position, and how often?

How will my management and leadership performances be measured? By whom?

Can this job, if done well, lead to other positions in the company? Which ones?

Can you describe the work environment?

What type of employee works here?

What kind of employee is successful here?

How empowered are employees?

What criteria determine who gets this job?

What do you like most about working for this company?

How would you describe your style of management?

How does the company support personal and professional growth?

What training opportunities are available?

Do you have any concerns about my skills, abilities, education, or experience?

Do you need anything else from me to have a complete picture of my qualifications and suitability for this job?

What is the next step in the interview process?

## 3. PREPARE AND REHEARSE YOUR RESPONSES

Another important step in preparing for an interview is to anticipate the questions you will be asked and how you will respond. Pages 328 to 332 show typical questions and responses.

## 4. CHOOSE WHAT TO WEAR

Dress is not just about receiving respect but also about conveying it. Your appearance at an interview reflects your personal presence in the context of a work culture, and it says a great deal about your work. Remember that the very first contact you have with people is visual.

Make that first impression a good one by taking the right steps to be dressed appropriately. For a cook's position, it is appropriate to wear casual business attire, as described here. For a Sous Chef or higher position, wear professional business dress, as also described below. Do not go to an interview in the Chef's clothes you wore to work that morning. If you are going to take a cooking test as part of the interview and therefore plan to wear your Chef's outfit, make sure everything is perfectly clean and pressed and that your shoes are polished. Here are guidelines for professional and casual business attire:

| Professional Business Dress: Men | Professional Business Dress: Women | Casual Business Dress: Men | Casual Business Dress: Women |
|---|---|---|---|
| Suit (navy blue, gray, black) | Suit (navy blue, gray, black) | Dress pants | Dark dress pants |
| Dress shirt | OR conservative dress | Dress shirt | OR dark skirt with blouse or sweater |
| Conservative tie | OR dark skirt with blouse or sweater | (jacket and tie not required but highly recommended) | Dress shoes |
| Dress shoes | Stockings | Dress shoes | Stockings (with skirt) |
| Dark socks to match shoes | Dress shoes | Dark socks to match shoes | |
| Matching belt and shoes | | Matching belt and shoes | |
| NO | NO | NO | NO |
| Loud ties | Miniskirts | Jeans | Jeans |
| White socks | Very high heels | Shorts | Shorts |
| Boots | Sandals | Boots | Miniskirts |
| | Low-cut clothing | T-shirts or polo shirts | Sandals |
| | | Loud ties | Low-cut clothing |
| | | White socks | |

The objective is to look reliable, not trendy. Don't wear clothes or accessories that draw attention *away* from you.

Avoid wearing lots of makeup, jewelry, perfume, or cologne, which can be distracting to the interviewer. Make sure your shoes are clean and polished, and check your personal hygiene—hair, fingernails, and so on.

Lastly, avoid last-minute clothing disasters by trying on your interviewing outfit a few days before the interview. Make sure it fits well, looks neat, and is clean and pressed. Also, plan for the unexpected. If you will be wearing stockings, make sure you have at least two pairs. If your shoes have shoelaces, get a spare pair in case they break.

## 5. CHOOSE WHAT TO BRING

You must take *some* things with you to the interview, but be sure to pack light. You don't need to lug a huge briefcase stuffed with lots of papers that have nothing to do with your interview. In many cases, a simple writing pad portfolio with a pocket for copies of your resume, references, and Interview Form, plus your calendar (paper or electronic) is enough. If you have a job portfolio in a loose-leaf binder, that's fine too. You must be able to immediately locate papers you want to share with the interviewer and refer to your list of questions. Have your calendar available in case you are asked to schedule another interview. If your cooking skills are going to be tested, bring your own set of knives to be most comfortable. If you carry these things in a slim, professional-looking briefcase, that's fine. Just make sure you have everything ready at least one to two days before the interview, and bring extra copies of your resume and references to hand out.

## 6. CALM DOWN ALREADY!

Most people are nervous when interviewing—but remember, you were asked to interview for the job because the employer believes you could be right for it. The interview is your chance to confirm that belief and establish rapport.

Also keep in mind that the interviewer is a little nervous too—nervous about selecting the wrong person! Employers often can't obtain a lot of feedback from your past employers, so they rely a lot on the interviewing process, which we all know doesn't always identify the best-qualified person.

To reduce nervousness, get a good night's sleep and maintain your usual routine. You might also call to mind some of your happiest memories or proudest moments before arriving for the interview. These relaxation techniques can also help.

◆ Take a deep, slow breath. Let the air come in through your nose and move deep into your lower stomach. Then breathe out through your mouth.

Repeat this for several minutes. Imagine that the air coming in carries peace and calm and that the air going out contains your tension.

◆ Slowly clench your fists. While keeping them clenched, pull your forearms tightly up against your upper arms. While keeping those muscles tense, tense all of the muscles in your legs. While keeping all those tense, clench your jaws and shut your eyes fairly tight. Now, while holding all your muscles tense, take a deep breath and hold it for five seconds. Then, let everything go all at once. Feel yourself letting go of your tensions.

A little bit of nervousness is okay. It will help you think clearly and concentrate.

# During the Interview

Now it is finally showtime! Because the interview is the first meeting between you and your prospective employer—and a relatively brief meeting, at that—your interviewer will have to base most decisions about you on first impressions. The manner in which you introduce yourself, your personal appearance, whether you maintain eye-to-eye contact with the interviewer throughout the conversation, the completeness and honesty of your answers to questions, whether you are on time—these factors will combine to form the interviewer's appraisal of you, both as a person and as a prospective employee.

## GETTING TO THE INTERVIEW

On the day of the interview, give yourself plenty of time to get ready and travel to the location. Plan to arrive 10 to 15 minutes early, even after allowing yourself extra driving time for traffic jams, roadwork, and other hazards. Consider taking a test drive or testing your public transportation route beforehand. You will be a lot more confident on the day of your interview if you know exactly where you are going.

Once you get there, find a restroom to check your appearance. Make sure to remove sunglasses, portable stereo, and chewing gum. Use a breath mint if needed. Then check in about five minutes early with the appropriate person. It's important to make a good impression from the moment you enter the reception area. Greet the receptionist cordially and try to appear confident. You never know what influence the receptionist has with your interviewer. If you are asked to fill out an application while you're waiting, be sure to do so completely. If you are instructed to sit down for a few minutes, look over your notes, read through company literature, and go over the major points you want to make in the interview. Keep smiling and be friendly!

## Table 17-2 General Interviewing Guidelines

- Turn off your cell phone or pager (or put it in silent mode) before you go into the interview.
- Don't chew gum or candy.
- Maintain good eye contact with the interviewer, especially when he or she is talking. This shows interest and self-confidence. Good eye contact does not mean staring; look away periodically.
- Use good body posture. Stand straight and sit correctly. Do not slouch.
- Show you are open and receptive by keeping your legs uncrossed. Don't cross your arms while you sit; it comes across as being defensive.
- Smile naturally at appropriate times.
- Check that you are not tapping your foot, running your hands through your hair, pulling on your jewelry, or using any other distracting mannerisms that show nervousness.
- Speak clearly and firmly. Don't talk too softly or too fast.
- Answer each question completely and directly. Be concise—most questions can be answered in 30 seconds to two or three minutes.
- Be specific when you answer questions. A good interviewer won't let you get away with being vague. Use specific examples to illustrate your points.
- It's okay to shed some modesty and brag a little about your accomplishments. Just don't overdo it, and don't even think about being arrogant!
- Talk about what you can do for the employer, not what the employer can do for you.
- Be honest.
- Never speak negatively about a former or current employer. It serves none of your purposes and will lower the interviewer's estimation of you. Put yourself in the place of the interviewer. If you speak poorly about one employer, what is to prevent you from speaking poorly about this employer if you get this job?
- Project enthusiasm about the prospect of working in this position.
- Be a good listener. Do not interrupt the interviewer. Instead of anticipating the interviewer's next question, concentrate on each question as it is being asked. Being a good listener is an excellent way to build a rapport with the interviewer.
- Pauses are a normal part of the interview process. It's okay to take a moment to put your thoughts together before answering tough questions.
- If you don't know the answer to a question, it's okay to say so. If you have never run a profit-and-loss account, don't fake it.
- Maintain your self-confidence throughout the interview.
- Let your personality come through.

*General guidelines of what to do and what not to do during interviews are listed in Tables 17-2 and 17-3. Read them over now, before the interview starts!*

## Table 17-3 What Not to Do During an Interview

- Be late.
- Dress informally.
- Have poor personal hygiene.
- Have bad breath.
- Tap your feet or fingers or click your pen.
- Let the interviewer pose all the questions.
- Be sarcastic.
- Be overbearing.
- Make negative statements about past supervisors or employers.
- Know it all.
- Be overassertive.
- Interrupt the interview constantly.
- Express yourself in an unclear manner.
- Overuse phrases or words such as "I guess," "yeah," and "like."
- Ask too many questions about salary and benefits.
- Ask any of the following questions.
    - Why do I have to fill out this job application when the information is on my resume?
    - Do I get compensation time for hours worked beyond 40 hours a week?
    - Can you tell me about the retirement plan?
    - Can I tape this interview?
    - I missed my lunch. Do you mind if I eat my sandwich while we talk?
    - Will this take long? My girlfriend is waiting for me outside.
    - When will I be eligible for my first vacation?
    - Is it possible to telecommute at all with this job?
    - Would I get an office or a cubicle?
    - Are you single?

## INTRODUCTORY PHASE

Make a favorable impression at once by smiling and greeting the interviewer by title (Mr. or Ms.) and name, then introducing yourself in a professional, self-confident manner. Never use the interviewer's first name unless you are invited to do so. Make eye contact and stand up straight. Be ready to shake hands if the interviewer extends a hand. Be sure your handshake is firm, but not firm enough to bruise. Sit down in the seat the interviewer indicates. Sit deep and comfortably in your seat. Take a deep breath. Try not to sit on the edge of the chair and look nervous.

The person conducting the interview will begin to form an opinion of you based on such things as the firmness of your handshake, the clearness of your voice, and whether you walk with purpose or shuffle along, so pay attention to what you are doing. This is not the time to try out the latest slang expressions or to move in low gear.

## THE HEART OF THE INTERVIEW

After introductions, the interviewer will probably discuss the company and describe the job. He or she will then ask questions meant to gauge how well you would fill the position. Many employers use resumes as guides, asking for additional details during the interview. In addition to finding out more information, they are observing how well you communicate and interact.

Some jobseekers are so focused on specific answers that they forget to relax and connect with the interviewer. An interview should be conversational, with the normal exchanges and pauses. It's okay to pause—for example, to stop and consider an answer to a difficult or unexpected question.

Certain questions will show up in many of your interviews, so it's a good idea to think about them ahead of time and plan how you will respond. Most of your responses will take from 30 seconds to two to three minutes. Your answers should be concise, but don't be afraid to adequately describe your skills, abilities, and accomplishments. Interviewers want to hear examples of how you use your knowledge, skills, and abilities. Their attitude can be summed up in two words: Show me. This is where your portfolio comes in. Use it to your advantage.

Interviewers recommend rehearsing your answers in front of a mirror or with a friend to gain confidence and poise. You may even want to videotape a mock interview to see how you really look and act. The goal is to become comfortable speaking with an interviewer about your education, experience, skills, abilities, achievement, and goals. Whatever you do, do not memorize your responses. The worst thing is to come across as if you are reading from a script.

### *"Tell me about yourself."*

This is a huge question. Don't make the mistake of giving a huge answer. What you want to do here is sum up your education and experience, then end with a statement about "how my background leads me to your company today to interview for this position." You might even start out with a mention of when you knew you were interested in the culinary field. Here's an example.

*I have been interested in working as a Chef since I worked summers on the New Jersey shore in my uncle's seafood restaurant. I worked in every position in that restaurant but loved being a line cook the most. Once I graduated from high school, I went straight to Middlesex County College to get my culinary degree. While completing my degree, I worked as a line cook at the Auberge, a fine dining restaurant. I learned a lot about à la carte dining, station setup, and mise en place. About five years ago, I got my degree and took a job with the Marriot Hotel in Princeton, where I am today. I've worked all the stations in the kitchen there, including working with banquets and catering, and was promoted to Sous Chef two years ago. I am also an ACF-certified Sous Chef. This is the background that leads me to this interview today.*

### "What are your strongest points?"

This question is a gift, so use it wisely. Think about your knowledge, skills, abilities, experience, personality, motivation, and so on. Mention four or five strengths, and give a specific, brief example to illustrate each. For example:

*I work well under pressure. For instance, last week the water main outside our building broke, and we had no water. Using our emergency plans, we were still able to feed all our guests satisfactorily until the water was turned back on that night.*

### "What are your major weaknesses?"

You can take two approaches to this classic question. First, you can mention something that is actually a strength, such as:

- *I'm something of a perfectionist.*
- *I'm a stickler for punctuality.*
- *I'm tenacious.*

Second, you can mention a weakness you can easily overcome, such as:

- *I need more computer training.*
- *I need to learn more about nutritious cooking methods. I've signed up for an online course about it.*
- *I need more experience doing public speaking.*

### "What do you hope to be doing five years from now?"

The interviewer is not only looking for information about your ambitions but is also seeing if your expectations for advancement match what the employer can offer. It's okay to want to continue climbing the career ladder; just be reasonable about how long it takes to do so. Here is one possible answer:

*I hope I will still be working here and have increased my level of responsibility based on my performance and abilities.*

Avoid citing specific time frames. Talk about what you enjoy doing and about realistic opportunities.

### "What do you know about our company? Why do you want to work here?"

This is where your research on the company will come in handy. Describe any encounters you have had with the company and offer positive feedback you have heard from customers or employees.

*Your company is a leader in your field and taking on new accounts every day. You run many of the college foodservices in this area, and I know, just from some friends, that you're doing wonderfully in your accounts. I would like to work for, and learn from, an industry leader.*

You might try to get the interviewer to give you additional information about the company by saying you are interested in learning more about the company objectives. This will help you focus your response on relevant areas.

### "What was your greatest accomplishment in your current or last job?"

Give a specific illustration from your previous or current job where the accomplishment was totally your doing and had a positive impact. If you have just graduated from college, try to find some accomplishment from your school-work, part-time jobs, or extracurricular activities. Don't exaggerate your achievements, and be sure to mention if others helped.

*When I started my current job, there was no catering menu, so we reinvented the wheel every time we did a catering affair. So I developed a standard catering menu, which our customers have enjoyed using. I have also gotten positive feedback from the cooks on it. It makes their jobs a little bit easier and more predictable.*

Use the technique shown in this example when you are asked to describe accomplishments: describe the problem, then the action you took, and the results of your action.

### "Why should we hire you?"

Cast your background in light of the company's current needs. Give compelling examples. If you don't have much experience, talk about how your education and training prepared you for the job.

*From our discussion, I think you would agree with me that I have the qualifications and experience to contribute to your company. I am also excited about this position and feel I would fit in well. I am sure I can expand your clientele as I did at my last job.*

### "Why do you want to leave your current job?"

This is not the time to mention that you can't stand your boss—although that may be true. It is generally expected that if you are looking for a new job you are looking for more money, a bigger challenge, a better shot at advancement, or simply a new environment. Make sure you point out why this job will provide you those things. Never complain, gossip, or whine about a current or past boss as this is not professional behavior.

*I want to develop my potential. I have never worked in a hotel foodservice and would like to get some experience doing catering and banquets. Also, this operation is a lot bigger than the one I left.*

### "Tell me about a problem you had in your last job and how you resolved it."

The employer wants to assess your analytical skills and see if you are a team player. Focus on the solution. Select a problem from your last job and explain how you solved it.

### "Describe a time you failed."

This may not sound like a question you want to hear, but you can use it to your advantage. We have all had times when our ideas didn't work. Think of a situation when you goofed, but the mistake didn't cause major problems and you learned a valuable lesson.

*One day I forgot to tell Maintenance about problems the baker was having with the floor mixer. The baker was furious at me the next morning because the mixer was still not working right. I ran to Maintenance and luckily got someone who fixed the problem temporarily until a regular repair person came in. Now I know the value of preventive maintenance and have put all the baker's equipment on a preventive maintenance schedule with our Maintenance department. I also learned to keep a pad and pen in my pocket at all times to write things down.*

### "How would you approach this job?"

This would be much easier to answer once you are in the job for a few weeks, but you're not that lucky! The interviewer wants to get an idea of what types of actions you will take and if those actions are appropriate. Mention that you will need time to observe and survey the operation before you take action. Name a couple of ideas you might implement after learning enough to do so.

*First, I would like to get to know the people in the kitchen and observe them. From what you have said, it seems that the room service area needs immediate attention. After seeing the menu and how the food is prepared and delivered, and after talking with the employees, I am sure to come up with solutions to the time-liness problem.*

### "Describe your management style."

This question probes how you work with people. Are you a participative manager? Do you like to empower or delegate to employees? Which do you value more: people or production? In the best situation, you have an idea of how this company treats their employees and whether or not your style matches it.

*My employees would say that I am a very participative manager. I try hard to listen because they are on the front line every day taking care of our guests. You've got to take good care of your employees to keep your turnover low and guest satisfaction high.*

### *"What is your philosophy of cooking?"*

Here you need to describe the guiding principles that drive you and your cooking, including your philosophy of foods and cooking, your work ethic, management philosophy, and so on. Your cooking philosophy may be, in brief, to emphasize local, organic foods in simple meals, or to blend traditional with contemporary cooking. In a healthcare setting, your cooking philosophy may be to provide home-style, attractive meals that patients enjoy.

You may find the interviewer asking questions that are not job-related. It is inappropriate for an interviewer to ask about your age, race, religion, or marital status. What can you do if you are asked such a question? Take a moment to evaluate the situation and respond in a way that is comfortable for you. For example, if you are asked about your age, be succinct and try to move the conversation back to an examination of your skills and abilities. Or you might say, "I'm in my forties, and I have a wealth of experience that would be an asset to your company." If you are not sure you want to answer the question, ask for a clarification of how it relates to your qualifications for the job. You may decide to answer if the explanation is reasonable. If you feel there is no justification for the question, you might say that you do not see the relationship between the question and your qualifications for the job and you prefer not to answer it.

## DON'T FORGET TO ASK YOUR QUESTIONS

Make sure to ask your list of questions. Just as the interviewer is evaluating you, you need to evaluate the job and the employer.

As the interviewer answers your questions, you may want to write down key points. Be sure to ask the interviewer ahead of time for permission to take notes. Asking permission shows that you are polite and respectful. You can phrase the question this way: "Do you mind if I jot down some notes about our discussion? Taking notes helps me organize all the wonderful information I am learning about your company and this job."

## ASK FOR A TOUR OF THE OPERATION

Most interviewers will want you to see the foodservice operation. We have prepared a list of points for you to investigate when you take your tour. Not all the items in Table 17-4 are set in stone, but we feel that a "yes" answer to most or all of these points indicates a quality place of employment. Keep in mind that, contingent on the type of operation, there are many ways to deal with most of these items. We have listed them to make you think about your potential workplace and how it is managed. After all, you may be spending more waking time there than anywhere else, so the place should live up to your expectations.

### Table 17-4 What to Look at During Your Tour of the Operation

- Uniforms clean and in good condition.
- Adequate and safe locker and changing facilities.
- Kitchen clean and orderly. No standing water, burned-out lights, or accumulated grease.
- Food production areas neat and orderly.
- Garbage area in good order.
- Staff has hats.
- Plastic gloves present.
- Break area in good shape. (This may be part of the restaurant in stand-alone facilities; that is acceptable as long as the management recognizes that breaks and nourishment are part of your workday.)
- Corners, walls, and ceilings clean.
- Refrigeration temperatures correct and temperature logs present.
- Employees happy and well directed.
- Facility maintained and in good repair.
- The kitchen stocked with needed equipment and smallwares.
- Food products wrapped, dated, labeled, and stored in proper containers.
- Temperature of the kitchen reasonable and the air fresh.
- General sanitation apparently correct.
- Hoods clean.
- China stored in an organized fashion.
- Chef certified at Chef de Cuisine or higher.
- Chef recently certified in sanitation.
- Chef's shoes clean and polished.
- Cooks taste what they are preparing.

## WHEN AND HOW TO DISCUSS MONEY AND BENEFITS

The right time to discuss money depends on whether you are applying for an hourly job or a salaried job. For an hourly job, it is appropriate to bring up the topic during your initial interview. Often you will know the hourly rate before the interview.

For salaried positions such as managerial jobs, it is risky to bring up salary issues during the interviewing process unless the interviewer does. It is best not to discuss your specific compensation package, especially salary, with the employer until you are offered the job, at which point you are in a much better position to discuss and negotiate salary. Remember: *He who mentions money first loses.*

If an interviewer asks what your salary requirements are, say you have a range that depends on the whole compensation package of salary, bonus, and benefits. If pushed, have a range in mind from your minimum salary requirement to 15 to 20% above that figure. Keep in mind that employers know you are looking to make more money than your current or last job, so put your minimum salary requirement above, but not outrageously beyond, your current or most recent salary. You also need to have a handle on the going rate in your locale for the type of position you want. Sources of salary information appear in Table 17-5.

## CONCLUDE THE INTERVIEW

Be sensitive enough to tell when the interview is over and it is time to leave. The interviewer may make one of the following statements to hint that the interview is coming to a close.

- I think that pretty much covers it.
- We've covered a lot of ground today.
- I really need to wrap this up.

Instead of saying something, the interviewer might look at the clock or an appointment book, or simply start shuffling papers.

Before the interview is over, be sure to find out what the next step will be. Are you to contact the interviewer, or is the interviewer to contact you? How long will it take for the interviewer to reach a decision? Should you contact the interviewer by phone or by email? If another interview is to be scheduled, get the necessary information. It is important to find out how you are supposed to follow up and then to follow the instructions.

Be sure to make your closing statement to the interviewer a positive one. You went into the interview expecting to land this job; it is hoped that you have reason to maintain this attitude throughout the interview. Now you want to leave the interviewer with the same positive feelings about you that you have presented throughout your meeting. In your closing statement, tell the interviewer that:

- You are very interested in the position.
- You are sure you would do the job well.
- You would enjoy working for the employer.

Also, don't forget to thank the interviewer for his or her time as you say good-bye.

## Table 17-5 Sources of Salary Information

INTERNET

jobstar.org/   Jobstar.org has lots of salary review information. Click on Salary Info.

www.salary.com   Try the Salary Wizard for salary information.

www.salaryexpert.com   This website can give you local salary information for many culinary jobs.

www.bls.gov/   The Bureau of Labor Statistics has tons of salary information; just be sure to check the date.

>  On the homepage for the Bureau of Labor Statistics, look under Occupations and click on *Occupational Outlook Handbook.* Next, click on A–Z Index. This book can tell you about earnings as well as the nature of the work, working conditions, employment, advancement, and job outlook for many occupations. Look under any of these three categories in the Index.

>  B—Bakers

>  C—Chefs, cooks, and food preparation workers

>  F—Food and beverage serving and related workers; foodservice managers

Make sure you are looking at the most current edition of the *Occupational Outlook Handbook.*

stats.bls.gov/oco/cg/cgindex.htm   This site is the index to the Bureau of Labor Statistics' *Career Guide to Industries.* It contains salary information about jobs in eating and drinking places, hotels and other lodgings, and health services. Some of the jobs covered are Chefs and Head Cooks, Restaurant Cooks, and Foodservice Managers.

www.bls.gov/   The Bureau of Labor Statistics posts salary information by state and metropolitan areas. On the home page, look under Wages, Earnings, and Benefits and click on Wages by Area and Occupation. Next, under State Wage Data, click on By State. You will now see a U.S. map. Click on the state you want, then click on the occupation you want, such as Food Preparation and Serving Related Occupations. Then you can click on 35-0000, Food Preparation and Serving Related Occupations, to get state salary data. At the bottom of this page are links to salary information for metropolitan areas in that state.

TRADE ASSOCIATIONS

TRADE PUBLICATIONS

COLLEGE/UNIVERSITY CAREER SERVICES OFFICE

YOUR NETWORK

YOUR PAST EXPERIENCE

# After the Interview

## EVALUATE THE INTERVIEW

As soon as possible after the interview, use your Interview Form (Figure 17-1) to help you go over it. Make sure to write down the names of people you met, along with their titles and any thoughts you have about them. Perhaps the Director of Human Resources was bossy and curt while the person who would be your supervisor was easy to get along with. Write down additional impressions from your interview such as which questions you answered well or not so well, what was appealing and not appealing about the job, the people, the employer, or the work environment, and so on.

Next, fill in the Follow-up box with your instructions about the next step(s) in the process. Make sure to write down the appropriate names and dates; note the dates in your calendar as well.

### *CONTACT INFORMATION*

Employer Name _____

Address/Phone/Website_____

Contacts (People you know, the interviewer, people you meet during the interview)

Name/Phone #/Title _____

Name/Phone#/Title _____

Name/Phone#/Title _____

### *FAST FACTS*

Headquarters: _____ Public or Private: _____

Number of Units: _____ Location of Units: _____

Number of Employees: _____ Annual Revenue/Sales:_____

Services/Products/Areas of Expertise_____

Interesting Statistics _____

Competitors _____

Company Strengths _____

Company Challenges _____

**Figure 17-1**
Interview form

## MY INTERVIEW QUESTIONS

1. _____
2. _____
3. _____
4. _____
5. _____
6. _____
7. _____
8. _____

## INTERVIEW IMPRESSIONS

## FOLLOW-UP

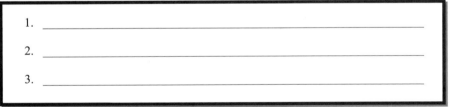

1. _____
2. _____
3. _____

## GOALS FOR NEXT INTERVIEW

1. _____

2 _____

Last, fill in the Goals for Next Interview box. Every interview is a little different, and each offers you opportunities to improve your interviewing skills. After you review the interview, set one or two goals for the next. Perhaps you have photographs of plated desserts you missed the chance to show the hiring manager. Maybe you need to work on talking at a slower pace.

## SEND A THANK-YOU NOTE

Even after the interview is over, your task is not complete. Secure a good impression by sending a thank-you letter to the interviewer. It is best to send the letter on the same day.

Thank-you letters should be brief—shorter than one page—and may be handwritten or word processed. Their purpose is to express your appreciation for the interviewer's time and to state again your interest in the job. Most thank-you letters have three main paragraphs (see Figure 17-2).

**Figure 17-2**
Sample Thank-You Letter

15 Spring Road
Hamlet, LS 41112
561-848-9487


April 15, 2005


Mr. Thomas Atkins
Executive Chef
Hilton Hotels
455 East Greenbush Avenue
Pittsburgh, PA 18944

Dear Mr. Atkins:

Thank you for the opportunity to interview with you yesterday afternoon. I am very interested in the Sous Chef position you described.

My culinary education and work experience as a Sous Chef in another local hotel have prepared me well for the open position. I am especially interested in expanding the banquet business, as we discussed. I would welcome the opportunity to contribute to that effort.

I enjoyed meeting you and your staff and look forward to hearing from you soon. If I can provide any additional information, please call me at 561-848-9487. Thank you again for your time and consideration.

Sincerely,


Peter Gates

**1.** The first paragraph is your chance to thank the interviewer again for meeting with you and to show enthusiasm for the job. Refresh the interviewer's memory by mentioning the date of the interview and the position for which you applied.

**2.** The second paragraph is for you to briefly repeat the skills that make you well suited for the job. You might also note a topic from the interview that was especially interesting to you. Include any important information you forgot to mention during the interview.

**3.** The third paragraph is where you thank the interviewer again, give your phone number, and state that you look forward to hearing from him or her.

Write or type the letter on solid white, off-white, or gray stationery. Use a standard business format. Put a colon after the interviewer's name and a space after each paragraph. And don't forget to sign your first and last name.

Many employers say an emailed thank-you letter is acceptable if email correspondence was exchanged between the interviewer and the candidate. Otherwise, an email message does not substitute for standard mail in most situations.

Be sure to proofread the letter and make sure you spell the interviewer's name correctly. If a group interviewed you, write to each person on the panel or to the person who led and coordinated the interview, mentioning the other people you met. Interviewers tell tales of misspelled, misused words written in thank-you letters that wreck the image of an otherwise impressive candidate. As you write your thank-you note, remind yourself that you might be writing to your next supervisor.

# Follow-up

Follow-up is *crucial* to your success. Job candidates stand out when they make the extra effort to reiterate their strong interest in a position and the company. Although you may think you are being a nuisance, you're not. You are being graded on initiative and persistence. This is how you can make an impression and stand out from other candidates. Contact the employer in the manner you were instructed to—phone, email, or in person. Repeat your interest in the job and ask if you were successful in obtaining a job offer.

# Employment Testing

Some employers use tests or other assessment tools as part of their screening process. This section discusses tests you may encounter.

## COOKING TESTS

Many of today's kitchens require practical cooking tests. Surprisingly for many, this can be the most intimidating part of the interviewing process. Most of the formats are truly not difficult in themselves; the pressure comes from what's riding on your performance. In most cases, the evaluators will be looking for skills relevant to the position you are seeking. For example, prep cook applicants may be tested on their ability to produce vegetable cuts, while a Sous Chef may be asked to prepare finished dishes. A common practice for the levels above cook is to ask the candidate to prepare a meal from what is called a market or mystery basket. In our opinion, the market basket is far too intimidating to the average Chef and, more important, not an effective evaluation tool. We heartily suggest that all candidates be fully informed as to the foods available or even asked to order the goods they want to use, with plenty of advance notice so candidates may plan their cooking. This scenario allows the employer to evaluate the candidate's planning and organizational skills as well as cooking skills.

Regardless of the venue or format, as cooks we all should have the same fundamental approach to any given task as well as the same basic concerns about accomplishing the work at hand. Those concerns are as follows.

### You as a Cook

Just as you wear a suit to the interview, you wear a perfectly clean, well-fitting uniform to the cook test. Your tools are spotless, your shoes shine, and your manners are impeccable. It is not foolish to buy a new uniform especially for this occasion. The uniform can reflect your personal style to some extent, but unless you really know the style of the Chef, keep it reasonably traditional. Understated style is rarely incorrect. Make sure your knives are sharp. Also, make sure to include gloves in your toolkit as well as special spices that may not be available. Tuck into your pockets some Band-Aids so you don't waste time running around looking for one if you need it.

Think about the foods you are most comfortable with, the dishes you know best and you know will work. This is not the time to experiment or try something new to show your creativity. Do now what you do best.

When you arrive, have a list of questions to ask so you are not constantly stopping and asking, "Where is this?" and "Do you have that?" Before beginning the cook's test, ask if you can tour the kitchen to familiarize yourself with the area. (As a side note, be sure not to smoke at any time while in this establishment. They can find out you smoke after you are hired.) You should set up your station, follow an organized game plan, and work quietly with a calm, controlled intensity, all the while taking time to chat a bit with the people around you. Most kitchens these days do want a cook to be personable and in good spirits.

### Sanitation

In most, if not all modern kitchens, sanitation and clean work habits are on the top of the list. You are expected to be certified in sanitation, or at the very least able to complete your tasks without violating the local sanitation code. Your workstation should be neat and clean at all times, and you should display an understanding of the importance of why we as cooks are concerned with this subject at every turn.

### Basic Skills

Every level of cooking calls for a skill set that applicants are expected to have. At each higher career level, that skill set expands. This applies to cooking methods as well as knife cuts and fundamental product processes such as roux, beurre manié, and the other starches used to give texture and visual appeal to every sauce. You should be able to demonstrate the skills commensurate with the position and type of establishment to which you are applying.

### Organization

A cook is basically a creature of habit and is therefore expected to be able to work in an organized fashion. This applies, for example, to managing small demonstration trays of cuts, the waste of various vegetables manicured for use in a three-course meal. Prepare food as if you were in your own kitchen. Separate garbage from trim that can be utilized later. For example, don't throw out chicken trimmings if you would normally use them in stock.

### Flavor Development

If the test is for a lower-level position and no finished dishes will be prepared, flavor development is not a factor. However, when dishes are to be formulated, executed, and eaten, you'd better believe they need to taste good! The unfortunate truth is that all too often the cook is so overcome with all the other aspects of the test that he or she forgets his reason for being—that is, to apply heat to edible substances to make them palatable and tasty. I would, in all cases, make this my first order of business when deciding what type of dish to make and how to prepare it. Do not forget all the wonderful herbs, spices, liquids, and the other flavor enhancers available to us. And don't be afraid to use ingredients such as salt, butter, cream, and oil in your cooking, as long as you execute to standard. When used correctly, these treasured ingredients do enhance food flavor.

### Craftsmanship

Craftsmanship reflects the level at which you aspire to be employed and should grow with your tenure as a Chef. By no means should a Chef ever be without craftsmanship. This is what the skills that reside in our hands and head do to the food: the way we cut a vegetable, the smooth removal of the flesh from a fish, the single motion of peeling a shrimp, and the slicing of a perfectly cooked roast. These processes require much practice and repetition in order to be done proficiently.

### Visual Appeal

Simplicity is the buzzword here. It is the way a cook lays a chicken breast next to a bundle of green beans, neatly drizzles a sauce on half of the breast and part of the plate, then tumbles three nicely browned potatoes into a group, and finishes the whole with a bit of caramelized mushrooms and shallots. This is the quiet skill that reflects a cook's ability to place food on a plate in an appetizing fashion. Don't sell the difficulty short; few culinarians can do this readily without a great deal of thought. Remember that everyone eats with their eyes first.

### Nutritional Understanding

Our basic responsibility is to nourish our patrons, so we need to have a prudent understanding of nutrients and how they affect the body. We need to know the good and bad of the foods we wish to serve. This is not to say we must cook in a low-calorie style or must direct our customers to what they ought to eat. However, it is our responsibility to use enhancing substances (such as salt) moderately, meaning to a level that is necessary and not obsessive. For instance, infuse more flavor into a product before you go and add salt.

### Culinary Integrity

All things in nature have many uses. The cook's challenge is to discover what each edible substance does best. At the same time, the cook must work to uncover how best to manicure, dismantle, or otherwise ready the product for the cooking process. This involves making the correct and best use of all the secondary parts, which too many of our colleagues irresponsibly refer to as garbage. Don't throw out usable by-products. As examples, you can make bread from the inside of a zucchini, braise hearts of celery, use the insides of tomatoes in sauces and then strain the seeds out, add flavor to stock with chicken trimmings, use meat trimmings for stews, and add artichoke trimmings to stock for a great soup.

### Interpersonal Skills

A kitchen is generally hot and busy, and you are going to spend a great deal of time next to your coworkers in high-stress situations. You don't get a cubicle. You need to rely on your coworkers and they on you, so wherever you are working, act as though you are there forever with your coworkers at your side. Treat each person with respect, and almost inevitably it will be returned. Even if it is not, you will never be the worse for the effort. You cannot be a good cook without the rest of the kitchen brigade.

## INTEGRITY TESTS

To help ensure they hire honest employees, employers may administer integrity tests. These usually involve two types of questions. The first type is about illegal or dishonest behaviors you may have exhibited in the past; for example, you might be asked if you have ever walked out of a restaurant without paying the bill. The second type asks about your attitudes toward dishonest behavior; for example, you could be asked about your views on punishing shoplifters. On an integrity test, you also might be asked questions about past involvement with drugs or alcohol.

## MEDICAL EXAMINATIONS

Medical exams are given to determine whether you have a physical condition that would prevent you from performing the job. It is illegal to give a preemployment physical exam or to ask about disabilities on the application. Physical exams, however, may be given after the job offer is made. The Americans with Disabilities Act (ADA) gives people with disabilities rights that prevent them from being unjustly rejected for a job. If you have a disability or medical condition that you think may pose barriers to your being hired, your state Vocational Rehabilitation Agency can offer assistance.

## DRUG TESTS

Drug tests indicate the presence of illegal drugs. An increasing number of companies use drug tests to screen candidates for all job categories, including managers and professionals. You should be aware that some medications, and even some foods, can produce a positive reading even though you have used no illegal drugs. It is important to inform the employer of any such medications you have taken recently. Be aware that drug tests may not be completely accurate. If you are told that your sample indicated drug use but you know you haven't used any illegal substances, ask if a formal appeals process is available. Tell the employer you would like to take the test again. Perhaps you can ask if there is a more sophisticated test you can take.

# Tips for Special Situations

## SCREENING INTERVIEWS WITH HUMAN RESOURCES

If your first interview is with a human resources representative, keep these few thoughts in mind. First, human resources interviewers can't offer you the job you want but they *can* reject your application, so you must be as attentive in this interview as in any other. Next, the interviewer is not likely to know everything about the job you're interested in but can tell you lots about the employer and work environment. He or she is checking your qualifications against a list, so make sure you present yourself well. The human resources interviewer also may ask questions about inconsistencies on your resume, such as work history gaps, and questions to reveal what kind of person you are. Last, be sure to make the human resources interviewer your ally.

Here are additional tips for interviewing with human resources representatives.

◆ Be very interested in the employer and the job.
◆ Be upbeat and positive.
◆ Be confident—but not overconfident.
◆ Treat the interviewer with respect.
◆ Listen carefully to what the interviewer says.
◆ Tell the interviewer why you are qualified to do the job. Give an example or two to back up the information in your resume.
◆ Keep your answers straightforward.
◆ Ask the interviewer about his or her experiences working for this employer and what attracted him or her to this employer.
◆ Gather information about the position and who you would be working with.

Keep in mind that human resources interviewers are often the most skilled interviewers you will meet. They are good at finding out if you should go on to the next interview.

## TELEPHONE INTERVIEWS

As mentioned, the telephone is often used for screening interviews to trim time and travel expenses. The procedure for this type of interview is the same as for face-to-face interviews: You need to prepare, and you need to follow up. Although a phone interview sounds informal, don't be caught off guard. You have to succeed in this interview to get to the next step, and you won't have body language to help you interpret how the interview is going. Your only clue is from the voice on the telephone.

When you are scheduling this interview, ask the employer for the names of the interviewers. More than one person may be involved. Make sure you know their names in advance; this makes it easier to identify who is talking during the interview.

Here are tips for telephone interviews:

◆ Use a phone in a quiet location where the interviewer won't hear distracting background noises. Avoid using a cell phone because the sound is better on land lines and you won't have to worry about the battery dying!
◆ Imagine that the other person is in the room and talk directly to him or her.
◆ Your tone says a lot, so try to sound animated and enthusiastic. It helps if you smile, too!

- Have a copy of your resume on hand to help you answer questions and give you confidence.
- Have paper and pen handy.
- Take the opportunity at the close of the interview to persuade the interviewer that you are the ideal candidate for the position.
- Don't rush—it's not your phone bill!
- Some people find it helpful to wear interview clothes so they can focus and project better.
- At the end, say thank you. Don't forget to send a thank-you note!

## PANEL INTERVIEWS

You may be asked to interview with a panel of two or more people who each have an interest in who fills the position. For example, you may be interviewed by the person who will be your supervisor, his or her supervisor, a human resources representative, and one or two potential coworkers. Panels are commonly used to interview candidates for teaching positions. The so-called search committee comprises mostly teachers who interview the candidates together.

Here are tips for handling panel interviews:

- Greet each person by name. Once you sit down, try to make a quick seating chart to help you remember each panel member's name.
- Each interviewer comes to the meeting with a different point of view and set of questions for you. Don't expect the interviewers to have a uniform front; they won't. Treat each one as an individual with his or her own quirks and questions.
- The interviewer who talks the most may not have the most say about whether you get the job. Conversely, the interviewer who says the least may have the most say. You don't know, so ignore no one and treat everyone respectfully.
- As in other interviews, ask questions when appropriate.
- Maintain eye contact with all panel members.
- At the end of the interview, thank the group for having you.

# If You Get a Job Offer

If you get a phone call extending an offer, what do you do? Do not immediately accept, even if this is your dream job. It is perfectly acceptable to ask for one day to consider the offer, including the compensation package. However, do not negotiate salary or anything else until you have had time to think over the offer. In any case, it is better to negotiate in person than on the phone.

## DO YOU WANT THIS JOB?

For each job you're offered, list the pros and cons and evaluate the offers using the following criteria.

- Specifics about the position: Duties, position level, hours of work, working conditions, travel requirements, and so on.
- Potential for growth and promotion; time frame for performance evaluations and rate increases.
- Salary and benefits:
  - Starting salary
  - Overtime/compensation time policy
  - Bonus plans
  - Vacation policy
  - Sick day policy
  - Personal day policy
  - Holiday schedule
  - Health insurance, including employee contribution
  - Dental insurance
  - Life insurance
  - Retirement savings plans
  - College tuition reimbursement plans
  - Stock options
- Training programs
- Provision and laundering of uniforms
- Moving expense benefit (if applicable)
- The company: Growth, success, reputation, management
- Your supervisors: Qualifications of Chef and Sous Chef, personalities, interactions, expectations, and so on
- Turnover rate
- Location: Housing availability and costs, recreation, quality of schools, and so on
- Any other factors you consider important

So how does this job rate? Only you can answer that question.

There will be times when you receive a job offer that is perfect... except for one thing. Rather than turning down the offer, consider negotiating with the employer.

## NEGOTIATION STRATEGIES

Negotiation is a nonadversarial type of communication in which two parties work together to come to an acceptable agreement. Only serious issues based on realistic expectations should be negotiated. Negotiable items include salary and benefits such as vacation time, sick leave, health insurance, life insurance, and tuition reimbursement. Your base salary and performance-based raises are probably the most negotiable parts of your compensation package.

Be aware that many companies offer a cafeteria of benefits whereby you select from a number of benefit options based on a total monetary value. In other words, the company spends a certain amount of money on each employee for benefits, and employees have some flexibility in selecting how those dollars are spent. For example, employees with children might select childcare reimbursement benefits, while employees interested in going back to school might choose tuition reimbursement.

When negotiating your compensation package, it is important to keep in mind the total package. Make sure to consider all the benefits the company has to offer, not just salary. Before you begin negotiating your compensation, decide which benefits are most important to you so you are ready to talk to the employer.

Here are a few things to keep in mind when entering into negotiations with an employer:

- Most employers expect to renegotiate some aspect of your compensation package.
- Negotiate only after an offer is made. Remember, *He who mentions money first loses.*
- Negotiate salary before benefits. If you can't get the salary you want, you might be able to make up for some of it with benefits.
- Realize that your current earnings usually provide the starting point for salary negotiations.
- Figure out the absolute lowest salary you will accept.
- Find out the employer's salary range for the job.
- Know what you are worth. Be familiar with salary ranges and typical benefits in the area where the job is located. See Table 17-5 for help in finding regional salary information.
- Present a realistic salary range that demonstrates your knowledge of the local market.
- Anticipate objections the employer might be able to raise (such as other employees at the same level earning less or lack of budget) and be prepared to justify your cost effectiveness.
- Negotiate based on your qualifications, skills, and experience. Demonstrate the benefit to the employer in paying you more.
- Ask questions such as "Your offer seems a bit modest. What would it take to get to a higher level within the pay scale?"
- Don't bring your personal needs to the discussion.
- Don't overnegotiate.
- Be a good listener. Make eye contact.
- Make the negotiation a friendly experience, because if you decide to accept the offer, the person you're negotiating with is likely to be your boss.
- Act professionally throughout the negotiations.

If you can't get your requested salary, ask for a performance review within a certain number of months so you will be eligible for an increase sooner. Alternatively, ask for a one-time sign-on bonus. Work with the employer to look at other avenues.

## EXERCISES

1. How would you handle yourself differently in each of the following types of interviews: screening, selection, and confirmation?

2. The following website offers an excellent tutorial and tips on how to research companies online. Use the website to do the following assignment. http://www.learnwebskills.com/company/index.html

   Pick an employer for whom you might one day want to work and use the resources noted in this chapter to find out at least six of these points of information: location of headquarters, public or private, number of units, number of employees, annual revenue, history, competitors, strengths, challenges.

3. Find the annual report for a managed service company. What are the major sections of the report? Identify five pieces of information in the report that would be useful to know for a job interview.

4. What would you wear to a job interview with a professional dress code? A casual dress code? Would you need to buy new clothes?

5. Develop responses to five of the questions interviewers like to use.

6. Describe how you would start an interview and how you would close an interview.

7. Describe interviews you have had. How are interviews different for hourly versus salaried positions? Have you ever been offered a job you didn't take? Why did you decline the offer?

8. For a salaried position, what should you do if the interviewer asks how much you want to earn?

9. Use the sources in Table 17-5 to get an idea of how much restaurant Chefs make in your area.

10. Much of the interview process revolves around whether you are the right person to fill the job. We also emphasized in this chapter that the interview is your opportunity to evaluate the employer. What questions can you ask and what actions can you take to see if this is a place you would like to work?

11. How would you handle yourself in each of these interviews: screening interview with a human resources representative, telephone interview, and panel interview?

12. Go to the following website, which contains links to many excellent resources on interviewing. Read one of the resources and write a paragraph on what you learned. http://www.acinet.org/acinet/library.asp?category=2.1

13. Plan a cook test menu and timetable.

**CHAPTER**

**18**

# Advancing Your Career

## Introduction

**ADVANCEMENT OPPORTUNITIES FOR COOKS AND CHEFS** depend on their training, work experience, ability to perform increasingly more responsible and sophisticated tasks, leadership and management skills, and level and type of education. Large establishments and managed service companies usually offer excellent advancement opportunities. Chefs and cooks who demonstrate an eagerness to learn new cooking skills and to accept greater responsibility can move up within the kitchen and take on responsibility for food purchasing, menu development, and supervision. Others may advance by moving from one kitchen or restaurant to another.

During the early part of your career, it's a good idea to stay in each job at least one year, or as long as it takes to learn everything you can there. If you are no longer challenged, perhaps it's time to look for another job to get experience with a different type of cuisine, work with a new chef, or simply work in another foodservice segment.

As Chefs improve their culinary skills, their opportunities for professional recognition and higher earnings increase. Chefs may advance to Executive Chef positions and oversee several kitchens within a foodservice operation, open their own restaurants as Chef-proprietors, or move into training positions as teachers or educators. Your career path is your own unique creation. This chapter discusses ways for you to advance.

## Succeeding in a Professional Kitchen

Whenever you start a new job, you receive some type of training. The most common type is on-the-job training, in which more experienced employees guide you. Large establishments and regional offices of nationwide chain or franchise operations increasingly use video and satellite TV training programs to educate newly hired staff. This type of corporate training generally covers the restaurant's history, menu, organizational philosophy, and daily operational standards. Nationwide chains often operate their own school for prospective managers so they can attend training seminars before acquiring additional responsibilities. Whatever type of training you receive, be sure to pay attention, take notes, and ask questions.

Following are additional guidelines for succeeding in a professional kitchen.

◆ Listen carefully to the Chefs and other supervisors.
◆ Be polite and don't use derogatory or demeaning terms.
◆ Get to work on time and be prepared to work long hours.
◆ Work quickly, efficiently, and neatly, always keeping in mind quality, safety, and sanitation.
◆ Ask questions such as "Would you please show me how to...?" You want to learn as much as possible in each kitchen you work.
◆ Keep a positive attitude.
◆ Spend time getting to know your supervisors, coworkers, and subordinates. This effort will most likely make work more enjoyable.
◆ When the Chef is looking for volunteers, step forward.
◆ Be honest.
◆ Have a sense of humor.
◆ Complete tasks on time.
◆ Follow the rules.
◆ Show respect for and support other team members.
◆ Remember that there is no substitute for work experience.
◆ Get the basics down before you experiment and innovate.

In sum, remember that the world is filled with variables and that you must be flexible, accommodating, and, most of all, patient. Hard work and effort mixed with dedication and loyalty is the recipe for success. In our industry, those qualities are not overlooked; they are commended and rewarded.

## Setting Career Goals

It is important that you construct your own career growth plan, since only you can decide where you want to go in the culinary field. Although developing this plan is your responsibility, it is helpful to enlist the guidance and assistance of others, such as your supervisor. Your goals should represent what you ultimately hope to accomplish within a given period of time. Most people set short-term goals (covering from three to five years) and long-term goals (such as ten years).

Your goals may be to obtain a specific position or to work in a specific segment of the culinary field. When you set your goals, don't get too hung up on job titles, as they are not always accurate. Concentrate instead on the knowledge, skills, and abilities you use on the job. Set a reasonable time frame indicating when you would like to reach your career goals. However, keep in mind that your career goals should be realistic and attainable ones that are reachable through your ongoing efforts to gain experience and education.

Develop a plan of activities to reach your goal. Think of this plan as a step-by-step statement of the specific activities needed to enhance your education, skills, knowledge, or experience. The activities should be measurable and tailored to achieve your specific career goals. You must be able to recognize when you are working toward your goal and when your goal has been accomplished. Be specific and set dates.

You should be prepared to commit a portion of your own time and effort to accomplishing this plan. Completing your planned work experience and/or training activities is your responsibility. You'll need to seek help when necessary.

Today's career paths don't necessarily progress straight up the organizational chart as they used to. At times, you may make lateral moves, possibly into different segments of the foodservice industry, instead of going up. Be flexible. You may wind up down a totally different road and love it!

## When You Leave a Job

When you leave a job, always do it on good terms. Even if you were not happy in the job, act professionally and give from two to four weeks' notice. The exact amount of notice depends on the level of your position and the employer's policies. You will burn bridges behind you if you leave without giving sufficient notice or if you decide to have it out with your boss or another employee just before you leave. If you leave on poor terms, a lot more people than you think will hear about it, and you may one day be asking one of them for a job.

Before you leave any job, it's a good idea to ask your employer to give you a written document listing your job title and dates of employment. See Figure 18-1 for an employment verification form. This document comes in handy when, for example, the restaurant you worked at goes out of business or is sold by your old boss. Be sure to get the form returned on or before your last day of work.

## Mentors

Chef apprentice programs use the mentor model to train new Chefs. A mentor is a person who helps someone, usually a subordinate, grow professionally. Mentors serve as role models and teach, guide, coach, and counsel their mentees (the individuals they mentor).

Date: _____


Dear Sir/Madam,

This will certify that _____

was employed at _____

from _____ to _____.

The job title was _____.

He/she supervised _____ personnel and worked _____ hours/week.


Signature _____

Title _____

Address _____

_____

Phone Number _____


Description of establishment: (full-service restaurant, hospital foodservice, etc.)

_____

_____

**Figure 18-1** Employment Verification Form

Although the mentee clearly benefits from the mentor relationship, the mentor gains advantages too, including the personal satisfaction of passing on his or her successes and the recognition for being a positive role model. Being a mentor also means being able to continually practice culinary, management, and interpersonal skills. Mentors are true professionals in the sense that they directly help others. Besides giving invaluable training, mentors boost the confidence of mentees and provide guidance about career moves.

Mentoring may be formal or relatively informal. Informal mentoring occurs all the time in the workplace. It happens when one person simply helps someone, usually a subordinate, and a career-helping relationship develops. Formal mentoring is found in structured programs in which a company or organization matches mentors with mentees. A good match is one in which the mentor gets along well with the mentee, has the expertise the mentee desires, and is a good teacher.

Regardless of whether mentoring is formal or informal, the following guidelines will help both parties get the most out of the relationship.

1. Have an initial meeting to discuss the expectations of both the mentor and the mentee.
2. Plan to commit to a partnership for six months to one year and discuss a no-fault termination in which either party can back out for any reason.
3. Identify mentee goals and make an action plan. Accept that these goals may change.
4. Set up how often the mentor and mentee will meet to discuss progress.
5. The mentee must be willing to accept constructive feedback, try new things, and take risks.
6. The mentor will use listening, coaching, guidance, career advising, and other techniques to help the mentee reach his or her goals.
7. Many mentoring relationships continue long past the initial time commitment, especially as the partners often become friends. When the mentor and mentee decide their work together is completed, they should go over the original action plan and discuss the progress and results. They should give each other constructive feedback that may help in future mentoring relationships.

In many cases, culinary professionals have to find their own mentors, often by networking, and be willing to work hard once they find a good mentoring relationship.

# Professional Organizations

Being active in professional organizations is key to job advancement and your career. Professional organizations give you lots of opportunities to network with other culinary professionals as well as these additional benefits (which vary by organization):

- Industry magazines and newsletters that keep you up to date on the culinary industry.
- Annual meetings/conferences to upgrade and update your knowledge and skills.
- Educational seminars to upgrade and update your knowledge and skills.
- Leadership opportunities.
- Industry contacts—for example, with representatives of foodservice equipment manufacturers.
- Job announcements.
- Recognition awards.

The Appendix gives detailed information about each organization.

The American Culinary Federation (ACF) is the largest and most prestigious organization dedicated to professional Chefs in the United States today. Their mission is as follows.

*It is our goal to make a positive difference for culinarians through education, apprenticeship, and certification, while creating a fraternal bond of respect and integrity among culinarians everywhere.*

ACF offers its members many opportunities to keep their knowledge and skills current through its monthly publications (*National Culinary Review* and *Center of the Plate*) and through seminars, workshops, national conventions, and regional conferences. ACF sanctions U.S. culinary competitions and oversees international competitions that take place in the United States. ACF accredits culinary programs at the secondary and postsecondary levels. Local chapters of ACF offer members opportunities to network with nearby culinary professionals. Also, ACF members are simultaneously enrolled in the World Association of Cooks Societies (WACS), which represents 54 countries.

## Certifications

Getting certified in the culinary field is a tremendous asset and can help you advance your career. ACF offers the only comprehensive certification program for Chefs in the United States. ACF certification is a valuable credential awarded to Cooks, Chefs, Pastry Cooks, and Pastry Chefs after a rigorous evaluation of industry experience, professional education, and detailed testing. Table 18-1 describes the certifications ACF offers. Table 18-2 lists other certifications in the culinary/foodservice field.

If you intend to take a certification test, give yourself at least several months to prepare for it so you can pass on your first try. Once you are certified, you will need to meet continuing education requirements to maintain your certification.

## Table 18-1 Certifications of the American Culinary Federation

### Certified Master Chef

The consummate Chef possesses the highest degree of professional culinary knowledge and skill. These Chefs teach and supervise their entire crew as well as provide leadership and serve as role models to ACF apprentices. A separate application is required. Certification as a CEC, CEPC, or CPEC is prerequisite.

### Certified Master Pastry Chef (CMPC)

The consummate Chef possesses the highest degree of professional culinary knowledge and skill. These Chefs teach and supervise their entire crew as well as provide leadership and serve as role models to ACF apprentices. A separate application is required. Certification as a CEC, CEPC, or CPEC is prerequisite.

### Certified Executive Chef (CEC)

An Executive Chef is the department head responsible for all culinary units in a restaurant, hotel, club, hospital, or other foodservice establishment. He or she might also be the owner of a foodservice operation. The Executive Chef must supervise a minimum of five full-time persons in the production of food, maintain a safe and sanitary work environment for all employees, and ensure that all kitchens provide nutritious, safe, eye-appealing, and properly flavored food. The Executive Chef is a department head and responsible to management, which coordinates responsibilities and activities with other departments. Other duties include menu planning, budget preparation, payroll, maintenance, controlling food costs, and maintaining financial and inventory records. The Executive Chef possesses basic knowledge of food safety and sanitation, culinary nutrition, and supervisory management.

### Certified Executive Pastry Chef (CEPC)

An Executive Pastry Chef is a department head, usually responsible to the Executive Chef of a food operation or to the management of his or her employing research or pastry specialty firm. The Executive Pastry Chef must maintain a safe and sanitary work environment for all employees and ensure that all bakery and pastry kitchens provide nutritious, safe, eye-appealing, and properly flavored food. The Executive Pastry Chef possesses a basic knowledge of food safety and sanitation, culinary nutrition, and supervisory management.

### Certified Culinary Administrator (CCA)

An executive-level Chef who is responsible for the administrative functions of running a professional foodservice operation. Must demonstrate proficiency in culinary knowledge, human resources, operational management, and business planning skills. This position supervises at least 10 full-time equivalent employees and reports directly to the owner, general manager, or corporate office.

### Personal Certified Executive Chef (PCEC)

A Chef with advanced culinary skills and a minimum of six years of professional cooking experience with a minimum of two full years as a personal Chef; provides cooking services on a cook-for-hire basis to a variety of clients; responsible for menu planning and development, marketing, financial management and operational decisions. The Personal Executive Chef provides nutritious, safe, eye-appealing, and properly flavored foods.

### Certified Culinary Educator (CCE)

An advanced-degreed culinary professional who is working as an educator in an accredited postsecondary institution or military training facility responsible for the development, implementation, administration, evaluation, and maintenance of a culinary arts or foodservice management curriculum. The Culinary Educator possesses superior culinary experience and expertise equivalent to that of the Certified Chef de Cuisine or Certified Working Pastry Chef and a superior knowledge of culinary arts, food safety and sanitation, and culinary nutrition.

## Table 18-1 Certifications of the American Culinary Federation (continued)

**Certified Secondary Culinary Educator (CSCE)**

An advanced-degreed culinary professional who is working as an educator in an accredited secondary or vocational institution responsible for the development, implementation, administration, evaluation, and maintenance of a culinary arts or foodservice management curriculum. The Secondary Culinary Educator possesses a thorough knowledge of culinary arts, educational development, food safety and sanitation, and culinary nutrition.

**Certified Chef de Cuisine (CCC)**

A Chef de Cuisine is in charge of food production in a foodservice operation, whether a single unit of a multi-unit operation or a freestanding operation. He or she is, in essence, the Chef of this operation with the final decision-making power as it relates to culinary operations. The Chef de Cuisine must supervise a minimum of three full-time people in the production of food. Normally, in large operations, the Chef de Cuisine is responsible to the Executive Chef and possesses a basic knowledge of food safety and sanitation, culinary nutrition, and supervisory management.

**Certified Working Pastry Chef (CWPC)**

The Working Pastry Chef is responsible for a pastry section or a shift within a foodservice operation with considerable responsibility for the preparation and production of pies, cookies, cakes, breads, rolls, deserts, confections, and other baked goods. The Working Pastry Chef possesses a thorough knowledge of food safety and sanitation, culinary nutrition, and supervisory management.

**Certified Sous Chef (CSC)**

A Sous Chef supervises a shift, station, or stations in a foodservice operation. A Sous Chef must supervise a minimum of two full-time people in the preparation of food. Job titles that qualify for this designation include Sous Chef, Banquet Chef, Garde Manger, First Cook, AM Sous Chef, and PM Sous Chef. The Sous Chef employed in a small operation is responsible to the Executive Chef and in a larger operation to the Chef de Cuisine. The Sous Chef possesses a thorough knowledge of food safety and sanitation, culinary nutrition, and supervisory management.

**Personal Certified Chef (PCC)**

A Chef with at least two full years employed as a Personal Chef engaged in the preparation, cooking, serving, and sorting of foods on a cook-for-hire basis. The Personal Chef is responsible for menu planning and development, marketing, financial management, and operational decisions of private business; provides cooking services to a variety of clients; and possesses a thorough knowledge of food safety and sanitation and culinary nutrition.

**Certified Pastry Culinarian (CPC)**

An entry-level culinary professional in the baking/pastry area of a foodservice operation is responsible for the preparation and production of pies, cookies, cakes, breads, rolls, desserts, and other baked goods. The Pastry Culinarian possesses a basic knowledge of food safety and sanitation, culinary nutrition, and supervisory management.

**Certified Culinarian (CC)**

An entry-level culinary professional within a commercial foodservice operation responsible for preparing and cooking sauces, cold foods, fish, soups and stocks, meats, vegetables, eggs, and other food items. The Culinarian possesses a basic knowledge of food safety and sanitation, culinary nutrition, and supervisory management.

## Table 18-2 Additional Certifications

**American Correctional Food Service Association**

Certified Correctional Food Service Professional (CCFP)

Certified Correctional Food Systems Manager (CFSM)

**School Nutrition Association**

School Foodservice and Nutrition Specialist (SFNS)

**Club Managers Association of America**

Certified Club Manager (CCM)

Master Club Manager (MCM)

**Dietary Managers Association**

Certified Dietary Manager (CDM)

Certified Food Protection Professional (CFPP)

**Educational Institute of the American Hotel and Lodging Association**

Certified Food and Beverage Executive (CFBE)

Certified Hospitality Educator (CHE)

**Foodservice Educators Network**

Certified Culinary Instructor (CCI)

**International Association of Culinary Professionals**

Certified Culinary Professional (CCP)

**International Food Service Executives Association**

Master Certified Food Executive (MCFE)

Certified Food Executive (CFE)

Certified Food Manager (CFM)

**Military Hospitality Alliance**

Registered Military Culinarian (RMC)

Registered Military Hospitality Manager (RMHM)

**National Association of Foodservice Equipment Manufacturers (NAFEM)**

Certified Food Service Professional (CFSP)

**National Restaurant Association Educational Foundation**

Foodservice Management Professional (FMP)

**National Ice Carving Association**

Certified Ice Carver

Certified Competition Ice Carver

Certified Professional Ice Carver

Certified Master Ice Carver

**Research Chefs Association**

Certified Culinary Scientist (CCS)

Certified Research Chef (CRC)

**Retailer's Bakery Association**

Certified Journey Baker (CJB)

Certified Baker (CB)

Certified Decorator (CD)

Certified Bread Baker (CBB)

Certified Master Baker (CMB)

# American Academy of Chefs

The American Academy of Chefs (AAC) is the honor society of the American Culinary Federation. AAC recognizes culinary professionals whose contributions positively affect the culinary industry. To apply for membership, a Chef's local ACF chapter must request an application from the AAC national office as well as sponsor the applicant. Some of the mandatory requirements for membership consideration in the AAC are as follows.

1. Must be certified as an ACF Master Chef, Master Pastry Chef, Executive Chef, Executive Pastry Chef, or Culinary Educator for not less than two years. Must continue to renew certification.

2. Must be in the culinary profession for not less than 15 years. Ten of the 15 years must have been at the Executive Chef or culinary educator level.

3. Must be a member of ACF at least ten years.

4. Must have attended any combination of four ACF- or ACF-approved regional conferences or national conventions within a ten-year period.

In addition to the mandatory requirements, candidates must complete at least ten additional requirements such as being a chapter president for a full term or being a chair or co-chair of a culinary art salon or hot food competition approved by ACF or WACS.

# Lifelong Learning

Lifelong learning, also called continuing education, refers to the continuous development of the skills, knowledge, and understanding essential to maintaining employment as well as to meeting personal needs. Chefs must be aware of the latest in cooking techniques, cuisines, purchasing, and much more. How do you keep up? Just check out any of the following:

◆ Professional associations such as the ACF present many opportunities for continuing education. The ACF and its chapters offer publications, seminars, workshops, online courses, and more.

◆ Many cooking schools and colleges offer continuing education classes and resources such as books and DVDs so working culinary professionals can learn about current practices. Some schools, including the Culinary Institute of America, offer online courses as well (www.ciaproChef.com).

◆ Culinary conferences and conventions always include seminars and workshops.

◆ Culinary industry magazines provide current information on trends, etc.

Lifelong learning is vital if you want to advance in the profession.

# EXERCISES

1. Pick out a certification you may want to have one day. Go to the appropriate website and find out what you need to do to obtain that certification. Find out how often you have to recertify (such as every five years) and what is required to recertify.

2. Go to the American Culinary Federation website (www.acfChefs.org) and find out the requirements for membership in the American Academy of Chefs.

3. Using the Career and Education Paths given in this book, plot your career path for the next five years. Your instructor will give you a worksheet.

4. Find three websites that offer culinary lifelong learning opportunities. Which site is most attractive to you and why?

5. Interview a culinary professional who has experience as either a mentor or a mentee. Was the experience worthwhile? How did it work?

# Culinary Professional Organizations

## AMERICAN CULINARY FEDERATION (ACF)

### Who Are They?

The American Culinary Federation (ACF) is the largest and most prestigious organization dedicated to professional Chefs in the United States today. Their mission is as follows.

It is our goal to make a positive difference for culinarians through education, apprenticeship, and certification, while creating a fraternal bond of respect and integrity among culinarians everywhere.

ACF offers its members many opportunities to keep up to date with the latest in knowledge and skills through its journals, seminars, workshops, national conventions, and regional conferences. ACF also sanctions U.S. culinary competitions and oversees international competitions that take place in the United States. ACF accredits culinary programs at the secondary and post-secondary levels. Local chapters of ACF offer members opportunities to network with nearby culinary professionals. As a member of ACF, you are simultaneously enrolled in the World Association of Cooks Societies (WACS), which represents 54 countries.

ACF offers the only comprehensive certification program for Chefs in the United States. ACF certification is a valuable credential awarded to Cooks and Chefs and Pastry Cooks and Pastry Chefs after a rigorous evaluation of industrial experience, professional education, and detailed testing.

### Where Are They?

180 Center Place Way
St. Augustine, FL 32095
800-624-9458
www.acfchefs.org

PUBLICATIONS
Monthly magazine: *The National Culinary Review*
Monthly newsletter: *Center of the Plate*

CERTIFICATIONS
Certified Master Chef (CMC)
Certified Master Pastry Chef (CMPC)
Certified Executive Chef (CEC)
Certified Executive Pastry Chef (CEPC)
Personal Certified Executive Chef (PCEC)
Certified Culinary Administrator (CCA)
Certified Culinary Educator (CCE)
Certified Secondary Culinary Educator (CSCE)
Certified Working Pastry Chef (CWPC)
Certified Sous Chef (CSC)
Personal Certified Chef (PCC)
Certified Pastry Culinarian (CPC)
Certified Culinarian (CC)

## AMERICAN DIETETIC ASSOCIATION (ADA)

### Who Are They?

The American Dietetic Association (ADA) is the largest and most visible group of professionals in the nutrition field. Most members are Registered Dietitians (RDs). Individuals with the RD credential have specialized education in human anatomy and physiology, chemistry, medical nutrition therapy, foods and food science, the behavioral sciences, and foodservice management. Registered dietitians must complete at least a bachelor's degree at an accredited college or university, a program of college-level dietetics courses accredited by the Commission on Accreditation for Dietetics Education, a supervised practice experience, and a qualifying examination. Continuing education is required to maintain RD status. Most RDs are licensed or certified by the state in which they live. They work in hospitals and other healthcare settings, private practice, sales, marketing, research, government, restaurants, fitness, and food companies.

### Where Are They?

120 South Riverside Plaza, Suite 2000
Chicago IL 60606
800-877-1600
www.eatright.org

PUBLICATIONS
Monthly journal: *Journal of the American Dietetic Association*
Bimonthly newsletter: *ADA Times*

REGISTRATIONS
Registered Dietitian (RD)
Registered Dietetic Technician (DTR)

## AMERICAN HOTEL AND LODGING ASSOCIATION (AHLA)

## EDUCATIONAL INSTITUTE OF THE AMERICAN HOTEL AND LODGING ASSOCIATION (EI)

### Who Are They?

The American Hotel and Lodging Association (AHLA) is a 93-year-old federation of state hotel lodging associations representing about 11,000 property members worldwide. AHLA provides its members with assistance in operations, education, and communications, and lobbying in Washington, DC, to encourage a climate in which hotels prosper.

The Educational Institute of the AHLA (EI) is the leading source of quality hospitality education, training, and professional certification for hospitality schools and colleges and industries around the world.

**Where Are They?**
American Hotel and Lodging Association
1201 New York Avenue NW #600
Washington, DC 20005
202-289-3100
www.ahma.com
American Hotel and Lodging Association Educational
Institute
800 North Magnolia Avenue, Suite 1800
Orlando, FL 32803
800-752-4567
American Hotel and Lodging Association Educational
Institute
2113 North High Street
Lansing, MI 48906
517-372-8800
www.ei-ahla.org

PUBLICATION
Monthly magazine: *Lodging*

CERTIFICATION
The Educational Institute offers a number of professional certifications, including Certified Food and Beverage Executive. The food and beverage manager must be an expert at providing quality service and have excellent leadership and organizational skills, technical proficiency, and a commitment to high standards. Certification lasts five years, during which requirements for recertification must be fulfilled. Certificate holders must maintain a qualifying position within the industry and earn points by carrying out various activities. The categories in which points are awarded include:

> professional experience
> professional development activities/seminars
> industry involvement
> educational services

EI provides a portfolio to help track professional development activities.

**AMERICAN INSTITUTE OF BAKING (AIB)**

**Who Are They?**

The American Institute of Baking (AIB) provides educational and other programs, products, and services to baking and general food production industries around the world. Its members range from international food ingredient and foodservice companies to small single-unit traditional and artisan retail bakeries. AIB is a resource for bakers looking for information and expertise in baking production, experimental baking, cereal science, nutrition, food safety and hygiene, occupational safety, and maintenance engineering.

**Where Are They?**
1213 Bakers Way
P.O. Box 3999
Manhattan, KS 66505-3999
800-633-5137
www.aibonline.org

PUBLICATIONS
Monthly magazine: *AIB Research Technical Bulletin*
Bimonthly magazine: *AIB Maintenance Engineering Bulletin*
Quarterly newsletter: *Bakers Way*

CERTIFICATIONS
Certified Baker—Bread and Rolls
Certified Baker—Cakes and Sweet Goods
Certified Baker—Cookies and Crackers

**AMERICAN INSTITUTE OF WINE AND FOOD (AIWF)**

**Who Are They?**

The American Institute of Wine and Food (AIWF) is an organization devoted to improving the appreciation, understanding, and accessibility of food and drink. With over 30 U.S. chapters, its members include restaurateurs, food industry professionals, food educators, nutritionists, Chefs, wine professionals, and dedicated food and wine enthusiasts.

**Where Are They?**
304 West Liberty Street, Suite 201
Louisville, KY 40202
800-274-2493
www.aiwf.org

PUBLICATION
Bimonthly newsletter: *American Wine and Food*

**AMERICAN PERSONAL CHEF ASSOCIATION (APCA)**

**Who Are They?**

The American Personal Chef Association (APCA) has been training, supporting, and representing successful personal Chefs since 1995.

**Where Are They?**
4572 Delaware Street
San Diego, CA 92116
800-644-8389
www.personalchef.com

CERTIFICATIONS (through APCA and ACF)
Personal Certified Executive Chef (PCEC)
Personal Certified Chef (PCC)

## AMERICAN SOCIETY FOR HEALTHCARE FOOD SERVICE ADMINISTRATORS (ASHFSA)

### Who Are They?

The American Society for Healthcare Food Service Administrators is an affiliate of the American Hospital Association. Members include food and nutrition service management professionals in hospitals, continuing care retirement communities, nursing homes, and other healthcare facilities. Members include Director of Food and Nutrition Services, Director of Dining Services, café/catering/vending managers, clinical nutrition managers, and dietitians. ASHFSA welcomes food and nutrition service professionals from both independent and contract operations.

### Where Are They?

304 West Liberty Street, Suite 201
Louisville, KY 40202
800-620-6422
www.ashfsa.org

PUBLICATION
Quarterly magazine: *Healthcare Food Service Trends*

CERTIFICATIONS/RECOGNITIONS
ASHFSA has a professional recognition program called APEX, which stands for Actions for Professional Excellence. The three levels of achievement are:
Level 1. Accomplished Health Care Foodservice Administrator (AHCFA)
Level 2. Distinguished Health Care Foodservice Administrator (DHCFA)
Level 3. Fellow Health Care Foodservice Administrator (FHCFA)
The ASHFSA Professional Recognition Program is designed to recognize those factors that are indispensable to true professionalism: basic and continuing education, experience, and participation in professional and society activities. Levels of recognition are achieved successively. Each level requires additional education, work experience, and participation in society activities. Members may put the acronym for the level they have accomplished after their name (for example, John Hall, DHCFA).

## ASSOCIATION FOR CAREER AND TECHNICAL EDUCATION (ACTE)

### Who Are They?

The Association for Career and Technical Education (ACTE) is dedicated to the advancement of education that prepares youth and adults for successful careers. ACTE provides resources to enhance the job performance and satisfaction of its members, increases public awareness of career and technical programs, and works on growth in funding for these programs. Its members include over 30,000 teachers, counselors, and administrators at the middle, secondary, and postsecondary school levels.

### Where Are They?

1410 King Street
Alexandria, VA 22314
800-826-9972
www.acteonline.org

PUBLICATIONS
Monthly magazine: *Techniques*
Email newsletter: *Career Tech Update*

## BLACK CULINARIAN ALLIANCE (BCA)

### Who Are They?

The Black Culinarian Alliance (BCA) is an educational and networking association of hospitality and foodservice professionals founded in 1993.

### Where Are They?

55 West 116th Street, Suite 234
New York, NY 10026
800-308-8188
www.blackculinarians.com

## BREAD BAKER'S GUILD OF AMERICA

### Who Are They?

The Bread Baker's Guild of America provides education resources for members and fosters the growth of artisan baking and the production of high-quality bread products. Its members include professional bakers, baking educators, home bakers, vendors, and others who share common goals.

### Where Are They?

3203 Maryland Avenue
North Versailles, PA 15137-1629
412-823-2080
www.bbga.org

PUBLICATION
Newsletter: *Bread Bakers Guild*

## CLUB MANAGERS ASSOCIATION OF AMERICA (CMAA)

### Who Are They?

Members of the Club Managers Association of America (CMAA) manage more than 3,000 country, city, athletic, faculty, yacht, town, and military clubs. CMAA provides its members with professional development programs, networking opportunitites, publications, and certification programs.

**Where Are They?**

1733 King Street
Alexandria, VA 22314-2720
703-739-9500
www.cmaa.org

PUBLICATION
Monthly magazine: *Club Management*

CERTIFICATIONS
Certified Club Manager (CCM)
Master Club Manager (MCM)

## CONFRÉRIE DE LA CHAÎNE DES RÔTISSEURS

**Who Are They?**

The Confrérie de la Chaîne des Rôtisseurs brings together professional and amateur gastronomes in an organization that celebrates the pleasures of food, wine, and spirits and encourages the development of young professionals. The organization is based on the traditions and standards of the medieval French guild of rôtisseurs, or "meat roasters." Members are in more than 70 countries around the world. The United States has its own chapter.

**Where Are They?**

Chaîne House
Fairleigh Dickinson University
285 Madison Avenue
Madison, NJ 07940-1099
973-360-9200
www.chaineus.org

PUBLICATION
Three times yearly: *Gastronome*

## DIETARY MANAGERS ASSOCIATION (DMA)

**Who Are They?**

The Dietary Managers Association is a national association with over 15,000 professionals dedicated to the mission of "providing optimum nutritional care through food service management." Dietary managers work in nursing homes and other long-term care facilities, hospitals, schools, correctional facilities, and other settings. Responsibilities may include directing and controlling menu planning, food purchasing, food production and service, financial management, employee hiring and training, supervision, nutritional assessment, and clinical care. Dietary managers who have earned the Certified Dietary Manager (CDM), Certified Food Protection Professional (CFPP) credential are also specially trained in food safety and sanitation. Dietary managers may work as foodservice directors, assistant foodservice directors, supervisors, and in other positions.

**Where Are They?**

406 Surrey Woods Drive
St. Charles, IL 60174
800-323-1908
www.dmaonline.org

PUBLICATION
Monthly magazine: *Dietary Manager*

CERTIFICATIONS
Certified Dietary Manager (CDM)
Certified Food Protection Professional (CFPP)

## FOODSERVICE CONSULTANTS SOCIETY INTERNATIONAL (FCSI)

**Who Are They?**

The Foodservice Consultants Society International (FCSI) promotes professionalism in foodservice and hospitality consulting and helps its members by supplying networking and educational opportunities, professional recognition, and other services. FCSI has members who work in layout and design, planning, research, training, technology, operations, and management.

**Where Are They?**

304 West Liberty Street, Suite 201
Louisville, KY 40202
502-583-3783
www.fcsi.org

PUBLICATIONS
Quarterly magazine: *The Consultant*
Monthly email newsletter: *The Forum*

## FOODSERVICE EDUCATORS NETWORK INTERNATIONAL (FENI)

**Who Are They?**

Foodservice Educators Network International (FENI) is a group of foodservice educators who work in high schools and postsecondary programs. FENI works with educators to help them advance their professional growth. A key element of FENI is to facilitate the means for culinary educators to share teaching techniques and other information with colleagues and industry partners to enhance high standards of culinary education.

**Where Are They?**

20 West Kinzie, 12th Floor
Chicago, IL 60610
312-849-2220
www.feni.org

PUBLICATION
Quarterly magazine: *Chef Educator Today*

CERTIFICATION
Certified Culinary Instructor (CCI)

## INSTITUTE OF FOOD TECHNOLOGISTS (IFT)

**Who Are They?**

The Institute of Food Technologists (IFT) has members who work in food science, food technology, and related professions in industry, academia, and government.

**Where Are They?**

525 West Van Buren, Suite 1000
Chicago, IL 60607
800-438-3663
www.ift.org

PUBLICATION
Monthly magazine: *Food Technology*

## INTERNATIONAL ASSOCIATION OF CULINARY PROFESSIONALS (IACP)

**Who Are They?**

The International Association of Culinary Professionals (IACP) is a group of approximately 4,000 food professionals from over 35 countries. IACP provides continuing education, networking, and information exchange for its members, who work in culinary education, communication, or the preparation of food and drink. Its mission is to "help its members achieve career success ethically, responsibly, and professionally." Many of its members are cooking school instructors, food writers, cookbook authors, Chefs, and food stylists.

**Where Are They?**

304 West Liberty Street, Suite 201
Louisville, KY 40202
502-581-9786
www.iacp.com

PUBLICATION
Quarterly magazine: *Food Forum*

CERTIFICATIONS
Certified Culinary Professional (CCP)

## INTERNATIONAL CATERERS ASSOCIATION (ICA)

**Who Are They?**

The International Caterers Association (ICA) includes both off-premises and on-premises caterers from around the world. ICA provides education, mentoring, and other services for professional caterers, and promotes the profession of catering to the public, vendors, and others.

**Where Are They?**

1200 17th Street NW
Washington, DC 20036
888-604-5844
www.icacater.org

PUBLICATIONS
Bimonthly newsletter: *CommuniCater*
Membership includes complimentary subscriptions to *Catering Magazine, Event Solutions,* and *Special Events Magazine.*

## INTERNATIONAL COUNCIL OF CRUISE LINES (ICCL)

**Who Are They?**

Members of the International Council of Cruise Lines (ICCL) includes the largest passenger cruise lines that call on ports in the United States and elsewhere, as well as individuals who work in the industry. ICCL offers many services to its members and is dedicated to ensure a safe and caring shipboard environment for both passengers and crew. ICCL advocates its positions to lawmakers, industry partners, and regulatory organizations.

**Where Are They?**

2111 Wilson Boulevard, 8th Floor
Arlington, VA 22201
800-595-9338
www.iccl.org

PUBLICATIONS
Monthly email: *Fast Facts*
Quarterly newsletter: *Even Keel*

## INTERNATIONAL COUNCIL ON HOTEL, RESTAURANT, AND INSTITUTIONAL EDUCATION (I-CHRIE)

**Who Are They?**

I-CHRIE is the advocate for schools, colleges, and universities that offer programs in hotel and restaurant management, food service management, and culinary arts.

**Where Are They?**

2613 North Parham Road
Richmond, VA 23294
804-346-4800
www.chrie.org

PUBLICATION
Monthly journal: *Journal of Hospitality and Tourism Education*

## INTERNATIONAL FOOD SERVICE EXECUTIVES ASSOCIATION (IFSEA)

**Who Are They?**

The International Food Service Executives Association (IFSEA) offers its diverse membership opportunities for personal development, networking, mentoring, and community service. IFSEA's membership includes professionals from many areas of the foodservice industry.

**Where Are They?**

2609 Surfwood Drive
Las Vegas, NV 89128
702-838-8821
www.ifsea.com

PUBLICATION
Magazine: *Hotline*

CERTIFICATIONS
Master Certified Food Executive (MCFE)
Certified Food Executive (CFE)
Certified Food Manager (CFM)

## INTERNATIONAL FOODSERVICE MANUFACTURERS ASSOCIATION (IFMA)

**Who Are They?**

The International Foodservice Manufacturers Association (IFMA) represents manufacturers of foodservice equipment. IFMA provides its members many services, training, networking, visibility, and marketing information.

**Where Are They?**

Two Prudential Plaza
180 North Stetson Avenue, Suite 4400
Chicago, IL 60601
312-540-4400
www.ifmaworld.com

PUBLICATION
Bimonthly magazine: *IFMA World*

## INTERNATIONAL INFLIGHT FOOD SERVICE ASSOCIATION (IFSA)

**Who Are They?**

The International Inflight Food Service Association (IFSA) serves the needs and interests of airline and railway personnel as well as airline and rail caterers who are responsible for providing passenger foodservice on regularly scheduled travel routes. IFSA's membership is dedicated to the advancement of the art and science of this segment.

**Where Are They?**

5775 Peachtree-Dunwoody Road
Building G, Suite 500
Atlanta, GA 30342
404-252-3663
www.ifsanet.com

PUBLICATION
Monthly email newsletter: *Onboard IFSA*

## LES DAMES D'ESCOFFIER

**Who Are They?**

Les Dames D'Escoffier is an invitation-only culinary

organization of successful women leaders. The organization mentors young women, educates the public about the pleasures of the table, awards scholarships, and supports food-related charities. Members include Chefs, restaurateurs, cookbook authors, food journalists and historians, wine professionals, food publicists, culinary educators, and hospitality executives.

**Where Are They?**

212-867-3929
www.ldei.org

## MILITARY HOSPITALITY ALLIANCE (MHA)

**Who Are They?**

The Military Hospitality Alliance (MHA) is the affiliate of the International Food Service Executives Association that focuses on the needs of the military. Projects include a military culinary competition, Enlisted Aide of the Year awards, and more.

**Where Are They?**

836 San Bruno Avenue
Henderson, NV 89015
888-234-3732
www.mhaifsea.com

CERTIFICATIONS
Registered Military Culinarian (RMC)
Registered Military Hospitality Manager (RMHM)

## NATIONAL ASSOCIATION FOR THE SPECIALTY FOOD TRADE (NASFT)

**Who Are They?**

The National Association for the Specialty Food Trade (NASFT) is an international organization of domestic and foreign manufacturers, importers, distributors, brokers, retailers, restaurateurs, caterers and others in the specialty foods business.

**Where Are They?**

120 Wall Street, 27th Floor
New York, NY 10005
212-482-6440
www.specialtyfood.com

PUBLICATION
Monthly magazine: *Specialty Food*

## NATIONAL ASSOCIATION OF COLLEGE AND UNIVERSITY FOOD SERVICES (NACUFS)

**Who Are They?**

The National Association of College and University Food Services (NACUFS) is the trade association for foodservice professionals at nearly 650 institutions of higher education in the United States, Canada, and abroad.

## Where Are They?

1405 South Harrison Road, Suite 305
Manly Miles Building
Michigan State University
East Lansing, MI 48824-5242
517-332-2494
Fax: 517-332-8144
www.nacufs.org

PUBLICATION
Quarterly magazine: *Campus Dining Today*

## NATIONAL ASSOCIATION OF FOODSERVICE EQUIPMENT MANUFACTURERS (NAFEM)

### Who Are They?

The North American Association of Food Equipment Manufacturers (NAFEM) represents companies throughout the United States, Canada, and Mexico that manufacture commercial foodservice equipment and supplies.

### Where Are They?

161 North Clark Street, Suite 2020
Chicago, IL 60601
312-821-0201
www.nafem.org

PUBLICATION
Quarterly magazine: *NAFEM in Print*

CERTIFICATION
Certified Food Service Professional (CFSP)

## NATIONAL FOOD PROCESSORS ASSOCIATION (NFPA)

### Who Are They?

The National Food Processors Association provides foodscience and technical expertise for the food processing industry on topics such as food safety, nutrition, and technical and regulatory matters. NFPA members process and package, fruit, vegetables, meat, fish, and other foods.

### Where Are They?

1350 I Street NW, Suite 300
Washington, DC 20005
800-355-0983

PUBLICATION
Monthly journal: *NFPA Journal*

## NATIONAL ICE CARVING ASSOCIATION (NICA)

### Who Are They?

The National Ice Carving Association (NICA) is the only organization in the United States devoted solely to promoting the art of ice sculpture across the country and around the world. With nearly 500 members, NICA sanctions and organizes ice carving competitions in North America, providing standardized guidelines for judging the quality of ice sculptures in competition.

### Where Are They?

P.O. Box 3593
Oak Brook, IL 60522-3593
630-871-8431
www.nica.org

PUBLICATION
Newsletter: *On Ice*

CERTIFICATIONS
Certified Master Ice Carver
Certified Professional Ice Carver
Certified Competition Ice Carver
Certified Ice Carver

## NATIONAL RESTAURANT ASSOCIATION (NRA) THE NATIONAL RESTAURANT ASSOCIATION EDUCATIONAL FOUNDATION (NRAEF)

### Who Are They?

The National Restaurant Association (NRA) is the leading business association for the restaurant industry. Together with the National Restaurant Association Educational Foundation, the Association's mission is to represent, promote, and educate (through reports, publications, research, training materials, networking, etc.) the restaurant and foodservice industry.

The National Restaurant Association Educational Foundation (NRAEF) is the not-for-profit organization dedicated to fulfilling the educational mission of the National Restaurant Association. The NRAEF is the premier provider of education resources, materials, and programs to attract, develop, and retain the industry's workforce. Examples of NRAEF programs include ServSafe® food safety certification, the ProStart® School-to-Career program, the ProMgmt.® program, and the Foodservice Management Professional (FMP) certification program.

### Where Are They?

National Restaurant Association
1200 17th Street NW
Washington, DC 20036
202-331-5900
www.restaurant.org
National Restaurant Association Educational Foundation
175 West Jackson Boulevard, Suite 1500
Chicago, IL 60604-2814
800-765-2122
312-715-1010 (in Chicago)

PUBLICATION (through the National Restaurant Association)
Monthly online magazine: *Restaurants USA Online*

CERTIFICATION (through Educational Foundation)
Foodservice Management Professional (FMP)

## NATIONAL SOCIETY FOR HEALTHCARE FOOD SERVICE MANAGEMENT (HFM)

### Who Are They?

The National Society for Healthcare Foodservice Management (HFM) is a professional association representing healthcare foodservice operators and their suppliers. HFM accepts only members who operate independent operations and are not contracted. HFM offers advocacy for independent healthcare foodservices as well as management tools to decrease costs, increase patient and staff satisfaction, and define successful operational performance. Members are mostly from hospitals.

### Where Are They?

204 E Street NE
Washington, DC 20002
202-546-7236
www.hfm.org

PUBLICATION(S)
Monthly magazine: *Innovator*

## RESEARCH CHEFS ASSOCIATION (RCA)

### Who Are They?

The Research Chefs Association (RCA) brings together Chefs, food scientists, and others who work in food research and development in restaurants, food companies, and many other businesses.

### Where Are They?

5775 Peachtree-Dunwoody Road
Building G, Suite 500
Atlanta, GA 30342
404-252-3663
www.culinology.com

PUBLICATIONS
Magazine: *Culinology*
Quarterly newletter: *Culinology Currents*

CERTIFICATIONS
Certified Research Chef (CRC)
Certified Culinary Scientist (CCS)

## RETAILER'S BAKERY ASSOCIATION (RBA)

### Who Are They?

The Retailer's Bakery Association (RBA) is a trade association of independent retail bakeries. The associa-tion works to improve the operations and profitability of its members by offering training, networking, conventions, communications, and meetings.

### Where Are They?

14239 Park Center Drive
Laurel, MD 20707-5261
800-638-0924
www.rbanet.com

CERTIFICATIONS
Certified Journey Baker (CJB)
Certified Baker (CB)
Certuified Decorator (CD)
Certified BreadBaker (CBB)
Certified Master Baker (CMB)

## SCHOOL NUTRITION ASSOCIATION (formerly the American School Food Service Association)

### Who Are They?

The School Nutrition Association (SNA) has over 55,000 members involved in some way with the National School Lunch Program. SNA works on making sure all children have access to healthful, tasty school meals and nutrition education. SNA does this by providing education and training, setting standards, and educating members on legislative, industry, nutritional, and other issues.

### Where Are They?

700 South Washington Street, Suite 300
Alexandria, VA 22314
703-739-3900
www.schoolnutrition.org

PUBLICATIONS
Monthly magazine: *School Foodservice and Nutrition*
Semiannual journal: *Journal of Child Nutrition and Management*

CREDENTIAL
School Foodservice and Nutrition Specialist (SFNS)

## SOCIÉTÉ CULINAIRE PHILANTHROPIQUE

### Who Are They?

The Société Culinaire Philanthropique is the oldest association of Chefs and Cooks in the United States. Founded by a group of French Chefs, its members organize the annual Salon of Culinary Arts in New York City.

### Where Are They?

305 East 47th Street, Suite 11B
New York, NY 10017
212-308-0628
www.societeculinaire.com

## SOCIETY FOR FOODSERVICE MANAGEMENT (SFM)

### Who Are They?

The Society for Foodservice Management (SFM) serves the needs and interests of executives in the on-site foodservice industry (predominantly B&I). SFM provides member interaction, continuing education, and professional development via information and research.

### Where Are They?

304 West Liberty Street, Suite 201
Louisville, KY 40202
502-583-3783
www.sfm-online.org

PUBLICATION
Monthly email newsletter: *FastFacts*

## UNITED STATES PERSONAL CHEF ASSOCIATION (USPCA)

### Who Are They?

The United States Personal Chef Association serves the needs and interests of personal Chefs. USPCA offers training, certification, personal Chef software, and other services.

### Where Are They?

481 Rio Rancho Boulevard NE
Rio Rancho, NM 87124
800-995-2138
www.uspca.com

PUBLICATION
Monthly magazine: Personal Chef

CERTIFICATION
Certified Personal Chef (CPC)

## WOMEN CHEFS AND RESTAURATEURS (WCR)

### Who Are They?

The Women Chefs and Restaurateurs (WCR) promote the education and advancement of women in the restaurant industry. Their goals are "exchange, education, enhancement, equality, empowerment, entitlement, environment, and excellence." WCR offers a variety of networking, professional, and support services.

### Where Are They?

304 West Liberty Street, Suite 201
Louisville, KY 40202
877-927-7787
www.womenchefs.org

PUBLICATION
Quarterly newsletter: *Entrez!*

## WOMEN'S FOODSERVICE FORUM (WFF)

### Who Are They?

The Women's Foodservice Forum was founded in 1989 to develop leadership talent and ensure career advancement of executive women in the foodservice industry. WFF currently offers three types of mentor programs and many other resources to help women move into senior-level positions. Its membership comes from restaurant operations, manufacturing, distribution, publishing, and consulting.

### Where Are They?

One General Mills Boulevard, MSW05D
Minneapolis, MN 55426
866-368-8008
www.womensfoodserviceforum.com

PUBLICATION
Monthly newsletter: *Open Doors*

# About the CD-ROM

## INTRODUCTION

This appendix provides you with information on the contents of the CD that accompanies this book. For the latest and greatest information, please refer to the ReadMe file located at the root of the CD.

## SYSTEM REQUIREMENTS

- A computer with a processor running at 120 Mhz or faster
- At least 32 MB of total RAM installed on your computer; for best performance, we recommend at least 64 MB
- A CD-ROM drive

NOTE: Many popular word processing programs are capable of reading Microsoft Word files. However, users should be aware that a slight amount of formatting might be lost when using a program other than Microsoft Word.

## USING THE CD WITH WINDOWS

To access the items from the CD, follow these steps:

1. Insert the CD into your computer's CD-ROM drive.
2. The CD-ROM interface will appear. The interface provides a simple point-and-click way to explore the contents of the CD.

If the opening screen of the CD-ROM does not appear automatically, follow these steps to access the CD:

1. Click the Start button on the left end of the taskbar and then choose Run from the menu that pops up.
2. In the dialog box that appears, type **d:\setup.exe.** (If your CD-ROM drive is not drive d, fill in the appropriate letter in place of d.) This brings up the CD Interface described in the preceding set of steps.

## WHAT'S ON THE CD

The following sections provide a summary of the software and other materials you'll find on the CD.

### Content

Any material from the book, including forms, slides, and lesson plans if available, are in the folder named "Content." The CD accompanying this book contains: checklists, templates, sample resumes, portfolio planning sheets, sample portfolios, cover letter templates and networking contact sheets. These documents are provided to aid the reader in their career planning and job search.

### Microsoft Word Documents

The documents are presented as figures in this book. These documents are provided for your reference and use in completing career planning and job search activities.

### Applications

The following applications are on the CD:

*Word Viewer*

Microsoft Word Viewer is a freeware viewer that allows you to view, but not edit, most Microsoft Word files. Certain features of Microsoft Word documents may not display as expected from within Word Viewer.

*Shareware programs* are fully functional, trial versions of copyrighted programs. If you like particular programs, register with their authors for a nominal fee and receive licenses, enhanced versions, and technical support.

*Freeware programs* are copyrighted games, applications, and utilities that are free for personal use. Unlike shareware, these programs do not require a fee or provide technical support.

*GNU software* is governed by its own license, which is included inside the folder of the GNU product. See the GNU license for more details.

*Trial, demo, or evaluation versions* are usually limited either by time or functionality (such as being unable to save projects). Some trial versions are very sensitive to system date changes. If you alter your computer's date, the programs will "time out" and no longer be functional.

## CUSTOMER CARE

If you have trouble with the CD-ROM, please call the Wiley Product Technical Support phone number at (800) 762-2974. Outside the United States, call 1(317) 572-3994. You can also contact Wiley Product Technical Support at http://www.wiley.com/techsupport. John Wiley & Sons will provide technical support only for installation and other general quality control items. For technical support on the applications themselves, consult the program's vendor or author.

To place additional orders or to request information about other Wiley products, please call (877) 762-2974.

# Index